Physics of Gas–Liquid Flows

Presenting tools for understanding the behavior of gas–liquid flows based on the ways large-scale behavior relates to small-scale interactions, this text is ideal for engineers seeking to enhance the safety and efficiency of natural gas pipelines, water-cooled nuclear reactors, absorbers, distillation columns, and gas lift pumps. The review of advanced concepts in fluid mechanics enables both graduate students and practicing engineers to tackle the scientific literature and engage in advanced research.

The text focuses on gas–liquid flow in pipes as a simple system with meaningful experimental data. This unified theory develops design equations for predicting drop size, frictional pressure losses, and slug frequency, which can be used to determine flow regimes, the effects of pipe diameter, liquid viscosity, and gas density. It describes the effect of wavy boundaries and temporal oscillations on turbulent flows, and explains transition between flow regimes, which is key to understanding the behavior of gas–liquid flows.

Thomas J. Hanratty is Professor Emeritus at the University of Illinois at Urbana-Champaign, and was a leader in establishing industrially important multiphase flow as a new academic discipline, by relating macroscopic behavior to small-scale interactions. His research has been recognized by nine awards from the American Institute of Chemical Engineers (AIChE), the American Society for Engineering Education, Ohio State University, Villanova University, and University of Illinois. He was the inaugural winner of the International Multiphase Flow Prize. Hanratty was named as one of the influential chemical engineers of the modern era at the AIChE centennial celebration in 2008. He has been elected to the National Academy of Engineering, the National Academy of Sciences, and the American Academy of Arts and Sciences.

"This authoritative and impressive monograph, written by a widely acclaimed pioneer in the field, is an excellent resource for new students as well as seasoned practitioner. It develops the concepts systematically and packages many decades worth of literature in this field lucidly, giving the readers a chance to understand appreciate the evolution of this field."

Sankaran Sundaresan, Princeton University

"Physics of gas-liquid flows" is a must read for graduate students, researchers and engineers seeking a solid basis or wanting to update their knowledge in the dynamics of gas-liquid systems. The book is an authoritative reference, mainly built around the results of the author, a leading expert in the field during several decades. It presents our current experimental and theoretical understanding at both the local and global scales with an original contribution to wave phenomena as they appear in film stratified and annular flows that may be a source of inspiration for researchers and teachers in the years to come."

Jean Fabre, Institut National Polytechnique, Toulouse

"A lifetime of probing research and deep thinking about gas-liquid flows is enclosed between the covers of this book. Starting from simple analyses – the style of which will be familiar to many undergraduates – the author moves gradually to more advanced topics building a succinct yet exhaustive picture of the present understanding of these important flows. Practicing engineers and researchers alike will find many gems in this book."

Andrea Prosperetti, Johns Hopkins University

Physics of Gas–Liquid Flows

THOMAS J. HANRATTY
University of Illinois at Urbana-Champaign

CAMBRIDGE
UNIVERSITY PRESS

CAMBRIDGE
UNIVERSITY PRESS

Shaftesbury Road, Cambridge CB2 8EA, United Kingdom

One Liberty Plaza, 20th Floor, New York, NY 10006, USA

477 Williamstown Road, Port Melbourne, VIC 3207, Australia

314–321, 3rd Floor, Plot 3, Splendor Forum, Jasola District Centre, New Delhi – 110025, India

103 Penang Road, #05–06/07, Visioncrest Commercial, Singapore 238467

Cambridge University Press is part of Cambridge University Press & Assessment, a department of the University of Cambridge.

We share the University's mission to contribute to society through the pursuit of education, learning and research at the highest international levels of excellence.

www.cambridge.org
Information on this title: www.cambridge.org/9781107041202

First published 2013

A catalogue record for this publication is available from the British Library

Library of Congress Cataloging-in-Publication data
Hanratty, Thomas J.
Physics of gas–liquid flows / Thomas J. Hanratty, University of Illinois at Urbana-Champaign.
pages cm
ISBN 978-1-107-04120-2 (hardback)
1. Multiphase flow. 2. Gas–liquid interfaces. I. Title.
TA357.5.M84H36 2013
532′.56–dc23

2013013373

ISBN 978-1-107-04120-2 Hardback

Contents

Preface

Gas–liquid flows are ubiquitous in industrial and environmental processes. Examples are the transportation of petroleum products, the cooling of nuclear reactors, the operation of absorbers, distillation columns, gas lift pumps. Quite often corrosion and process safety depend on the configuration of the phases. Thus, the interest in this area should not be surprising.

The goal of this book is to give an account of scientific tools needed to understand the behavior of gas–liquid systems and to read the scientific literature. Particular emphasis is given to flow in pipelines.

The following brief historical account is taken from a plenary lecture by the author at the Third International Conference on Multiphase Flow, Lyon, France, June 8–12, 1998. (*Int. J. Multiphase Flow* 26, 169–190, 2000):

> A symposium held at Exeter (P. M. C. Lacey) in 1965 brought together 160 people with a wide range of interests. Discussions at the 42 presentations indicated, to me, that something special was happening and that future directions of work on multiphase flow were being defined. This thrust was continued in conferences at Waterloo, Canada, in 1968 (E. Rhodes, D. S. Scott) and at Haifa, in 1971 (G. Hetsroni). Intellectual activity in ensuing years is exemplified by more focused conferences on Annular and Dispersed Flows held at Pisa, 1984 (S. Zanelli, P. Andreussi, T. J. Hanratty) and in Oxford, England, in 1987 (G. F. Hewitt, P. Whalley, B. Azzopardi), the Symposium on Measuring Techniques at Nancy (J. M. Delhaye, 1983) and the Conference on Gas Transfer at Heidelberg (Jähne, 1995). However, the 350 papers presented at the Second International Conference on Multiphase Flow in 1995 (A. Serizawa, Y. Tsuji) manifested a new level of activity.

A fair question is what happened between 1965 and 1995. My own assessment is that major successes came about, mainly through efforts that relate macroscopic properties of multiphase systems to small-scale behavior. An outcome of this approach is the possible emergence of a new field. This is evidenced in many ways, of which the establishment of the *International Journal of Multiphase Flow* (Gad Hetsroni, 1973) and the Japan Society of Multiphase Flow (A. Akagawa, T. Fukano, 1987) are examples. The following excerpt from a talk by R. T. Lahey at the inauguration of the Japan Society would indicate that my observations are not original: "I believe that this new field will become as widely accepted in the future as other emerging fields . . .".

Physics of Gas–Liquid Flows addresses both graduate students and practitioners. The treatment is based on a course, taught at the University of Illinois, which required only that the students had taken one undergraduate course in fluid dynamics. As a consequence, attention is given to topics that are usually bulwarks in graduate courses in fluid dynamics, such as ideal flow theory, the Navier–Stokes equations and interfacial waves.

In reporting the results from published research I have retained the units (metric or imperial) as used in the original work.

This work has a kinship to *One Dimensional Two-Phase Flow* by Graham Wallis, published by McGraw-Hill in 1969. One should recognize that much has happened since the publication of this book and that my way of presenting the material could be different from that of that of Dr. Wallis.

Physics of Gas–Liquid Flows leans heavily on contributions by researchers in my laboratory. A verbal summary of these works is contained in an account of the Research of Thomas J. Hanratty which was deposited with the University of Illinois in the Illinois Digital Environment for Access to Learning and Scholarship (IDEALS). It can be visited at http://hdl.handle.net/2142/9132. The illustrations were prepared by Dorothy Loudermilk. Taras V. Pogorelov provided a helping hand in transmitting the manuscripts to Cambridge University Press.

Cover

The photograph featured in the cover was obtained in the laboratory of Thomas J. Hanratty by James B. Young. It captures the trajectories of particles in a turbulent liquid flowing down a vertical pipe. Cross-sections at several locations were illuminated by thin sheets having different colors. Axial viewing photography was used to capture the paths of the particles. The color of a particle gives its axial location.

Symbols

A	Area
A	Measure of the thickness of the viscous wall layer, used by van Driest
A_G	Area of the gas space
A_L	Area of the liquid space
A_{LS}	Area of the liquid in a slug
A_{L1}	Area of the liquid layer in front of a slug
A_{Pi}	Acceleration of a particle in the i-direction
A_{L0}	Critical area for a liquid layer to sustain a stable or growing slug
A_t	Area of a tube or pipe
a	Amplitude of a wave
a_v	Interfacial area per unit volume
B	Height of a rectangular channel, or the height of the gas space
C	Concentration
C_0	Concentration at $y = 0$
C_B	Bulk concentration
C_D	Drag coefficient
C_L	Lift coefficient
C_P	Heat capacity at constant pressure
C_V	Heat capacity at constant volume
C_W	Concentration at the wall
c	Velocity of sound; molecular velocity
c	Complex wave velocity $= c_R + ic_I$
c_0	Wave velocity for stagnant fluids
c_F	Velocity of the liquid at the front of a slug
c_B	Velocity of the bubble behind a slug
c_g	Group velocity, the speed at which wave energy is transmitted
c_G	Sound velocity in the gas phase of a gas–liquid flow
c_{KW}	Kinematic wave velocity, defined by (7.51)
c_L	Sound velocity in the liquid phase of a gas–liquid flow
c^t	Turbulent concentration fluctuation
D	Molecular diffusion coefficient
d_m	Maximum drop diameter
d_t	Pipe diameter

d_h	Hydraulic diameter = $4A$/wetted perimeter $d_{v\mu}$
$d_{v\mu}$	Volume median diameter
d_{qm}	Drop diameter defined by Mugele & Evans
d_{10}	Number mean diameter
d_P	Diameter of a bubble or a drop
d_{32}	Sauter mean diameter
E	Entrainment
E	Energy associated with one wavelength = $P + T$
E_M	Maximum possible entrainment
$E(n)$	Spectral density function
e	Energy per unit mass
e^c	Capture efficiency
$\vec{F}$	Force
F	Mechanical energy lost per unit mass of fluid due to friction
F	Flux of particles
F_D	Resisting force on a particle due to fluid drag
F_D	Flux due to diffusion
F_B	Buoyancy force
$\vec{F}_L$	Lift force
f	Fanning friction factor
f	Factor that accounts for the volume of fluid dragged along by a particle
$f_n(d_P)$	Number distribution function
$f_v(d_P)$	Volume distribution function
f_S	Friction factor for a smooth surface
f_S	Frequency of slugging
f_i	Friction factor for gas flow over a gas–liquid interface
f_{Pi}	Force of the fluid on a particle
G	Mass velocity defined by (1.4)
G_C	Mass velocity at the choking condition
G_L	Mass velocity of the liquid
G_G	Mass velocity of the gas
G_{LE}	Mass velocity of drops entrained in the gas
$\vec{g}$	Acceleration of gravity
$\hat{g}$	Defined by (3.20) and (3.21)
g^+	Equal to $g\upsilon/v^*$
H	Submergence of injector in a gas-lift pump
h	Enthalpy
h	Height above a datum plane
h_H	Enthalpy of a gas–liquid mixture
h^L	Enthalpy of the liquid
h^G	Enthalpy of the gas
h_L	Height of liquid layer in a channel
h_{L1}	Height of the liquid layer in front of a slug

h_L	In a pipe, the length of the bisector of the liquid layer
h_L^+	Dimensionless height of the liquid layer $= h_L v^*/\upsilon$
h_G	Height of the gas layer in a channel
h_{L0}	Critical height of the liquid layer needed for slugs to appear
h_L^c	Location of the centroid of an area
h_{LB}	Height of the liquid at the bottom of a horizontal pipe
h_W	Distance of the top of the waves from the wall
h_{LS}	Height of the liquid layer if the flow is steady
Δh	Wave height
I	Intermittency, fraction of the time that disturbance waves are present
I_1	Integral defined by (12.63)
I_2	Integral defined by (12.64)
i	Internal energy per unit mass
k	Wave number
k_M	Wave number at which wave growth is a maximum
k_S	Sand roughness
k_D	Deposition coefficient
L	Length of pipe or channel
L	Length through which liquid is lifted in a gas lift pup
L_D	Pipe length needed for slugs to develop
L_S	Slug length
L_U	Length of a region of unstable stratified flow
l	Length in the flow direction
ℓ	Mixing length; characteristic viscous length
M	Molecular weight
M_{LG}	Mass transfer rate per unit area from the liquid to the gas
M_{GL}	Mass transfer rate per unit area from the gas to the liquid
m	Average height of the liquid layer around the circumference
m_P	Mass of a particle
m^+	Dimensionless height based on the friction velocity and the viscosity
m_G^+	Ratio of the film height to a gas-phase length scale
m_c^+	Dimensionless film height, based on v_c^* and the viscosity of the liquid
N	Local rate of mass transfer
n	Frequency, cycles per second
P	Perimeter
P	Potential energy associated with one wavelength
P_G	Length of the pipe perimeter in contact with the gas
P_L	Length of the pipe perimeter in contact with the liquid
P_S	Pressure imposed on a solid or liquid surface by a flowing fluid
$\tilde{P}$	Pressure made dimensionless with $\rho_L g d_t$
p	Pressure
p_a	Ambient pressure
p_G	Pressure along a sinusoidal wavy interface $= p_{GR} + i p_{Gi}$

p_{GR}	Pressure component which has a minimum at the wave crest
p_{GI}	Pressure component which is in phase with the wave slope
p_{Gi}	Gas pressure at the interface
p_{Li}	Liquid pressure at the interface
Q_{A}	Volumetric rate at which a liquid layer is atomizing
Q_{G}	Volumetric flow of gas
Q_{L}	Volumetric flow of the liquid
Q_{sh}	Volumetric flow at which slugs shed liquid
q	Velocity of the fastest-moving particle on the interface of a wave
q	Magnitude of the velocity
q	Volumetric flow per unit breadth $\approx hu_{\mathrm{a}}$
q^2	Equal to $u_x^2 + u_y^2 + u_z^2$
$\dot{q}$	Time rate at which heat is added to a control volume
q_{W}	Rate of heat addition at a wall per unit area
$\dot{q}_{\mathrm{rev}}$	Heat transfer rate if changes are occurring reversibly
q_{WL}	Heat transfer rate per unit area from a wall to a liquid
q_{WG}	Heat transfer rate per unit area from a wall to a gas
q_{GL}	Heat transfer per unit area from a gas to a liquid
q_{LG}	Heat transfer per unit area from a liquid to a gas
R	Wave resistance defined by (4.116) and (4.117)
R	Molar gas constant
R	Radius of curvature of a surface
R_{D}	Mass rate of deposition of particles per unit area
R_{A}	Mass rate of atomization of the wall layer per unit area
$R_{1,2}$	Principal radii of curvature of a surface
R_{ij}	The Reynolds stress divided by the density
R_i^{path}	Correlation coefficient of fluid velocity fluctuations seen by a particle as it moves around in a turbulent field
R_i^{L}	Lagrangian correlation coefficient for turbulent fluid particles
$R_{\mathrm{P}i}^{\mathrm{L}}$	Lagrangian correlation coefficient for particles in a turbulent field
R^{E}	Eulerian correlation coefficient
r_{t}	Radius of a tube or a pipe
r_{P}	Particle or bubble radius
r_{ij}	Wave-induced variation of the R_{ij}
S	Slip ratio, equal to gas velocity divided by the liquid velocity
S	Projected area of a particle
S_{i}	Length of the interface in an idealized stratified flow
s	Entropy
s	Sheltering coefficient, defined by Jeffreys
T	Temperature
T	Kinetic energy in one wavelength
T_{S}	Tangential stress imposed on the interface by a flowing fluid
t	Time

t_C	Time constant reflecting the rate of growth of waves
U_P	Particle velocity
U_A	Velocity of atomizing drops in the flow direction
U_D	Velocity of depositing drops in the flow direction
U_B	Velocity of a bubble in a stationary liquid; bulk velocity
$U_{S\infty}$	Rise velocity of a bubble in an infinite stagnant liquid
U_1	Convection velocity of a homogeneous, isotropic field
U_C	Settling velocity of a suspension of particles
$\vec{U}_S$	Relative velocity between a particle (or a bubble) and a fluid
u	One-dimensional velocity
$\vec{u}$	Fluid velocity vector
u_c	Fluid velocity at the center of a pipe or channel
u_t^t	Turbulent velocity component
u^+	Velocity made dimensionless using the friction velocity
u_n	Component of the velocity normal to a surface
u_t	Component of the velocity tangent to a surface
u_H	Velocity of a homogeneous mixture, defined by (1.48)
u_L	Velocity of the liquid
u_{L1}	Velocity of the stratified layer in front of a slug
u_{L3}	Velocity of the liquid in the body of a slug
u_{slug}	Velocity of a slug
u_G	Gas velocity
u_{Gc}	Critical gas velocity for the initiation of atomization
u_{GM}	Maximum value of u_G
u_m	Mixture velocity = $u_{GS} + u_{LS}$
u_a	Spatially averaged liquid velocity
u_S	Liquid velocity at the interface
u_{ga}	Spatially averaged gas velocity
u_{GSt}	Critical superficial gas velocity at which Kelvin–Helmholtz waves appear
u_{GS}	Superficial gas velocity = volumetric flow of gas divided by A
u_{LS}	Superficial liquid velocity = volumetric flow of liquid divided by A
u_o	Rise velocity of bubbles
$u_{G\infty}$	Rise velocity of single bubbles in infinite media
V	Volume
V_D	Mean velocity of particles striking a wall
V_{DL}	Drift velocity associated with a lift force
V_{Drift}	Drift velocity, defined by (8.29)
$\overline{V}_D$	Mean velocity with which particles are depositing
V_W	Average velocity with which particles strike a boundary
V_P	Volume of a particle or a bubble
V_T	Terminal free-fall velocity
$\overline{V}_R$	Average velocity of entrained particle in the radial direction
V_i^0	Velocity with which particles enter the field

V_{tp}	Turbophoretic velocity
V^*_{GS}	Dimensionless group defining the initiation of annular flow (11.1)
v	Specific volume
v_G	Specific volume of a gas
v_L	Specific volume of a liquid
v^*	Friction velocity, using the shear stress at the wall or the interface
v^*_c	Critical frictional velocity
v^*_G	Friction velocity based on gas density and stress at the interface
v^*_L	Friction velocity based on the liquid density and the stress at the wall
v^*_c	Friction velocity based on liquid density and the characteristic stress
$\vec{v}$	Velocity of a fluid particle
$\vec{v}_P$	Velocity of an entrained particle or bubble
v^t_{Pi}	Turbulent velocity fluctuation
W	Weight rate of flow
W_L	Weight flow rate of liquid
W_{LP}	Mass flow of liquid in the "pool" at the bottom of the pipe
W_G	Weight flow rate of gas
W_W	Mass flow of liquid on the wall
W_{LE}	Mass flow of entrained drops
W_{LF}	Mass flow rate in the wall layer
W_{LFC}	Critical flow below which atomization does not occur
$\dot{w}$	Rate at which fluid in a volume is doing work on surroundings
w	Width of a control volume
$\overline{X^2_P}$	Mean square displacement of particles in a turbulent field
x	Mixture quality, mass fraction that is a gas or vapor
x	Coordinate in the flow direction in a Cartesian coordinate system
y	Coordinate that is perpendicular to a wall
y_{fl}	Location at which particles start a free-flight to the wall
y^+	Distance from the wall made dimensionless with the friction velocity
y_0	Average location of the interface
z	Coordinate in the direction of flow in a pipe flow
z	Distance from reference plane

Greek symbols

α	Volume fraction of gas in a flowing mixture
α	Ratio of the fastest moving particle at the interface to wave velocity
α_W	Ratio of wave height to what would be observed on a deep liquid
α	Parameter representing shape of velocity profile
α	Equals $\left(\tau_i^{\text{path}}\right)^{-1}$
α_P	Volume fraction of particles or bubbles
am^+	Dimensionless quantity defined by (3.28)
β	Reciprocal of the inertial time constant of an entrained particle

$\tilde{\beta}$	Reciprocal time constant for a particle in a liquid (10.50)
Γ	Gamma function
Γ	Volumetric flow per unit length in the spanwise direction
Γ_c	Critical film flow below which atomization does not occur
Γ_c^*	Modified critical film flow (12.74)
Γ_S	A shape factor for the velocity profile (7.6)
γ	Ratio of the heat capacities at constant pressure and constant volume
δ	Thickness of a viscous boundary layer
δ_{ij}	Delta function
ε	Rate of dissipation of mechanical energy per unit volume
ε	Lagrangian turbulent diffusivity of fluid particles
ε	Ratio of the fastest-moving particle in the interface to the wave velocity
ε_h	Diffusivity representing turbulent mixing in the wall layer
ε_L	Liquid holdup $= (1 - \alpha)$
ε_P	Lagrangian turbulent diffusivity of entrained particles
ε_P^E	Eulerian diffusivity of entrained particles
η	Displacement of the interface from its average location
θ	Inclination angle to the horizontal
θ	Phase for an oscillating quantity
θ	Angular location on the wall of a pipe, where $\theta = 0$ can be the top or bottom of the pipe
θ_g	Orientation angle of flow to the gravitational vector (Figure 1.2)
Λ	Lagrangian length scale
λ	Wavelength
λ_m	Wavelength which is growing the fastest
λ^T	Lagrangian micro-time-scale
μ	Viscosity
μ_G	Gas viscosity
μ_L	Liquid viscosity
μ_P	Viscosity of the fluid inside a drop or bubble
μ^t	Turbulent viscosity
ρ	Density
ρ_L	Liquid density
ρ_G	Gas density
ρ_P	Particle density
ρ_f	Fluid density
ρ_S	Density of a dissolving solid surface
ρ_H	Density of a homogeneous mixture
ρ_G^*	$\rho_G \coth(kh_G)$
ρ_L^*	$\rho_L \coth(kh_L)$
σ	Surface tension
σ_P	Root-mean-square of the turbulent velocity fluctuations of particles
σ	Root-mean square of a fluctuating quantity

τ	Magnitude of a shear stress
τ_c	Time constant characterizing the rate of growth of waves
τ_c	Characteristic stress in a liquid layer = $(2/3)\tau_W + (1/3)\tau_i$
τ_P	Inertial time constant of a particle
τ_P	Effective stress due to atomization and deposition
τ_i^{path}	Time constant characterizing R_i^{path}
τ^L	Lagrangian time constant
τ^E	Eulerian time-scale
τ_{PB}	Volume-average inertial time constant
τ_{PS}	Time constant for a particle obeying Stokes law
τ_W	Shear stress at a wall
τ_{WG}	Resisting stress at the wall on a gas
τ_{WL}	Resisting stress at the wall on a liquid
τ_i	Shear stress at an interface
τ_{ij}	Component of a stress tensor
τ_{ij}^t	Component of a turbulent stress tensor
τ_{DA}	Net momentum flux due to atomization and deposition
υ	Kinematic viscosity
υ_L	Kinematic viscosity of the liquid
υ_G	Kinematic viscosity of the gas
υ^t	Turbulent kinematic viscosity
ϕ	Potential function describes velocity for an irrotational field
ϕ_G^2	Dimensionless frictional pressure gradient defined by (1.100)
ϕ_L^2	Dimensionless frictional pressure gradient defined by (1.101)
ω	Circular frequency = kc
$\vec{\omega}_P$	Angular rotation of a particle or bubble
ψ	Stream function defined by (6.20)

Dimensionless groups

$\tilde{\beta}$	$= 3C_D\rho_f\lvert\vec{u} - \vec{v}_P\rvert/4r_P(2\rho_P + \rho_f)$
Bo	Bond number $= \dfrac{d_h}{[\sigma/(\rho_L - \rho_G)g]^{1/2}}$
F	$= \dfrac{\gamma(\text{Re}_{LF})}{\text{Re}_G^{0.9}}\dfrac{\nu_L}{\nu_G}\sqrt{\dfrac{\rho_L}{\rho_G}}$ is a parameter in (3.41)
F_H	F for a horizontal pipe (12.48)
Fr	Froude number = $u_{GS}/(gd_t)^{1/2}$
Ma	Mach number = ratio of the fluid velocity to the velocity of sound
Re	Reynolds number
Re_G	Gas-phase Reynolds number = $d_t W_G/A_t\mu_G$ for a pipe
Re_{LF}	Film Reynolds number = $4\Gamma/\upsilon_L$
Re_{L0}	Liquid Reynolds number based on the velocity at the interface

$\mathrm{Re_P}$	Particle Reynolds number $= d_P \rho U_S/\mu$
Sc	Schmidt number
θ	Number reflecting the effect of viscosity on wave growth
We	Weber number
X	Martinelli parameter

Other symbols

$\langle\ \rangle$	Spatial average, ensemble average or phase average
$(\ )_i$	Quantity evaluated at the interface
$\|\ \|$	Signifies an absolute value
$\dfrac{D}{Dt}$	Substantial derivative. Time change seen in a framework of a moving fluid particle (defined by equation (4.13))
$\overline{N}$	Overbar indicates a time-average
N'	Prime indicates a fluctuating quantity
N^t	A turbulent fluctuation
$\hat{N}$	Indicates complex amplitude of a fluctuating quantity induced by sinusoidal waves
$\tilde{N}$	Tilde indicates $\hat{N}$ divided by the wave amplitude, a
$\tilde{\tilde{N}}$	Difference between the phase-average and the time-average
$\hat{N}$	Defined by (3.20), (3.21)
$\left(\dfrac{dP}{dz}\right)_F$	Frictional contribution to the pressure gradient
$\left(\dfrac{dP}{dz}\right)_{GS}$	Frictional pressure gradient for gas flowing along a pipe
$\left(\dfrac{dP}{dz}\right)_{LS}$	Frictional pressure gradient for liquid flowing along a pipe

1 One-dimensional analysis

1.1 Introduction

The "simplest" models for gas–liquid flow systems are ones for which the velocity is uniform over a cross-section and unidirectional. This includes flows in a long straight pipe and steady flows in a nozzle.

A treatment of pipe flow with a constant cross-section is initiated by reviewing analyses of incompressible and compressible single-phase flows. A simple way to use these results is to describe gas–liquid flows with a homogeneous model that assumes the phases are uniformly distributed, that there is no slip between the phases and that the phases are in thermodynamic equilibrium. The volume fraction of the gas, α, is then directly related to the relative mass flows of the phases. However, the assumption of no slip, $S = 1$, can introduce considerable error. This has prompted a consideration of a separated flow model, where uniform flows of gas and liquid are pictured as moving parallel to one another with different velocities and to be in thermodynamic equilibrium.

The two-fluid model develops equations for the interaction of two interpenetrating streams. It does not require the specification of S or the assumption of equilibrium between the phases. However, it introduces several new variables.

An interesting feature of the single-phase analysis of a compressible fluid is the existence of the choking phenomenon whereby there is a maximum flow which can be realized for a system in which the pressure at the pipe inlet and the pressure in the receiver are controlled. The homogeneous model predicts much smaller choking velocities for gas–liquid flows than would exist for gas flowing alone.

The separated flow model requires equations for the stress at the wall, τ_W, and α (or the slip ratio). The homogeneous model requires only an equation for τ_W. The specification of these quantities is a continuing interest and is a focus in future chapters. A starting point for this pursuit is the widely used Lockhart–Martinelli analysis discussed in this chapter.

Steady flow in a nozzle is different from flow in a straight pipe in that changes in inertia override the importance of wall resistance and of gravity. Equations that use the separated flow model are developed in this chapter for gas–liquid flows in a nozzle.

The materials in Section 1.6 on flow in a nozzle and on the two-fluid model can be ignored in a first reading since they are not needed to follow the main narrative of this book.

1.2 Single-phase flow

1.2.1 Flow variables

When a single phase is flowing, the field is described by the variables pressure, p, velocity, $\vec{u}$, density, ρ and temperature, T. Four equations are needed: conservation of momentum, conservation of energy, conservation of mass, and a relation among state variables. Examples of the last are the perfect gas law and tables of thermodynamic properties.

1.2.2 Momentum and energy theorem

For flow systems, it is convenient to formulate Newton's second law of motion as a momentum theorem, which focuses on a fixed volume in space rather than on particles moving through a field. A sketch of an arbitrary volume is shown in Figure 1.1. Note that, on the boundary, the velocity vector, $\vec{u}$, can be represented by components perpendicular, u_{n}, and tangent, u_{t}, to the boundary. The theorem is stated as follows: "The time rate of change of momentum in a *fixed* volume in space plus the net flow of momentum out of the volume equals the net force acting on the volume."

$$\int \frac{\partial(\rho\vec{u})}{\partial t} dV + \int \rho u_{\mathrm{n}} \vec{u}\, dA = \vec{F} \tag{1.1}$$

Thermodynamic variables are defined for systems which are at equilibrium. The application of conservation of energy and the equation of state to a flowing system, therefore, usually involves the assumption that the adjustment of molecular properties to a change of the environment occurs much more rapidly than the change of the flow field. (Examples where this assumption might not be valid are flow through a shock wave or the stagnation of flow at a very small impact tube.) The energy of a flowing fluid is defined as

$$e = i + \frac{|u|^2}{2} \tag{1.2}$$

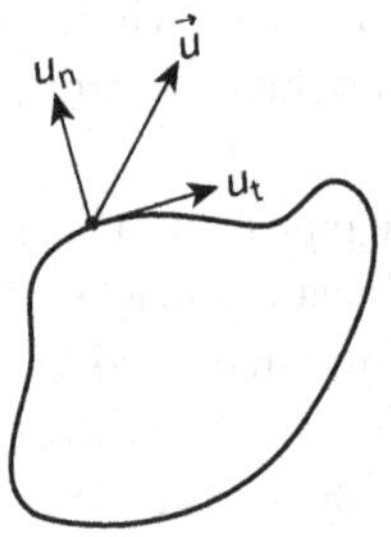

Figure 1.1 An arbitrary volume in space.

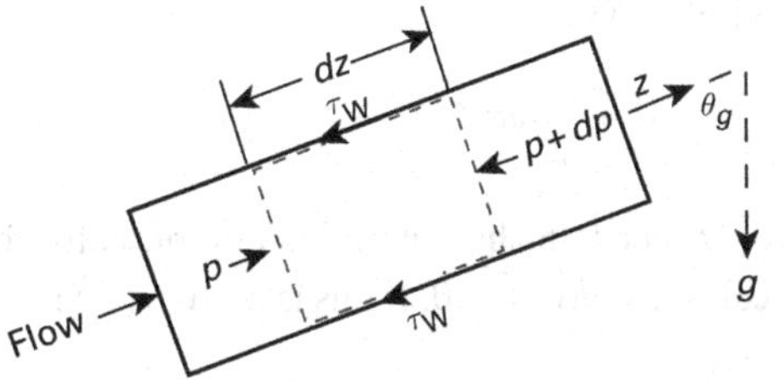

Figure 1.2 Control volume in a pipe.

where i is the internal energy per unit mass and $|u|^2/2$ is the kinetic energy per unit mass.

The first law of thermodynamics is formulated for a flow system as the energy theorem: "The time rate of change of energy in a *fixed volume* in space plus the net flow of energy out of that volume equals the rate at which heat is added, $\dot{q}$, minus the rate at which the fluid in the volume is doing work on the surroundings, $\dot{w}$."

$$\int \frac{\partial(\rho e)}{\partial t} dV + \int \rho u_n e \, dA = \dot{q} - \dot{w} \tag{1.3}$$

where ρe is the energy per unit volume.

1.2.3 Steady flow in a duct of constant area

The equations defining the steady flow in a pipe of constant area, A, can be derived by considering the fixed control volume shown in Figure 1.2. The pipe is inclined at an angle θ_g to the direction of the gravitational vector. The fluid is flowing in the z-direction with an average velocity, u. Equations are to be developed describing the changes of u, p, ρ, and T with z.

Under steady flow conditions, the mass flow, ρuA, does not vary along the pipe. The equation for conservation of mass is

$$AG = u\rho A = \text{constant} \tag{1.4}$$

where term G is called the mass velocity.

The flux of momentum along the pipe is given as GAu. Since the flow is steady, the time rate change in the control volume, indicated by the dashed lines in Figure 1.2, is zero so the momentum theorem states that the net flow of momentum out of a fixed volume equals the net force. The flows of momentum into and out of the control volume are given by $\rho u^2 A$ and $\rho u^2 A + d(\rho u^2 A)$, so the net flow of momentum out is $d(\rho u^2 A)$.

The forces due to pressure acting on the front and back faces are pA and $-(p + dp)A$, so the net force is $-(dp)A$. The force of gravity is $-\rho g \cos\theta_g Adz$. The wall resists the flow with a stress, τ_W, acting on the side walls of the control volume and the net force due to this stress is $-\tau_W Pdz$ where P is the perimeter of the duct. The momentum theorem gives

$$d(AGu) = -(dp)A - \tau_W Pdz - \rho Ag \cos\theta_g dz \tag{1.5}$$

The wall shear stress is usually represented as

$$\tau_W = \frac{1}{2}\rho u^2 f \tag{1.6}$$

where the Fanning friction factor, f, needs to be determined empirically for turbulent flows. Conservation of mass indicates that $d(GA)=0$. If this is used in (1.5), the following is obtained after substituting (1.6).

$$G\,du = -dp - \frac{fPG^2}{2A\rho}dz - \rho g \cos\theta_g dz \tag{1.7}$$

From conservation of mass, if A is constant, $du = G\,d(1/\rho)$ so that

$$G^2 d(1/\rho) = -dp - \frac{fPG^2}{2A\rho}dz - \rho g \cos\theta_g\, dz \tag{1.8}$$

This shows that pressure changes result from balancing the frictional drag, the gravitational force and the acceleration of the fluid.

Equation (1.8) can be solved directly for p if ρ is constant, that is, if there is no change in velocity in the direction of flow. If ρ is not constant the relation $\rho\ (p)$ is needed. This can be obtained from the energy theorem and the equation of state. The rate of energy addition to the control volume can be calculated from the rate of heat addition at the wall per unit area as $q_W P\,dz$. The rate at which the fluid in the control volume does work on the surroundings consists of two terms. The rate at which the fluid at z works to get into the control volume is puA. The rate at which the fluid at $z + dz$ does work to get out of the control volume is $(puA + d(puA))$. The net rate of work on the surroundings due to these effects is $d(pua)$. The fluid in the control volume works against gravity to lift the fluid through a distance $\cos\theta_g\, dz$. This contributes a rate of work of $GAg\cos\theta_g\, dz$.

Thus, for the steady flow depicted in Figure 1.2, the energy theorem gives

$$AG\,di + AG\,d\frac{|u^2|}{2} = q_W P\,dz - d(pAu) - GAg\ \cos\theta_g\, dz \tag{1.9}$$

The enthalpy per unit mass is given as

$$h = i + \frac{p}{\rho} \tag{1.10}$$

where $1/\rho$ is the volume per unit mass. The kinetic energy per unit mass is $|u^2|/2$. If (1.10) and $u = G/\rho$ are substituted into (1.9) and A is constant,

$$dh = \frac{1}{2}G^2 d\left(\frac{1}{\rho^2}\right) + g\cos\theta_g dz = \frac{q_W P}{AG}dz \tag{1.11}$$

Equation (1.11) defines the change of enthalpy. For a given fluid, the thermodynamic state is defined by any three state variables, so

$$h = f(p,\rho) \tag{1.12}$$

Therefore, a table or a diagram representing the thermodynamic variables provides a relationship between ρ and p if $h(z)$ is known. Equation (1.11) gives a relation between p and ρ along the z-axis. This can be used in (1.8) to eliminate either $p(z)$ or $\rho(z)$.

1.2.4 Choking

Consider a length of pipe for which the fluid enters at a pressure p_1 and discharges into a chamber whose pressure, p_2, is changed. Equation (1.8) can be solved to obtain G for a given p_2/p_1. The solution shows that G initially increases as p_2/p_1 decreases. Eventually it gives a maximum and the impossible result that G decreases with decreasing p_2/p_1. The maximum in G represents a choking condition for which further decreases of the pressure in the discharge chamber have no effect on the mass flow and the flow along the pipe remains unchanged. This can be understood when it is realized that the equations predict that a decrease in p_2 is accompanied by an increase in the velocity and a decrease in the density. These have opposite effects on the mass velocity since $G = \rho u$. Thus, for p_2/p_1 less than the critical value the equations predict that decreases in ρ offset the increases in u.

At choking, the fluid velocity equals the velocity with which small disturbances (sound waves) propagate in a fluid. Thus, information about the change in the pressure at the outlet cannot be transmitted upstream and the behavior in the pipeline is not affected by what is happening in the receiving chamber. The fluid discharging from the pipeline expands supersonically and forms shock waves in the receiver.

This interpretation is illustrated, in a direct way, by considering the momentum balance, equation (1.8). The term on the left side of (1.8) may be written as

$$G^2 d\left(\frac{1}{\rho}\right) = -\frac{G^2}{\rho^2} d\rho = -u^2 \frac{d\rho}{dp} dp \tag{1.13}$$

where $dp/d\rho$ is the change of pressure with density along the pipe. Substituting (1.13) into (1.8) gives

$$\frac{dp}{dz}\left(1 - \frac{u^2}{c^2}\right) = -\frac{fG^2 P}{2A\rho} - \rho g \cos\theta_g \tag{1.14}$$

where the velocity of sound, c, is given by

$$c^2 = \frac{dp}{d\rho} \tag{1.15}$$

It is noted that the coefficient of dp/dz changes sign when the Mach number, Ma $= u/c$, assumes values greater than unity. This suggests that the frictional pressure gradient changes from a negative to a positive value. A positive frictional pressure gradient is an impossibility so Ma cannot change from a value less than unity to a value greater than unity in a pipeline of constant area. The flow chokes when $u = c$ at the outlet.

The entropy of a fluid can be defined by the following thermodynamic relation

$$Tds = dh - \frac{dp}{\rho} \tag{1.16}$$

Thus, the change of entropy along the pipeline can be calculated if $h(z)$, $p(z)$ and $\rho(z)$ are known. This calculation shows that, for an adiabatic flow, s increases with z because of the irreversible effects associated with wall friction. At the choking condition $ds = 0$ so, from (1.15), the disturbance velocity is given as

$$c^2 = \left(\frac{\partial p}{\partial \rho}\right)_s \tag{1.17}$$

This is a thermodynamic quantity which is the classical definition of the velocity of sound. When the flow is not adiabatic, $q_W \neq 0$, this is not the case. For example, if the frictional heating is balanced by heat losses, the temperature might be kept approximately constant so that

$$c^2 = \left(\frac{\partial p}{\partial \rho}\right)_T \tag{1.18}$$

1.2.5 Flow of an ideal gas in a pipe of constant area

An ideal gas is defined with the equations

$$\frac{p}{\rho} = \frac{R}{M}T \tag{1.19}$$

$$di = C_V dT \tag{1.20}$$

$$dh = C_P dT \tag{1.21}$$

where R is the molar gas constant, M is the molecular weight, C_V is the heat capacity at constant volume and C_P is the heat capacity at constant pressure. Thus, for an ideal gas

$$\left(\frac{\partial p}{\partial \rho}\right)_s = \gamma\frac{p}{\rho} \tag{1.22}$$

$$\left(\frac{\partial p}{\partial \rho}\right)_T = \frac{R}{M}T \tag{1.23}$$

where $\gamma = C_P/C_V$ and

$$\frac{R}{M} = C_P - C_V \tag{1.24}$$

The application of the equations developed in Section 1.2.3 will be illustrated by considering the adiabatic flow of an ideal gas in a pipe of constant area. Since a gas is being considered, gravitational effects can be neglected and the momentum balance equation is written as

$$-\frac{\rho}{G^2}dp+\frac{d\rho}{\rho}-\frac{2f}{d_t}dz=0 \tag{1.25}$$

where d_t is the pipe diameter. This can be integrated if the dependency of p on ρ is known. Conservation of energy for $q_W = 0$ gives

$$dh+\frac{1}{2}Gd\left(\frac{1}{\rho^2}\right)=0 \tag{1.26}$$

Since conservation of mass gives G = constant, the equation can be integrated to give

$$(h-h_1)=\frac{G^2}{2\rho_1{}^2}-\frac{G^2}{2\rho^2} \tag{1.27}$$

where subscript 1 indicates inlet conditions. From (1.21),

$$(h-h_1)=C_P(T-T_1) \tag{1.28}$$

If (1.19) and (1.24) are used in (1.28)

$$h-h_1=\frac{R}{M}\left(\frac{\gamma}{\gamma-1}\right)(T-T_1)=\left(\frac{\gamma}{\gamma-1}\right)\left(\frac{p}{\rho}-\frac{p_1}{\rho_1}\right) \tag{1.29}$$

The following relation between p and ρ is obtained if (1.27) is substituted into (1.29):

$$\frac{p}{p_1}=-\frac{(\gamma-1)}{\gamma}\frac{G^2}{2p_1\rho_1}\frac{\rho_1}{\rho}+\frac{(\gamma-1)}{\gamma}\frac{G^2}{2p_1\rho_1}\frac{\rho}{\rho_1}+\frac{\rho}{\rho_1} \tag{1.30}$$

Equation (1.30) is used to eliminate p from (1.25). The following relation for ρ is obtained if f is assumed to be constant.

$$\frac{(\gamma+1)}{\gamma}\frac{1}{2}\ln\frac{\rho}{\rho_1}-\frac{(\gamma-1)}{\gamma}\frac{1}{4}\left(\frac{\rho^2}{\rho_1^2}-1\right)-\frac{p_1\rho_1}{2G^2}\left(\frac{\rho^2}{\rho_1^2}-1\right)-\frac{2f(z-z_1)}{d_t}=0 \tag{1.31}$$

If the density variation is calculated with (1.31), the pressure variation is obtained from (1.30) and the velocity can be obtained from conservation of mass

$$\frac{u}{u_1}=\frac{\rho_1}{\rho} \tag{1.32}$$

The temperature is obtained from (1.19).

$$\frac{T}{T_1}=\frac{p}{p_1}\frac{\rho_1}{\rho} \tag{1.33}$$

From (1.16), (1.11), (1.19) the following equation is obtained for the entropy

$$ds=\frac{R}{Mp}\left(\frac{G^2}{\rho^2}d\rho-dp\right) \tag{1.34}$$

This can be integrated to obtain $s-s_1$, using the equations for p and ρ developed above.

If an isothermal flow is assumed

$$\frac{p}{p_1} = \frac{\rho}{\rho_1} \tag{1.35}$$

This condition will hold for a high molecular weight gas for which C_V is quite large and $\gamma = (C_V + R)/C_V \approx 1$. If $\gamma = 1$ is substituted into (1.30), it is seen that (1.35) results. However, in general, isothermal conditions can be maintained only if heat is exchanged with the surroundings; that is, the flow would not be adiabatic.

Equation (1.25) can be integrated directly if (1.35) is substituted for ρ

$$-\frac{p_1\rho_1}{2G^2}\left[\left(\frac{p}{p_1}\right)^2 - 1\right] + \ln\frac{p}{p_1} - \frac{2f(z - z_1)}{d_t} = 0 \tag{1.36}$$

For an isothermal flow of an ideal gas, the enthalpy h is constant so the energy equation gives

$$\frac{q_W P}{AG}(z - z_1) = \frac{G^2}{2\rho^2} - \frac{G^2}{2{\rho_1}^2} \tag{1.37}$$

The heat that is added to the flow to maintain a constant temperature must increase with increases in kinetic energy. Since $dh = 0$, equation (1.16), describing the change in entropy, gives

$$ds = \frac{1}{T}\frac{dp}{\rho} \tag{1.38}$$

This can be integrated, using (1.19), to yield

$$s - s_1 = -\frac{R}{M}\ln\frac{p}{p_1} \tag{1.39}$$

The entropy will increase as long as p decreases. It need not attain a maximum at the value of p/p_1 for choking to occur. This difference from the adiabatic case arises since the system cannot be considered as isolated so that a reversible process need not correspond to one for which the entropy is constant.

1.2.6 Mechanical energy balance

The first law of thermodynamics defines a change in internal energy as

$$di = \frac{\dot{q}_{rev}P}{GA}dz - pd\left(\frac{1}{\rho}\right) \tag{1.40}$$

where $\dot{q}_{rev}$ is the heat needed for a reversible operation. If this is substituted into the energy balance, equation (1.9),

$$\frac{dp}{\rho} + d\left(\frac{u^2}{2}\right) + g\cos\theta_g dz = (q_W - \dot{q}_{rev})\frac{P}{AG}dz \tag{1.41}$$

Define F as the rate of mechanical energy loss per unit mass of fluid, where

$$dF = (\dot{q}_{\mathrm{rev}} - q_{\mathrm{W}}) \tag{1.42}$$

Thus (1.41) can be rewritten as

$$\frac{dp}{\rho} + d\left(\frac{u^2}{2}\right) + g\cos\theta_g dz = -dF \tag{1.43}$$

A comparison of (1.43) with the momentum balance equation gives

$$\frac{dF}{dz} = \frac{\tau_{\mathrm{W}} P}{\rho A} = \frac{4}{d_{\mathrm{t}}}\frac{\tau_{\mathrm{W}}}{\rho} \tag{1.44}$$

1.3 The homogeneous model for gas–liquid or vapor–liquid flow

1.3.1 Basic equations

The simplest approach to flows involving a gas and a liquid is to treat the mixture the same as a single phase. This requires assuming (1) zero slip between the phases, (2) uniform flow and (3) equilibrium between the phases (one can use thermodynamic tables when the vapor and liquid are in contact). Define the quality, x, as the mass fraction of the flowing mixture that is a gas or vapor and a void fraction, α, as the fraction of the cross-section that is occupied by the gas. For flows such as air–water, where phase changes are not occurring, x does not vary along the pipeline. For flows with a fluid pair such as steam–water, vaporization can occur, so x can change.

Define a mixture density as the ratio of the mass flow to the volume flow:

$$\rho_{\mathrm{H}} = \frac{GA}{AxGv_{\mathrm{G}} + (1-x)GAv_{\mathrm{L}}} \tag{1.45}$$

where

$$v_{\mathrm{G}} = \frac{1}{\rho_{\mathrm{G}}} \quad v_{\mathrm{L}} = \frac{1}{\rho_{\mathrm{L}}} \tag{1.46}$$

Thus

$$\frac{1}{\rho_{\mathrm{H}}} = xv_{\mathrm{G}} + (1-x)v_{\mathrm{L}} = \frac{x}{\rho_{\mathrm{G}}} + \frac{(1-x)}{\rho_{\mathrm{L}}} \tag{1.47}$$

A mixture velocity is defined as

$$u_{\mathrm{H}} = \frac{G}{\rho_{\mathrm{H}}} \tag{1.48}$$

For the condition $u_{\mathrm{G}} = u_{\mathrm{L}}$

$$\alpha = \frac{Q_{\mathrm{G}}}{Q_{\mathrm{G}} + Q_{\mathrm{L}}} = \frac{xGAv_{\mathrm{G}}}{xGAv_{\mathrm{G}} + (1-x)GAv_{\mathrm{L}}} = x\frac{v_{\mathrm{G}}}{v_{\mathrm{H}}} = x\frac{\rho_{\mathrm{H}}}{\rho_{\mathrm{G}}} \tag{1.49}$$

Also

$$\frac{\alpha}{1-\alpha} = \frac{x}{(1-x)} \frac{\rho_L}{\rho_G} \tag{1.50}$$

The momentum balance, equation (1.8), is rewritten as

$$-\frac{\rho_H}{G^2}dp + \frac{d\rho_H}{\rho_H} - \frac{\rho_H^2}{G^2}g\cos\theta_g dz - \frac{\tau_W \rho_H P}{G^2 A}dz = 0 \tag{1.51a}$$

or, alternatively, as

$$-\frac{dp}{dz} = G^2\frac{d\rho_H^{-1}}{dz} + \rho_H g\cos\theta_g + \frac{\tau_W P}{A} \tag{1.51b}$$

The conservation of energy equation (1.11) is given as

$$dh_H + \frac{1}{2}G^2 d\left(\frac{1}{\rho_H^2}\right) + g\cos\theta_g dz = \frac{q_W P}{AG}dz \tag{1.52}$$

where

$$h_H = xh^G + (1-x)h^L \tag{1.53}$$

The enthalpy of the gas, h^G, and the enthalpy of the liquid, h^L, are obtained from thermodynamic correlations.

For two-component systems such as air and water, the quality, x, does not vary along the pipeline and is fixed by the inlet conditions. The enthalpies, h^L and h^G, are fixed for a given p and ρ_H (which relate to ρ_G and ρ_L through equation (1.47)). For a single-component system, such as steam–water, the quality, x, and the densities, ρ_L and ρ_G, are fixed for a given pressure and enthalpy, if equilibrium is assumed between the phases. Thus, (1.52) can be used to provide a relation between ρ_H and p. This can be used to integrate (1.9). Equation (1.52) and thermodynamic data can be used to calculate the variation of p, ρ_H, G, x with distance along the pipe.

1.3.2 Choking

The momentum balance for a homogeneous flow can be rewritten as

$$-dp\left[1 - \frac{G^2(d\rho_H/dp)}{\rho_H^2}\right] - \frac{\rho_H^2}{G^2}g\cos\theta_g\, dz - \frac{\tau_W P}{A}dz = 0 \tag{1.54}$$

Choking occurs when the term in the brackets changes sign

$$\frac{1}{G_C^2} = -\frac{dv_H}{dp} \tag{1.55}$$

From (1.15), define

$$\frac{1}{c^2} = \frac{d\rho_H}{dp} = \frac{dv_H^{-1}}{dp} = -\rho_H^2\frac{dv_H}{dp} \tag{1.56}$$

Substitute (1.48) and (1.56) into (1.54) to get

$$-dp\left[1-\frac{u_{\mathrm{H}}^{2}}{c^{2}}\right]=\frac{\tau_{\mathrm{W}}P}{A}dz \tag{1.57}$$

Choking is predicted when $u_{\mathrm{H}}^{2}/c^{2}=1$.

Using (1.47) for v_{H}, one obtains

$$\frac{dv_{\mathrm{H}}}{dp}=x\left(\frac{dv_{\mathrm{G}}}{dp}\right)+(1-x)\frac{dv_{\mathrm{L}}}{dp}+(v_{\mathrm{G}}-v_{\mathrm{L}})\frac{dx}{dp} \tag{1.58}$$

From (1.56) and (1.58)

$$\frac{1}{\rho_{\mathrm{H}}^{2}c^{2}}=\frac{x}{\rho_{\mathrm{G}}^{2}c_{\mathrm{G}}^{2}}+\frac{(1-x)}{\rho_{\mathrm{L}}^{2}c_{\mathrm{L}}^{2}}+(v_{\mathrm{G}}-v_{\mathrm{L}})\frac{dx}{dp} \tag{1.59}$$

where c_{G} and c_{L} are the sound velocities in the gas and in the liquid

$$c_{\mathrm{G}}^{2}=-\frac{1}{\rho_{\mathrm{G}}^{2}}\left(\frac{dp}{dv_{\mathrm{G}}}\right) \tag{1.60}$$

$$c_{\mathrm{L}}^{2}=-\frac{1}{\rho_{\mathrm{L}}^{2}}\left(\frac{dp}{dv_{\mathrm{L}}}\right) \tag{1.61}$$

Since, for a homogeneous flow, $x=\alpha$

$$c^{-2}=[\alpha\rho_{\mathrm{H}}+(1-\alpha)\rho_{\mathrm{L}}]\left[\frac{\alpha}{\rho_{\mathrm{G}}c_{\mathrm{G}}^{2}}+\left(\frac{(1-\alpha)}{\rho_{\mathrm{L}}c_{\mathrm{L}}^{2}}\right)+\frac{(v_{\mathrm{G}}-v_{\mathrm{L}})}{v_{\mathrm{H}}}\frac{dx}{dp}\right] \tag{1.62}$$

Take a path of $dx/dp=0$. Note that

$$c^{2}=c_{\mathrm{G}}^{2}\qquad\alpha=1 \tag{1.63}$$

$$c^{2}=c_{\mathrm{L}}^{2}\qquad\alpha=0 \tag{1.64}$$

For $\alpha\neq 1$ and ρ_{L} much greater than ρ_{G}

$$c^{2}=\frac{\rho_{\mathrm{G}}}{\rho_{\mathrm{L}}}\frac{c_{\mathrm{G}}^{2}}{(\alpha)(1-\alpha)} \tag{1.65}$$

This has a minimum at $\alpha=1/2$. The important point to be made is that the homogeneous model predicts smaller sound velocities than would exist in the gas alone. Thus choking could occur at a smaller exit velocity for a gas–liquid mixture than for the pure gas. For example, Wallis (1969) points out that homogeneous theory predicts choking velocities as low as 2.1 m/s for air–water flows. This is to be compared with a value of about 340 m/s in air at room temperature.

1.3.3 Calculation of wall shear stress

The usual practice in making calculations with a homogeneous model is to define a friction factor based on ρ_H.

$$\tau_W = \frac{f}{2}\rho_H u_H^2 \tag{1.66}$$

Single-phase results are used to obtain f. This requires a definition of viscosity which is dependent on the viscosities of both the gas and the liquid.

This approach does not have a sound basis. The use of friction factors defined by (1.66) for flows which have drops entrained in a gas flow can produce predictions of the pressure drop which show an effect of liquid flow which is stronger than observed. Thus, Henstock & Hanratty (1976) defined friction factors for annular flows based on the gas density. Likewise, the prediction of pressure drops in the bubbly flow that constitutes the body of a slug (Fan & Hanratty, 1993) are best represented by using the density of the liquid, rather than ρ_H, in (1.66).

1.3.4 Comments about the reliability of the homogeneous model

There are a number of problems in using the homogeneous model. The assumption of a homogeneous distribution of the phases is sometimes a poor approximation. The use of single-phase relations for τ_W with ρ_H substituted for ρ and a viscosity which is a combination of the liquid and gas viscosities is not substantiated. An appreciable slip can exist between the phases, so the equating of α to x can be incorrect and lead to errors in calculating the contribution of the pressure drop associated with hydrostatic head. Also, the assumption of equilibrium between vapor and liquid phases can be in error.

1.4 Separated flow model for gas–liquid flow

1.4.1 Basic equations

The separated flow model differs from the homogeneous model by recognizing that the velocities of the two phases are usually different. It is a special one-dimensional version of the two-fluid model, which pictures the phases to be two separate streams.

A slip ratio between the phases is defined as

$$S = \frac{u_G}{u_L} \tag{1.67}$$

and the quality of the mixture, x, retains its definition as the ratio of the mass flow of the gas to the mass flow of the mixture. The volume flows of the gas and the liquid are given as Gxv_G and $G(1-x)v_L$ so the velocities are

$$u_G = \frac{GAx}{\rho_G A\alpha} \tag{1.68}$$

$$u_L = \frac{GA(1-x)}{\rho_L A(1-\alpha)} \tag{1.69}$$

where α and $(1-\alpha)$ are the fractions of the cross-section, A, occupied by the gas and the liquid. From (1.67), (1.68) and (1.69) the slip ratio is given as

$$S = \frac{x}{(1-x)}\frac{(1-\alpha)}{\alpha}\frac{\rho_L}{\rho_G} \tag{1.70}$$

Thus, the relation between x and α for a homogeneous model, equation (1.50), is modified as indicated in (1.70).

The momentum flux includes the contributions of both the gas and liquid flows

$$\text{Momentum flux} = GAxu_G + GA(1-x)u_L \tag{1.71}$$

The substitution of (1.68) and (1.69) into (1.71) gives

$$\text{Momentum flux} = GAx\frac{Gx}{\rho_G\alpha} + GA(1-x)\frac{G(1-x)}{\rho_L(1-\alpha)} \tag{1.72}$$

As in Section 1.2.2, the momentum theorem is applied to a differential length along the pipe. The net flow of momentum out of the control volume, $(dz)A$, equals

$$\frac{d}{dz}\left(\frac{G^2Ax^2}{\alpha\rho_G} + \frac{G^2A(1-x)^2}{(1-\alpha)\rho_L}\right)dz$$

The force of gravity on the control volume equals

$$-g\cos\theta_g[\rho_L(1-\alpha) + \rho_G\alpha]A\,dz$$

In a liquid–vapor mixture, an increase in x is associated with an increase in the momentum of the vapor and a decrease in the momentum of the liquid given by

$$AG(u_G - u_L)dx$$

The momentum balance for the separated flow model, therefore, gives

$$d\left[\frac{G^2Ax^2}{\alpha\rho_G} + \frac{G^2A(1-x)^2}{(1-\alpha)\rho_L}\right] = -Adp - g\cos\theta_g[\rho_G\alpha + (1-\alpha)\rho_L]Adz - P\langle\tau_W\rangle dz + AG(u_G - u_L)dx \tag{1.73}$$

This equation states that the change in the momentum flux equals the pressure force, the force of gravity, the force of the wall on the fluid, where $\langle\tau_W\rangle$ is the average wall shear stress around the periphery, the change in the momentum associated with phase change. Unlike the momentum balance for homogeneous flow, (1.51a) and (1.51b), both x and α (or x and S) appear. Since u_G and u_L are given by (1.68) and (1.69), the last term in (1.73) can be subtracted from the momentum flux term to give

$$-\frac{dp}{dz} = \frac{P\langle\tau_W\rangle}{A} + G\left[x\frac{du_G}{dz} + (1-x)\frac{du_L}{dz}\right] + g\cos\theta_g[\rho_G\alpha + (1-\alpha)\rho_L] \quad (1.74)$$

The third term in (1.74) represents a weighted average of the accelerations of the gas and the liquid.

The flow of kinetic energy (KE) also has contributions from both the gas and liquid.

$$\begin{aligned}\frac{\text{Flow of KE}}{\text{Mass flow}} &= \frac{x}{2}\left(\frac{GAx}{\rho_G\alpha A}\right)^2 + \frac{(1-x)}{2}\left[\frac{GA(1-x)}{\rho_L(1-\alpha)A}\right]^2 \\ &= \frac{1}{2}\left[\frac{G^2x^3}{\alpha^2\rho_G^2} + \frac{G^2(1-x)^3}{(1-\alpha)^2\rho_L^2}\right]\end{aligned} \quad (1.75)$$

The energy theorem gives

$$\frac{d\langle h\rangle}{dz} + \frac{1}{2}\frac{d}{dz}\left[\frac{G^2x^3}{\alpha^2\rho_G^2} + \frac{G^2(1-x)^3}{(1-\alpha)^2\rho_L^2}\right] + g\cos\theta_g = \frac{q_W P}{AG} + \left(\frac{u_G^2}{2} - \frac{u_L^2}{2}\right)\frac{dx}{dz} \quad (1.76)$$

$$\langle h\rangle = (1-x)h^L + xh^G \quad (1.77)$$

The last term in (1.76) is the change in kinetic energy associated with phase change. Since u_G and u_L are given by (1.68) and (1.69),

$$\frac{d\langle h\rangle}{dz} + \left[x\frac{1}{2}\frac{du_G^2}{dz} + (1-x)\frac{1}{2}\frac{du_L^2}{dz}\right] + g\cos\theta_g = \frac{q_W P}{AG} \quad (1.78)$$

Note that the kinetic energy term represents the weighted average of the kinetic energy of the gas and the kinetic energy of the liquid.

1.4.2 Choking

The momentum equation (1.73) for a separated flow can be written as

$$-\frac{dp}{dz}\left[1 + \frac{d}{dp}\left(\frac{G^2x^2}{\alpha\rho_G} + \frac{G^2(1-x)^2}{(1-\alpha)\rho_L}\right)\right] = \frac{P\langle\tau_W\rangle}{A} + g\cos\theta_g[\rho_G\alpha + (1-\alpha)\rho_L] \quad (1.79)$$

since $d/dz = d/dp(dp/dz)^{-1}$. Again, choking is defined as the condition for which the term multiplying dp/dz changes sign.

$$\frac{1}{G_C^2} = -\frac{d}{dp}\left(\frac{x^2}{\alpha\rho_G} + \frac{(1-x)^2}{(1-\alpha)\rho_L}\right) \quad (1.80)$$

It is not convenient to use (1.80) to calculate G_C. A direct approach is more appropriate. Usually, the upstream conditions are fixed. Then, (1.73) and/or (1.76) are solved (quite often by making further approximations) to obtain the downstream pressure, p_2, for a given G. The ratio p_2/p_1 is decreased until a maximum value of G is calculated. This is G_C.

Usually, the balance equations are written with S, rather than α. This is done by using (1.70), that is,

$$\frac{1}{\alpha} = 1 + \frac{S(1-x)}{x}\frac{\rho_G}{\rho_L} \tag{1.81}$$

to eliminate α. The integration requires a relation between x, ρ_G, ρ_L and p. This can be obtained by assuming equilibrium between the phases and using thermodynamic relations. If S is assumed to be independent of z the void fraction is fixed if x is fixed. This simplification is called a separated flow with constant slip. The separated flow model probably gives a better approximation for the choking condition than the homogeneous model ($S = 1$) does, since it allows for an assessment of the effects of slip (which can be quite large).

1.5 The Lockhart–Martinelli analysis

1.5.1 Evaluation of homogeneous and separated flow models

The homogeneous and separated flow assumptions are not complete in that they do not give models for τ_W and α. Their chief merit is that they estimate the contributions of acceleration and kinetic energy to the pressure change and the choking condition. Empirical relations are needed for τ_W and α. The approach that has been taken is to use measurements of the pressure gradient and the void fraction under conditions that the flow is not changing significantly in the flow direction, so the contribution of acceleration to the pressure gradient can be ignored.

Empirical correlations of these measurements are aided by using dimensional analysis. An important step in this approach is to discover the relevant dimensionless groups. A widely used correlation is due to Lockhart & Martinelli (1949). It is now known that this approach provides only a rough estimate, but its simplicity makes it a useful tool in obtaining a first approximation.

1.5.2 Frictional pressure gradient

The momentum balance, equation (1.5), can be written in the following form for a passage with constant area:

$$-\frac{dp}{dz} = \frac{d(GA)}{dz} + \frac{\tau_W P}{A} + \rho g \cos\theta_g \tag{1.82}$$

This equation indicates that the pressure gradient is due to fluid acceleration, frictional effects associated with the wall drag and the overcoming of gravity. The frictional pressure gradient is given as

$$-\left(\frac{dp}{dz}\right)_F = \frac{\tau_W P}{A} \tag{1.83}$$

or, for a circular pipe, as

$$-\left(\frac{dp}{dz}\right)_F = \frac{4\tau_W}{d_t} \tag{1.84}$$

where d_t is the pipe diameter. Lockhart and Martinelli chose to develop correlations for the frictional pressure gradient rather than for τ_W. If τ_W is represented by the friction factor relation (1.6),

$$-\left(\frac{dp}{dz}\right)_F = \frac{2\rho u^2}{d_t} f \tag{1.85}$$

For a single-phase flow, the friction factor can be approximated (Knudsen & Katz, 1958) by

$$f = 0.046(\mathrm{Re})^{-0.2} \tag{1.86}$$

for a turbulent flow and by

$$f = 16(\mathrm{Re})^{-1} \tag{1.87}$$

for a laminar flow, where the Reynolds number is defined as

$$\mathrm{Re} = \frac{d_t u \rho}{\mu} \tag{1.88}$$

A similar result is obtained for gas–liquid flows. For example, (1.84) gives the frictional contribution to the pressure drop as $P\tau_W/A$. It is noted that the gravitational contribution $g \cos\theta_g[\rho_G \alpha + \rho_L(1-\alpha)]$ requires a knowledge of the void fraction, α.

1.5.3 Rationale for the Lockhart–Martinelli plots

A rationale for using the method of Lockhart and Martinelli is obtained by using a separated flow model which pictures the gas and liquid as flowing in parallel pipes of areas of αA and $(1-\alpha)A$, as shown in Figure 1.3. Liquid and gas are flowing in a pipeline with a diameter of d_t and an area of A. The mass velocity is given as

$$G = G_L + G_G \tag{1.89}$$

where G_L and G_G are the contributions by the liquid and the gas. The fictitious parallel pipes have areas $(1-\alpha)A$ and αA and diameters of d_G and d_L. Thus

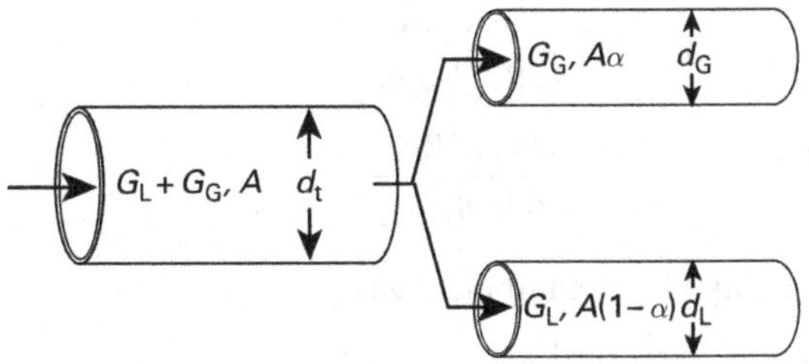

Figure 1.3 Rationale for using the method of Lockhart & Martinelli.

$$\alpha = \frac{d_G^2}{d_t^2} \qquad (1-\alpha) = \frac{d_L^2}{d_t^2} \tag{1.90}$$

From (1.85) and (1.86) the frictional pressure loss for a turbulent flow scales as

$$\left(\frac{dp}{dz}\right)_F \propto \frac{u^{1.8}}{d_t^{1.2}} \tag{1.91}$$

For the gas-filled pipe, the following is obtained if the flow is turbulent

$$\begin{aligned}\left(\frac{dp}{dz}\right)_F &= \left(\frac{dp}{dz}\right)_{GS}\left(\frac{u_G}{u_{GS}}\right)^{1.8}\left(\frac{d_t}{d_G}\right)^{1.2}\\ &= \left(\frac{dp}{dz}\right)_{GS}\alpha^{-1.8}\alpha^{-0.6}\end{aligned} \tag{1.92}$$

where $(dp/dz)_{GS}$ is the pressure gradient for the gas flowing alone at a velocity u_{GS} in a pipe with a diameter d_t, and d_G is defined in Figure 1.3. For a liquid-filled pipe

$$\left(\frac{dp}{dz}\right)_F = \left(\frac{dp}{dz}\right)_{LS}(1-\alpha)^{-2.4} \tag{1.93}$$

Since $(dp/dz)_F$ is presumed to be equal in the two fictitious pipes, it can be eliminated between (1.92) and (1.93) to give

$$\left(\frac{\alpha}{1-\alpha}\right)^{2.4} = \left(\frac{dp}{dz}\right)_{LS}\Big/\left(\frac{dp}{dz}\right)_{GS} = X_{tt}^2 \tag{1.94}$$

The term X_{tt} is called the flow factor for a turbulent liquid and a turbulent gas. This suggests that the void fraction is a function of X, which can be X_{tt}, X_{tv}, X_{vt}, X_{vv} depending on whether the flow is turbulent (Re > 1000) or laminar (Re < 1000). Thus,

$$X_{tt}^2 = \frac{0.046\rho_L u_L^2 \mathrm{Re}_L^{-0.2}}{0.046\rho_G u_G^2 \mathrm{Re}_G^{-0.2}} \tag{1.95}$$

$$X_{vt}^2 = \frac{16\mathrm{Re}_L^{-1}\rho_L u_L^2}{0.46\mathrm{Re}_G^{-0.2}\rho_G u_G^2} \tag{1.96}$$

$$X_{\mathrm{tv}}^2 = \frac{0.046\rho_{\mathrm{L}} u_{\mathrm{L}}^2 \mathrm{Re}_{\mathrm{L}}^{-0.2}}{16\mathrm{Re}_{\mathrm{G}}^{-1} \rho_{\mathrm{G}} u_{\mathrm{G}}^2} \tag{1.97}$$

$$X_{\mathrm{vv}}^2 = \frac{16\rho_{\mathrm{L}} u_{\mathrm{L}}^2 \mathrm{Re}_{\mathrm{L}}^{-1}}{16\rho_{\mathrm{G}} u_{\mathrm{G}}^2 \mathrm{Re}_{\mathrm{G}}^{-1}} \tag{1.98}$$

For example, for a turbulent liquid and a turbulent gas

$$X_{\mathrm{tt}} = \left(\frac{\mu_{\mathrm{L}}}{\mu_{\mathrm{G}}}\right)^{0.1} \left(\frac{\rho_{\mathrm{G}}}{\rho_{\mathrm{L}}}\right)^{0.5} \left(\frac{W_{\mathrm{L}}}{W_{\mathrm{G}}}\right)^{0.9} \tag{1.99}$$

where W_{L} and W_{G} are the weight rates of flow for the liquid and the gas. This model, thus, suggests that measurements of the void fraction, α, should be correlated as a function of X, which depends on whether the phases are turbulent or laminar.

A dimensionless frictional pressure gradient can be defined as

$$\phi_{\mathrm{G}}^2 = \frac{(dp_{\mathrm{F}}/dz)}{(dp/dz)_{\mathrm{GS}}} \tag{1.100}$$

or as

$$\phi_{\mathrm{L}}^2 = \frac{(dp_{\mathrm{F}}/dz)}{(dp/dz)_{\mathrm{LS}}} \tag{1.101}$$

From (1.92) and (1.93), ϕ_{G} and ϕ_{L} are functions of α. Thus, the above equations suggest that measured frictional pressure losses can be correlated by plotting ϕ_{G}^2 or ϕ_{L}^2 as a function of X.

1.5.4 Lockhart–Martinelli correlation

Lockhart & Martinelli (1949) have exploited the above correlations in a landmark paper. The application of this correlation has been thoroughly discussed by Hewitt (1978) and by Wallis (1969). A simple fit to their correlations by Chisholm (1967) is as follows:

$$\phi_{\mathrm{L}}^2 = 1 + \frac{C}{X} + \frac{1}{X^2} \tag{1.102}$$

$$\phi_{\mathrm{G}}^2 = 1 + CX + X^2 \tag{1.103}$$

where $C=20$ for X_{tt}, $C=12$ for X_{vt}, $C=10$ for X_{tv}, $C=5$ for X_{vv}. Flow factor X_{vt} refers to a viscous liquid and a turbulent gas. Note that X is zero and ϕ_{G}^2 is unity for a liquid flow that is zero. Both X and ϕ_{G} increase with increasing liquid flow. Note also that for zero gas flow $X=\infty$. Thus ϕ_{L} is unity under this condition. With increasing gas flow, X decreases and ϕ_{L} increases.

Wallis (1969) gives the following representation of the plot for void fraction developed by Lockhart & Martinelli (1949):

$$\alpha = \left(1 + X^{0.8}\right)^{-0.375} \tag{1.104}$$

where X can be defined for any of the four regimes.

The data represented by (1.102), (1.103) and (1.104) are mainly for gas–liquid flows. Martinelli & Nelson (1948) and Thom (1964) developed a modified approach for steam/water. Detailed discussions of these works are presented by Hewitt (1978) and by Wallis (1969).

1.6 Steady flow in a nozzle

1.6.1 Preliminary comments

Another situation where the one-dimensional analysis can be used is steady flow in a nozzle. This differs from steady flow in a pipe in that the cross-sectional area, A, is varying in the flow direction. Usually the changes in velocity are large enough that drag on the wall and gravitational effects can be ignored compared to inertial changes. The analysis presented here makes the simplifying assumption that the flow is uniform over a cross-section.

1.6.2 Single-phase flow

A differential control volume in a contracting flow is depicted in Figure 1.4.

Conservation of mass gives

$$\rho u A = \text{constant} \tag{1.105}$$

which can be written as

$$\frac{du}{u} + \frac{d\rho}{\rho} + \frac{dA}{A} = 0 \tag{1.106}$$

The momentum balance equation is obtained by applying the momentum theorem to the differential volume in Figure 1.4. The flow of momentum into and out of the volume are

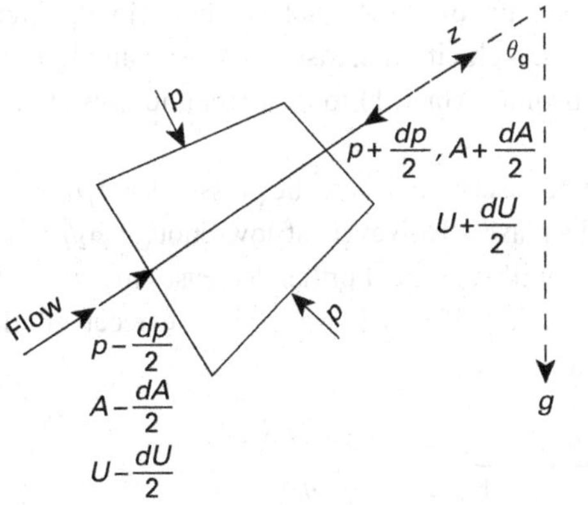

Figure 1.4 Differential control volume in a contracting flow.

ρuAu and $\rho uA(u+du)$ so the net flow of momentum out is $\rho uAdu$. The only force acting on the volume is due to pressure since gravity and the resisting force at the wall are neglected. The pressures on the front and back faces are pA and $-(p+dp)(A+dA)$. In addition there is a force acting on the slant face whose magnitude is $p|dA|$, where $|dA|$ is the magnitude of the projection of the area of the slant face on a plane perpendicular to the flow direction. For the volume in Figure 1.4, the pressure force on the slant face is negative, since dA is negative for a converging section. Thus, the force on the slant face is $p\,dA$.

When all of the pressure forces are added, the net force is found to be $-Adp$ if second-order terms in the differential quantities are ignored. Thus, not surprisingly, the momentum balance gives the Bernoulli equation

$$udu + \frac{dp}{\rho} = 0 \tag{1.107}$$

This can be rearranged to

$$\frac{d\rho}{\rho} = -\frac{du}{u}\left[\frac{u^2}{dp/d\rho}\right] \tag{1.108}$$

If $d\rho/\rho$ is eliminated by using mass balance equation (1.106)

$$-\frac{dA}{A} = \frac{du}{u}\left[1 - \mathrm{Ma}^2\right] \tag{1.109}$$

where

$$\mathrm{Ma}^2 = \frac{u^2}{c^2} \tag{1.110}$$

and c^2 is given as

$$c^2 = \frac{dp}{d\rho} \tag{1.111}$$

If Ma < 1, du/u has the opposite sign of dA/A, that is, the velocity increases in a converging section. For Ma > 1 the velocity increases in an expanding section. Thus, it is impossible to go from a subsonic (Ma < 1) to a supersonic flow in a converging section.

If a nozzle is located between two reservoirs where the pressures are p_1 and p_2, the flow will initially increase as p_2/p_1 decreases. However, at low enough p_2/p_1 the flow will choke and the velocity at the throat will be sonic. Further decreases in p_2/p_1 will result in a supersonic expansion and the creation of shock waves in the receiver. If (1.106) is combined with (1.107) one obtains

$$\frac{dp}{dz} = \frac{(G^2/2\rho A^2)(dA/dz)}{\left[1 - G^2(dv/dp)\right]} \tag{1.112}$$

where v is the specific volume. This gives a choking condition at the throat $dA/dz = 0$ when the mass velocity at the throat is

$$G_C^2 = \frac{dp}{dv} = \left(\frac{dv}{dp}\right)^{-1} \tag{1.113}$$

The mass flow under this condition is the product of G_C and the area of the throat.

The energy balance gives

$$dh + d\left(\frac{u^2}{2}\right) = \frac{q_W}{W} \tag{1.114}$$

where W is the weight rate of flow. If $(u^2/2)$ is eliminated between (1.107) and (1.114)

$$dh - \frac{dp}{\rho} = \frac{q_W}{W} \tag{1.115}$$

For $q_W = 0$ the flow is adiabatic and reversible; that is, it has a constant entropy. Thus

$$dh = \frac{dp}{\rho} \tag{1.116}$$

and

$$c^2 = \left(\frac{\partial p}{\partial \rho}\right)_S \tag{1.117}$$

It is noted that, for an adiabatic situation, changes of the flow in a converging section are defined by (1.107) and (1.114) with $q_W = 0$. The flow can be calculated either with the energy or the momentum balance equation and the condition of constant entropy.

For a perfect gas, (1.116) can be written as

$$C_P dT = \frac{dp}{\rho} \tag{1.118}$$

By using (1.19) and (1.24) to eliminate T from (1.118),

$$\frac{\gamma}{\gamma - 1} d\left(\frac{p}{\rho}\right) = \frac{dp}{\rho} \tag{1.119}$$

This can be written as

$$\frac{\gamma}{(\gamma - 1)} \frac{dp}{\rho} + \frac{\gamma}{(\gamma - 1)} p d\left(\frac{1}{\rho}\right) = \frac{dp}{\rho} \tag{1.120}$$

Multiplying by $(\gamma - 1)$ and rearranging

$$\frac{dp}{\rho} = -\gamma p d\left(\frac{1}{\rho}\right) \tag{1.121}$$

Integration gives

$$\frac{p}{\rho^{\gamma}} = \text{constant} \tag{1.122}$$

or

$$\frac{p}{p_1} = \left(\frac{\rho}{\rho_1}\right)^{\gamma} \tag{1.123}$$

This can be substituted into the momentum equation (1.107) to eliminate either p or ρ. Note that the relationship between p and ρ for a straight pipe, equation (1.31), differs from (1.123) because it is for an irreversible flow that includes effects of wall friction.

1.6.3 Two-phase flow in a nozzle

Two-phase flow in a nozzle is pertinent to a number of practical problems. For example, the calculation of choking is of particular interest in considering the performances of relief systems and of rocket nozzles.

The influence of wall drag and gravity are ignored so the application of the momentum theorem to the differential volume, shown in Figure 1.4, gives

$$-A\frac{dp}{dz} = \frac{d}{dz}\left[\frac{G^2Ax^2}{\alpha\rho_G} + \frac{G^2A(1-x)^2}{(1-\alpha)\rho_L}\right] \tag{1.124}$$

where from conservation of mass

$$GA = \text{constant} \tag{1.125}$$

Conservation of energy is given by

$$\frac{d\langle h\rangle}{dz} + \frac{1}{2}\frac{d}{dz}\left[\frac{G^2x^3}{\alpha^2\rho_G^2} + \frac{(1-x)^3G^2}{(1-\alpha)^2\rho_L^2}\right] = 0 \tag{1.126}$$

with $\langle h\rangle$ defined by (1.77). The velocities of the gas and liquid streams are given by (1.68) and (1.69). Equation (1.81) is used to eliminate α from (1.124) and (1.126) so that the terms in the brackets then contain the slip, S, and x rather than α and x.

By combining (1.125) and (1.124) one obtains

$$\frac{dp}{dz}\left\{1 + G^2\frac{d}{dp}\left[\frac{x^2}{\alpha\rho_G} + \frac{(1-x)^2}{(1-\alpha)\rho_L}\right]\right\} = G^2\frac{dA}{dz}\left[\frac{x^2}{\alpha\rho_G} + \frac{(1-x)^2}{(1-\alpha)\rho_L}\right] \tag{1.127}$$

Note that (1.127) predicts a change in the sign of dp/dz in a converging section (dA/dz has negative value) when the term multiplying dp/dz changes sign. This is an impossibility so the flow chokes at the throat ($dA/dz = 0$) where $G = G_C$, given by (1.80). This indicates that the choking for flow in a converging nozzle is given by the same condition as for flow in a long straight pipe.

As already discussed in Section 1.4.2, the calculation of G_C requires the solution of (1.124) and/or (1.127), with α eliminated using (1.81). These integrations are used to determine G as a function of the ratio of the pressures at the outlet and the inlet. Equilibrium is usually assumed to exist between the phases so that thermodynamic correlations can be used to calculate x. The slip ratio varies with z but, as an approximation, it is usually assumed to have a constant value. A maximum in G indicates that choking is occurring.

An extensive review of various approaches to the calculation of G_C has been given by Henry (1981), including the separated flow model that uses the assumptions of a constant slip ratio and equilibrium between the phases. From data for air–water he suggests that the use of $S = 3.2$ provides a good approximation.

An interesting solution has been presented by Moody (1965) for a fluid which is undergoing phase changes as it flows through a nozzle. At the inlet the fluid is a vapor–liquid mixture or a saturated vapor. The velocity is small enough for stagnation conditions ($p_o, \rho_o, h_o, s_o, u_o = 0$) to exist in the entering fluid. The two terms in the brackets in (1.126) are $u_G^2 x$ and $u_L^2(1-x)$. The integration of (1.126) from the inlet to the outlet, designated by the subscript 2, gives

$$\langle h_2 \rangle - \langle h_o \rangle + \frac{1}{2}\frac{G_2^2 x_2^3}{\alpha_2^2 \rho_{G2}^2} + \frac{1}{2}\frac{G_2^2 (1-x_2)^3}{(1-\alpha_2)^2 \rho_{L2}^2} = 0 \tag{1.128}$$

where α is eliminated by using an equation presented earlier

$$\frac{1}{\alpha} = 1 + \frac{S(1-x)}{x}\frac{\rho_G}{\rho_L} \tag{1.129}$$

Moody assumed the flow to be isentropic, so that $\langle s_2 \rangle = \langle s_o \rangle$. Thus, for a given p_2, thermodynamic relations give $\langle h_2 \rangle, x_2, \rho_{G2}, \rho_{L2}$. If p_o and ρ_o are known, $\langle h_o \rangle$ and $\langle s_o \rangle$ can be obtained from thermodynamic relations. Equations (1.128) and (1.129) give $\langle h_2 \rangle$ in terms of G_2, the slip ratio, S, and downstream conditions. Thus, if S is fixed, G_2 can be calculated for different values of p_2/p_o. The value of p_2/p_o at which G_2 is a maximum represents the choking condition.

This critical mass velocity depends on the choice of S, so a homogeneous model ($S = 1$) predicts lower G_C than does a separated flow model. Moody (1965) suggested that the maximum possible G_C, calculated from the separated flow model, is given by

$$S = \left(\frac{\rho_L}{\rho_G}\right)^{1/3} \tag{1.130}$$

He made calculations for the steam/water system using (1.130).

However, there is no physical reason to expect that this is the right value. The actual S which characterizes the evolution of a given system to a choking condition must be established either empirically or by a hydrodynamic model which depends on an assumed distribution of the phases. Clearly, there is a need for more research in this area.

1.6.4 One-dimensional two-fluid model

The assumption of equilibrium between the phases used in previous sections is often inaccurate. This is especially a concern in rapidly varying flows such as exist in choking situations. The application of a two-fluid model to represent such systems is presented by Boure (1978). Only a brief discussion is presented here.

This model represents the flow as consisting of interpenetrating gas and liquid streams. Separate equations are developed for the gas and the liquid. The velocity then emerges in a natural way so correlations for slip are not introduced. Thermodynamic relations are used for the separate phases but equilibrium between phases is not assumed. Thus, x is not obtained from thermodynamics and the temperatures of the two phases need not be equal. The equations presented here are for a simplified version of the model in that a one-dimensional flow is considered; that is, the flow at different locations in the cross-sections are assumed to be the same.

The equations of conservation of mass for steady flow in a duct with constant area, A, can be written as

$$\frac{d}{dz}(A\alpha\rho_{\mathrm{G}}u_{\mathrm{G}}) - M_{\mathrm{LG}}Aa_{\mathrm{v}} = 0 \tag{1.131}$$

$$\frac{d}{dz}[A(1-\alpha)\rho_{\mathrm{L}}u_{\mathrm{L}}] - M_{\mathrm{GL}}Aa_{\mathrm{v}} = 0 \tag{1.132}$$

where M_{GL} is the mass transfer rate per unit area from the gas to the liquid, M_{LG} is the mass transfer rate from the liquid to the gas and a_{v} is the interfacial area per unit volume. Thus

$$M_{\mathrm{LG}} = -M_{\mathrm{GL}} \tag{1.133}$$

and

$$M_{\mathrm{LG}} = GA\frac{dx}{dz} \tag{1.134}$$

By adding (1.131) and (1.132), one gets

$$\frac{dG}{dz} = 0 \quad \text{or} \quad G = \text{constant} \tag{1.135}$$

The momentum balance equations for the gas and for the liquid for the case of constant A and G are

$$\frac{d}{dz}(\rho_{\mathrm{G}}u_{\mathrm{G}}^2\alpha) = M_{\mathrm{LG}}a_{\mathrm{v}}u_{\mathrm{G}} - \alpha\frac{dp}{dz} - \rho_{\mathrm{G}}\alpha\cos\theta_{\mathrm{g}} - \frac{P\tau_{\mathrm{WG}}}{A} + a_{\mathrm{v}}\tau_{\mathrm{LG}} \tag{1.136}$$

$$\frac{d}{dz}\left[\rho_{\mathrm{L}}u_{\mathrm{L}}^2(1-\alpha)\right] = M_{\mathrm{GL}}a_{\mathrm{v}}u_{\mathrm{l}} - (1-\alpha)\frac{dp}{dz} - \rho_{\mathrm{L}}g(1-\alpha)\cos\theta_{\mathrm{g}} - \frac{P\tau_{\mathrm{WL}}}{A} - a_{\mathrm{v}}\tau_{\mathrm{LG}} \tag{1.137}$$

The liquid can exert a force on the gas at the interface. The component of this force in the flow direction is represented by $\tau_{\mathrm{LG}}a_{\mathrm{v}}Adz$ where τ_{LG} is an average effective stress and

$a_v A dz$ is the interfacial area in the control volume. Similarly, the force of the gas on the liquid is given by $\tau_{GL} a_v A dz$. The other stresses in (1.136) and (1.137), τ_{WG} and τ_{WL}, represent the resisting stresses at the wall on the gas and on the liquid.

If evaporation is occurring the term M_{LG} is positive so $M_{LG} a_v u_G$ and $M_{GL} a_v u_L$ represent increases and decreases of momentum associated with the phase change. The sum of these two terms is the net increase of the momentum of the flow system already discussed in Section 1.4.1. The sum of (1.136) and (1.137) gives (1.74).

Conservation of energy for the gas and for the liquid gives

$$\frac{d}{dz}\left(u_G \alpha \rho_G \frac{u_G^2}{2}\right) + \frac{d}{dz}\left(u_G \alpha \rho_G h^G\right) = M_{GL} a_v \frac{u_G^2}{2} + q_{LG} a_v - \rho_G u_G \alpha g \cos\theta_g + q_{WG}\frac{P}{A} \tag{1.138}$$

$$\begin{aligned}\frac{d}{dz}\left[u_L(1-\alpha)\rho_L \frac{u_L^2}{2}\right] + \frac{d}{dz}\left[u_L(1-\alpha)\rho_L h^L)\right] &= M_{GL} a_v \frac{u_L^2}{2} + q_{GL} a_v p \\ &\quad - \rho_L u_L (1-\alpha) g \cos\theta_g + q_{WL} a_v\end{aligned} \tag{1.139}$$

where h^G and h^L are the enthalpies of the gas and the liquid, q_{LG} and q_{GL} are the heat transfer rates per unit area from the liquid to the gas or from the gas to the liquid, q_{WG} and q_{WL} are the heat transfer rates per unit area from the wall to the gas or from the wall to the liquid. The first terms on the right side of (1.138) and (1.139) represent changes of the kinetic energy of the gas or the liquid. The sum of (1.138) and (1.139) gives (1.78).

The void fraction, α, can be calculated from conservation of mass if u_G and u_L are known and if G is fixed

$$GA = u_L(1-\alpha)A + u_G \alpha A \tag{1.140}$$

The two-fluid model thus avoids the need to specify a slip between the phases and the void fraction because the velocities of the two phases are calculated. Also, equilibrium between the phases need not be assumed.

However, new variables, which need to be evaluated, are introduced. These include the rate of mass transfer, M_{LG}, the rate of heat transfer, q_{LG}, between the phases and the interfacial drag, τ_{LG}. Also, a theory needs to be developed to predict the interfacial area per unit volume, a_v. This quantity introduces, in a direct way, new concepts into the one-dimensional balances, in that it requires a knowledge of the structure of the phases. For example, it might require a specification of a bubble size distribution.

Drew & Flaherty (1994) and others have shown that solutions of the one-dimensional two-fluid equations are unstable for the simple case of an adiabatic flow with no phase change. (This is not the case for the homogeneous or separated flow models.) This suggests that more attention needs to be given to the formulation of the two-fluid model and the method of solution. The use of finite difference methods to solve the differential equations introduces numerical viscosity which can artificially stabilize the equations.

Three-dimensional two-fluid equations have been successfully used to study adiabatic bubbly flows (Lance & Lopez de Bertodano, 1994) and solid–fluid flows.

References

Boure, J. A. 1978 Constitutive equations for two-phase flows. In *Two-phase Flows and Heat Transfer with Applications to Nuclear Reactor Design Problems*, ed. J. J. Ginoux, Washington, DC: Hemisphere, pp. 169–175.

Chisholm, D. 1967 A theoretical basis for the Lockhart–Martinelli correlation for two-phase flow. Nat. Eng. Lab., UK, Report no. 310 (also *Int. J. Heat Mass Transfer* 10, 1767–1778).

Drew, D. A. & Flaherty, J. E. 1994 Analysis of multiphase flow. *Multiphase Science and Technology* 8, 207–255.

Fan, Z. & Hanratty, T. J. 1993 Pressure profiles for slugs in horizontal pipelines. *Int. J. Multiphase Flow* 19, 421–437.

Henry, R. E. 1981 Calculational techniques for two-phase critical flow. In *Two-Phase Fluid Dynamics* (Papers presented at the Japan–US Seminar, held July 31–August 3, 1979, Kansei, Kobe, Japan), eds. A. E. Bergles & S. Ishigai, Washington, DC: Hemisphere, pp. 415–438.

Henstock, W. H. & Hanratty, T. J. 1976 The interfacial drag and the height of the wall layer in annular flows. *AIChE Jl* 22, 990–1000.

Hewitt, G. F. 1978 Simple momentum and energy balances and their related empirical correlations. In *Two–phase Flows and Heat Transfer with Applications to Nuclear Reactor Design Problems*, ed. J. J. Ginoux, Washington, DC: Hemisphere.

Knudsen, J. G. & Katz, D. L. 1958 *Fluid Dynamics and Heat Transfer*, New York: McGraw-Hill, pp. 89, 173.

Lance, M. & Lopez de Bertodano, M. 1994 Phase distribution phenomena wall effects in bubbly two-phase flows. *Multiphase Science and Technology* 8, 207–255.

Lockhart, R. W. & Martinelli, R. C. 1949 Proposed correlation of data for isothermal two-phase, two-component flow in pipes. *Chem. Eng. Progr.* 45, 39–48.

Martinelli, R. C. & Nelson, D.B. 1948 Prediction of pressure drop during forced-circulation boiling of water. *Trans. ASME* 70, 695–702.

Moody, F. J. 1965 Maximum flow rate of a single component, two-phase mixture. *J. Heat Transfer* 87, 134–142.

Thom, J. R. S. 1964 Prediction of pressure drop during forced circulation of boiling water. *Int. J. Heat Mass Transfer* 7, 709–724.

Wallis, G. B. 1969 *One-dimensional Two-phase Flow.* New York: McGraw-Hill.

2 Flow regimes

2.1 Need for a phenomenological understanding

The one-dimensional analysis and the correlations for frictional pressure drop and void fraction (presented in Chapter 1) have been widely used as a starting point for engineering designs. However, these correlations have the handicap that the structure of the phase boundaries is ignored. As a consequence, they often give results which are only a rough approximation and overlook phenomena which could be of first-order importance in understanding the behavior of a system.

It is now recognized that the central issue in developing a scientific approach to gas–liquid flows is the understanding of how the phases are distributed and of how the behavior of a multiphase system is related to this structure (Hanratty *et al.*, 2003). Of particular interest is the finding that macroscopic behavior is dependent on small-scale interactions. An example of this dependence is that the presence of small amounts of high molecular weight polymers can change an annular flow into a stratified flow by damping interfacial waves (Al-Sarkhi & Hanratty, 2001a).

A goal of this book is to develop an understanding of the basic scientific tools needed to describe gas–liquid flows. This chapter opens this discourse by describing flow patterns that have been defined. Detailed discussions of the theory will be presented in later chapters.

Four systems are considered: flow in horizontal pipes, flow in vertical pipes, microgravity flows, flow in capillaries and microchannels. Discussions of the flow regimes in horizontal pipes and of the mechanisms for transitions from one regime to another provide the motivation for a large portion of this book.

The Kelvin–Helmholtz (inviscid) instability of a stratified flow plays an important role in understanding gas–liquid flows. Therefore, it is discussed in detail in Chapter 4. The results of this theory are previewed in this chapter since they arise in much of the discussion of the behavior of gas–liquid flows. The theory is also of historic interest because of its early use to explain the effect of pipe diameter on the transition from a stratified flow to an intermittent (slug or plug) flow.

Chapter 5 shows that the orientation of a pipe can have a strong effect on the flow pattern because the role of the component of the gravitational force in the direction of flow cannot be ignored. The present chapter discusses only the patterns observed in horizontal flows and in vertical flows.

Changes in fluid properties can influence the flow configuration. The sections in this chapter on the effects of liquid viscosity and gas density address this issue. Increases in

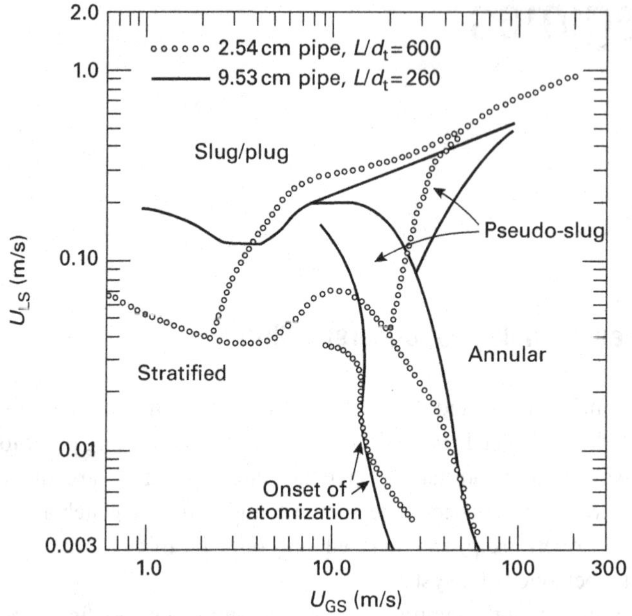

Figure 2.1 Flow regime map for air and water flowing in horizontal 2.54 cm and 9.53 cm pipes. Lin & Hanratty, 1987b.

gas density cause increases in the kinetic energy of the gas flow and of energy transfer from the gas flow to interfacial waves. The generation of waves is easier. Thus, at large enough densities, the appearance of slugs depends on their stability, rather than the stability of stratified flows.

2.2 Flow regimes in horizontal pipes

2.2.1 Flow regimes for air and a low-viscosity liquid

Figure 2.1 presents a Mandhane (Mandhane *et al.*, 1974) plot for air and water flowing in horizontal pipes with diameters of 2.54 cm and 9.53 cm and lengths of 15.5 m and 24.6 m under atmospheric conditions (Lin & Hanratty, 1987b). These observations were made at the downstream end of pipes, where the flow is approximately fully developed. The abscissa and ordinate are the superficial gas and liquid velocities, defined as the volumetric flows of the gas and liquid divided by the area of the pipe. A detailed consideration of this figure provides an introduction to the physical challenges that are presented in relating the behavior of a gas–liquid flow to interfacial structure.

2.2.2 Wavy stratified flow in air–water systems

Consider, first, the results for low gas and liquid velocities in a 9.53 cm pipe, for which the flow is stratified. In this pattern, all of the liquid moves along the bottom of the pipe.

At very low gas and liquid velocities the interface is smooth. However, at larger gas velocities waves appear.

Three types of waves have been identified: Jeffreys waves, Kelvin–Helmholtz waves and roll waves. The first interfacial disturbances observed in air–water flow with increasing gas velocity are Jeffreys waves (Jeffreys, 1925) for which a balance exists between the energy fed to the waves by the air flow and dissipation in the liquid (Hanratty & Engen, 1957; Cohen & Hanratty, 1965; Andritsos & Hanratty, 1987). This transition is observed to occur at u_{GS} = 4 m/s to 0.5 m/s as u_{LS} varies from 0.001 m/s to 0.1 m/s. Note that the area occupied by the gas decreases with increasing u_{LS}. Thus, the actual critical air velocity at transition is roughly equal to 4 m/s at all water velocities.

The presence of waves induces pressure variations along the interface which, for very small amplitude sinusoidal waves, can be represented as the sum of a component which is a maximum where the wave slope is a maximum, p_{GI}, and a component, p_{GR}, which is a minimum (a suction) where the wave height is a maximum. Energy is fed to Jeffreys waves through p_{GR}.

The magnitude of the induced gas phase pressure variations at the interface increases with increasing gas velocity. Irregular large-amplitude waves appear at large gas velocities for which the destabilizing effect of p_{GR} becomes large compared with the stabilizing effects of gravity and surface tension (Andritsos & Hanratty, 1987; Andreussi & Persen, 1987). For air–water flow, they are observed at $u_{GS} \approx 14$ m/s to 5 m/s as u_{LS} varies from 0.002 m/s to 0.1 m/s. These disturbances are called Kelvin–Helmholtz waves since they are initiated by a Kelvin–Helmholtz instability. At u_{GS} approximately twice that needed to initiate irregular waves, drops are observed to be ripped from the crests of waves (see Figure 2.1).

Roll waves (Hanratty & Engen, 1957; Hanratty & Hershman, 1961) appear over a range of conditions. These are large-wavelength disturbances with steep fronts and gradually sloping backs. They may be considered as surges of the flow in the liquid layer. The surface of the surges can be smooth (Figure 2.2b) or wavy (Figure 2.2d), depending on the interfacial condition of the stratified flow on which these waves are imposed.

2.2.3 Annular flow

The drops created by atomization of irregular waves can deposit at the top of the pipe. If the rate of deposition is large enough a turbulent film forms. This pattern is called an annular flow (Lin & Hanratty, 1987b), whereby part of the liquid flows along the wall and part flows as drops entrained in the gas (see Figure 2.2a and c). Waves aid in bringing liquid up the pipe wall so that the gas velocity needed to form an annular flow decreases with increasing u_{LS}. For air–water flow in a 9.53 cm pipe, an annular pattern is initiated at gas velocities of 30–50 m/s (see Figure 2.1). For very low u_{LS}, atomization does not occur in the air–water system, so an annular flow is not observed.

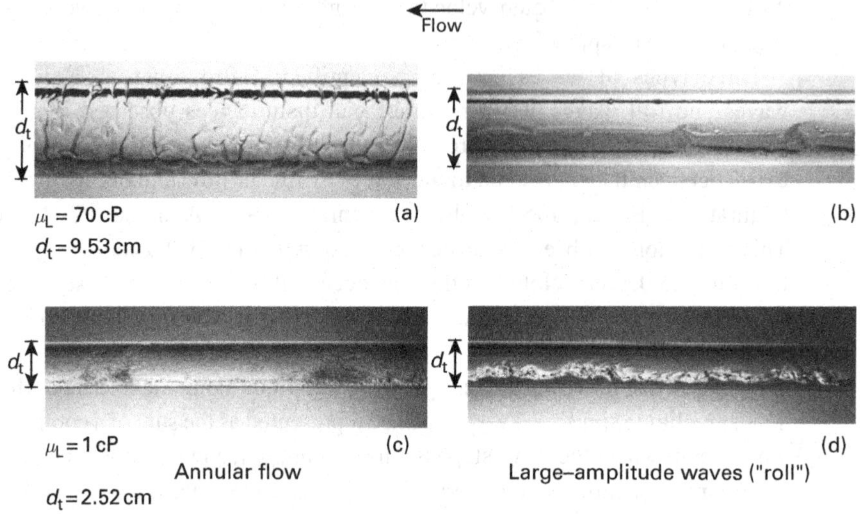

Figure 2.2 Annular flows and large roll waves for 1 cP and 70 cP liquids. Andritsos & Hanratty, 1987.

2.2.4 Plug and slug flow

At large enough u_{LS}, waves that touch the top of the pipe can form a liquid bridge. At $u_{SG} < \sim 0.6$ m/s and large u_{LS} a plug flow occurs whereby these bridges are part of a train of elongated bubbles which move along the top of the pipe (Ruder & Hanratty, 1990). At larger gas velocities, the bridges form liquid slugs, sketched in Figure 9.1. These aerated blocks of liquid move over a slowly flowing stratified layer at approximately the same velocity as the gas. Slugs can be quite long, so they can have large momenta and can cause vibrations when they impact on surfaces. Therefore, it is usually desirable to design a piping system so as to avoid slugging. Pictures of the front and back of a slug are given in Figure 2.3b, c.

The transition from a slug to a plug flow has been defined by Ruder *et al.* (1989) as the condition for which a sudden increase in liquid height (a hydraulic jump) in front of a slug cannot exist. The jump moves with a velocity c_F. An equation for this condition (see Figure 9.1) is derived by arguing that dissipation of mechanical energy accompanies the sudden increase in height. (See Chapter 9 for a derivation.) For a rectangular channel, this concept yields

$$\frac{(c_F - u_{L1})^2}{gB} > 1 \tag{2.1}$$

where h_{L1} and u_{L1} are the height and velocity of the liquid in front of the slug and B is the channel height. The term $(c_F - u_{L1})$ is the velocity of the slug relative to the velocity of the liquid layer in front of the slug. For a pipe with diameter d_t, the critical condition is predicted to be

$$\frac{(c_F - u_{L1})^2}{gd_t} > 0.95 - 1.21 \tag{2.2}$$

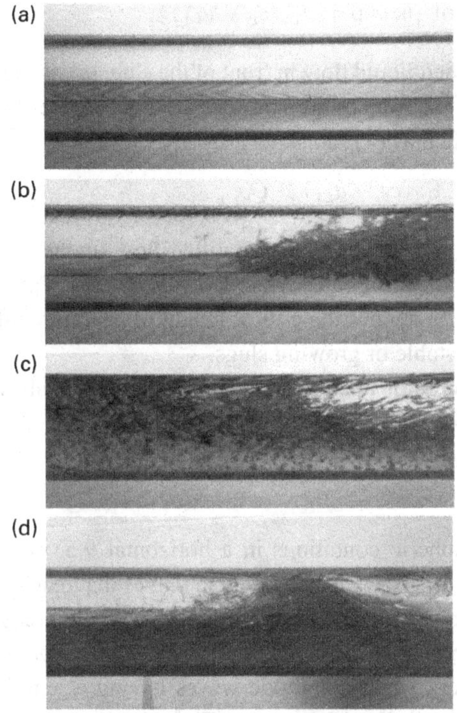

Figure 2.3 (a) Regular waves observed 180 diameters from the entry of a 9.53 cm pipe for $u_{LS} = 0.06$ m/s, $u_{GS} = 4.3$ m/s, $u_L = 20$ cP. (b) Front of a slug that was formed 100 diameters from the entry of a 9.53 cm pipe and observed at 180 diameters. (c) Tail of the slug in (b). (d) A slug that has just formed at 30 diameters from the entry of a 9.53 cm pipe for $u_L = 0.033$ m/s, $u_{GS} = 7$ m/s, $u_L = 100$ cP. Andritsos *et al.*, 1989.

where the term on the right depends on the value of h_{L1}/d_t. (See Ruder *et al.*, 1989, Table 1.) Ruder & Hanratty (1990) carried out visual studies for air and water flowing in a 9.53 cm pipe and defined the transition from plug to slug flow as occurring when

$$\frac{(c_F - u_{L1})^2}{g d_t} > 1.2 \tag{2.3}$$

Hurlburt & Hanratty (2002) showed how this condition can be used to predict the critical u_{GS}.

Barnea *et al.* (1980) have differentiated between plug and slug flow by considering whether the liquid blockage contains gas bubbles. This appears to be equivalent to the condition cited above for air and water flowing in horizontal pipes since the hydraulic jump in the front of the slug occludes gas at high gas velocities.

2.2.5 Stability of a slug

An understanding of the transition to intermittent (plug or slug) flow requires a consideration of the stability of a slug, as well as the stability of a stratified flow. Slugs moving along a stratified layer pick up liquid at their front at a rate given by

$$\text{Rate of pickup} = A_{L1}(c_F - u_{L1}) \tag{2.4}$$

where A_{L1} is the area of the stratified liquid flow in front of the slug, u_{L1} is the velocity in this layer and c_F is the velocity of the front of the slug. The volumetric rate at which slugs shed liquid at their back is specified as Q_{sh}, so slugs grow if

$$A_{L1}(c_F - u_{L1}) > Q_{sh} \tag{2.5}$$

Define A_{L0} and h_{L0} as the area and height of the stratified flow under conditions for which the pickup rate equals the shedding rate. When $h_{L1} > h_{L0}$ a slug will grow; when $h_{L1} < h_{L0}$, it will decay. Thus, (2.5) defines the critical height of a stratified flow, below which it is impossible to sustain stable or growing slugs.

A detailed derivation and a discussion about the implementation of the above ideas are given in Section 9.2 of the chapter on slug flow.

2.2.6 Transition to slug/plug flow

For air–water flow under atmospheric conditions in a horizontal 9.53 cm pipe, slugs are initiated at low gas velocities, $u_{GS} < \sim 5$ m/s, by regular waves that reach the top wall of the pipe. Jeffreys waves with a wavelength of approximately 8.5 cm are generated in the early part of the pipeline. These grow in height and increase in velocity to a point where they can resonate with very small amplitude waves having a length of 17 cm. Energy is fed from the 8.5 cm wave to the 17 cm wave. These larger wavelength waves can grow and tumble or, if the liquid flow is large enough, strike the top wall to form a slug. (See Fan *et al.*, 1993 and Figure 2.4.)

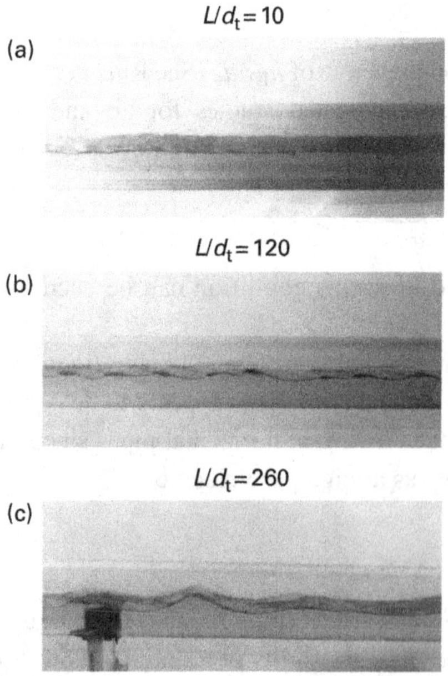

Figure 2.4 Development of waves on water, along a 9.53 cm pipe. Fan *et al.*, 1993.

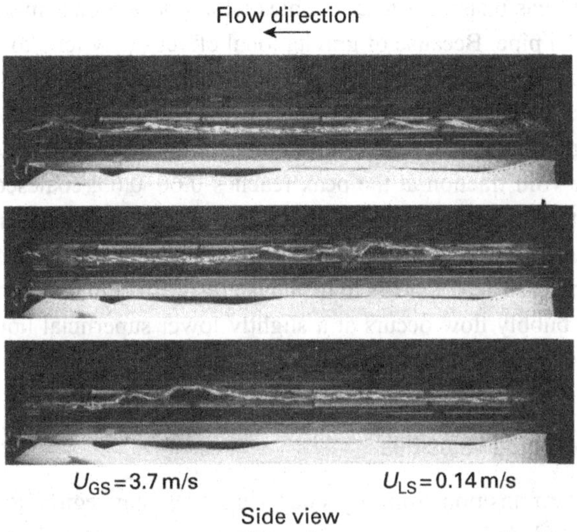

Figure 2.5 Formation of slugs by coalescing of roll waves in a 9.53 cm pipe. Lin & Hanratty, 1987b.

For large superficial gas velocities, irregular Kelvin–Helmholtz waves appear. These can coalesce to form a high enough wave that it touches the top wall (see Figure 2.5).

The height of the stratified layer decreases with increasing gas velocity so it can become too small to support a stable slug (Woods & Hanratty, 1996). Then, a consideration of the stability of slugs provides a prediction of the possibility of forming a slug flow.

It is possible for slugs to exist at sub-critical conditions (as defined by the stability of a stratified flow) if disturbances are introduced to the system and conditions are such that slugs are stable. (Terrain slugging is an example.)

2.2.7 Pseudo-slugs

At $u_{GS} > 9$ m/s a pseudo-slug region can exist in a 9.53 cm pipe carrying air and water (see Figure 2.1). The lower boundary of this region defines locations where large amplitude waves are initiated. These waves can coalesce and touch the top of the pipe to form a possible slug. The higher boundary of the pseudo-slug region is defined by the stability of a slug. Visual observations could, mistakenly, label these large-amplitude waves as slugs. However, they do not have the basic properties of slugs. That is, they do not have velocities which are approximately equal to the gas velocity and they are not coherent over a long length of pipe. Instrumental methods have been developed by Lin & Hanratty (1987a) to differentiate pseudo-slugs from slugs.

2.2.8 Bubble/foam regime

At very large liquid velocities and low gas velocities a bubble/foam regime exists. The transition (not shown in Figure 2.1) from a bubbly flow to a flow with elongated bubbles (plug flow) has been defined as occurring at $u_{LS} = 4$–5 m/s (Mandhane *et al.*, 1974). In this bubble region, the gas phase exists as bubbles which have Sauter mean diameters of 2–5 mm in a 5.03 cm pipe. Because of gravitational effects, they tend to move to the top of the pipe. A maximum in the void fraction occurs at about one bubble radius from the wall (Kocamustafaogullari & Wang, 1991). The distribution of void fraction is related to a balance of the gravitational force on the bubbles and forces due to liquid phase turbulence. When the void fraction at the peak reaches 0.60–0.65, coalescence ensues and a transition to intermittent flow occurs. Taitel & Dukler (1976) suggested that the transition occurs because a decrease in liquid flow reduces the effectiveness of turbulence in suspending the bubbles. There appears to be a hysteresis effect in that transition from a plug/slug flow to a bubbly flow occurs at a slightly lower superficial liquid velocity ($u_{LS} = \sim 2$–4 m/s) than transition from a bubbly flow to a plug/slug flow.

2.2.9 Flow at large gas and liquid velocities

The mechanism for the transition from a pseudo-slug or a slug regime to an annular regime is not well defined. At low gas velocities, slugs occlude gas at their front. The gas escapes at their back. The suggestion has been made that, at large gas velocities, gas is occluded at the rear and that the slug is eventually destroyed by blow-through. (However, this mechanism has not been substantiated.)

The flow pattern that exists at high gas and liquid velocities, also, is not understood. As one increases the liquid velocity in the annular regime, there is evidence for the appearance of large structures, formed by drop coalescence, which have many of the characteristics of slugs. These could be considered as slugs with a continuous gas phase (Dukler & Hubbard, 1975).

2.2.10 Effect of pipe diameter on flow regimes

The chief differences to be noted in comparing the mappings for air–water flow in a 2.54 cm pipe and in a 9.53 cm pipe are the strong effect of pipe size on the critical liquid velocity needed to generate slugs at low gas velocities, the increase in the size of the pseudo-slug region with a decrease in pipe diameter and the need for larger gas velocities to generate annular flow in large diameter pipes.

Some of these effects can be explained if the transition is predicted by an instability of stratified flow to a disturbance whose wavelength is large compared with the heights of the gas space and the liquid layer. The presence of these waves causes contractions and expansions of the gas space. These are accompanied by changes in the area of the gas space and, through the Bernoulli effect, by decreases in the pressure at the crest. This destabilizing effect, which is counterbalanced by the effect of gravity, increases with decreasing pipe size since the size of the gas space decreases. Thus, the initiation of roll waves occurs at smaller liquid flows in a smaller pipe.

Coalescence of waves can lead to the formation of a larger-amplitude wave that temporarily touches the top wall (a pseudo-slug), or to a slug flow. The ability of coalescing waves to reach the top wall is enhanced in smaller-diameter pipes, so the pseudo-slug region is larger. The transitions to slug flow for 2.54 cm and 9.53 cm pipes at large liquid velocities appears to be associated with the stability of slugs (equation (2.5)), so they are observed at approximately the same u_{LS} in the two pipes. Detailed discussions of this explanation have been given by Soleimani & Hanratty (2003) and by Hurlbert & Hanratty (2002).

As mentioned in the previous section, both deposition of drops and the transport of liquid around the circumference of the pipe by liquid phase turbulence associated with disturbance waves (shown in Figure 2.2c) are responsible for the initiation of annular flow. The ability of waves to wet the top of the pipe increases with decreasing pipe diameter so the transition to annular flow can be observed at smaller u_{GS} in a 2.54 cm pipe than in a 9.53 cm pipe.

The initiation of annular flow by deposition requires that the gas velocity is large enough for drops to form by atomizing the liquid layer. However, it also requires that the gas phase turbulence is sufficient to carry drops from the bottom of the pipe, where a thick liquid layer exists, to the top of the pipe. Drop mixing might become a more important factor in pipes with very large diameters. Thus, from a consideration of the data reported by Wu *et al.* (1987) for a natural gas/condensate pipeline with a diameter of 20.3 cm, Baik & Hanratty (2003a) point out that larger u_{GS} (than would be expected from experiments summarized in Figure 2.1) are needed to facilitate a transition to annular flow. This topic is covered in greater detail in Chapter 12.

2.3 Large-wavelength Kelvin–Helmholtz waves / viscous large-wavelength instability

Since theoretical considerations of the initiation of slugs/plugs at low gas velocities has centered on the stability of a stratified flow to large-wavelength disturbances, a brief historical account is appropriate. More detailed discussions of the theory are given in later chapters and in the paper by Hurlburt & Hanratty (2002).

Kordyban & Ranov (1970) and Wallis & Dobson (1973) explored an inviscid stability mechanism whereby the slugs arise from the growth of infinitesimal disturbances at the interface of a stratified flow. Since the spacing between slugs is large it was natural to restrict the analysis to wavelengths that are large compared with the height of the gas space. If viscous effects are neglected and the waves are considered to have small amplitudes, the following relation between the wave velocity and wave number (which is derived in Chapter 4, equation (4.73)) is obtained for a rectangular channel of height B inclined at an angle θ to the horizontal

$$k\rho_L(u_L - c)^2 \coth kh_L + k\rho_G(u_G - c)^2 \coth kh_G = g\cos\theta(\rho_L - \rho_G) + \sigma k^2 \qquad (2.6)$$

In this equation, k is the wave number, σ is the surface tension, u_G is the gas velocity, u_L is the liquid velocity, h_G is the height of the gas space, h_L is the height of the liquid space, ρ_G is the gas density, ρ_L is the liquid density and g is the acceleration of gravity. The wave number is defined as $k = 2\,\pi/\lambda$, where λ is the wavelength.

If (2.6) is solved for the wave velocity, it is found, for large enough u_G, that c is complex, thus indicating an instability. The critical condition at which this occurs is defined by the following equations:

$$(u_G - u_L)^2 = \frac{h_G}{\rho_G}\left[g \cos\theta(\rho_L - \rho_G) + k^2\sigma\right]\left[\frac{\tanh(kh_G)}{kh_G} + \frac{\rho_G}{\rho_L}\frac{\tanh(kh_L)}{kh_L}\right] \tag{2.7}$$

$$c = \frac{u_G\rho_G h_L + u_L\rho_L h_G}{\rho_L h_G + \rho_G h_L} \tag{2.8}$$

For long waves ($kh_L \ll 1$, $kh_G \ll 1$), surface tension effects can be ignored. Thus, stability conditions (2.7) can be written as

$$(u_G - u_L)^2 = \frac{h_G}{\rho_G}(g\rho_L \cos\theta) \tag{2.9}$$

Taitel & Dukler (1976) reformulated (2.9) and (2.8) as follows for a pipe with diameter d_t, area A_t and $\rho_G h_L/\rho_L h_G$ small:

$$\rho_G(u_G - c)^2 > \rho_L g \cos\theta \frac{A_G}{S_i} \tag{2.10}$$

$$c = u_L \tag{2.11}$$

where A_G is the area occupied by the gas and S_i is the length of the interface; that is,

$$\left(\frac{S_i}{d_t}\right)^2 = 1 - \left(2\frac{h_L}{d_t} - 1\right)^2 \tag{2.12}$$

where h_L is the height of the liquid along a vertical diameter.

Equations (2.9) and (2.10) define critical h_L/B or h_L/d_t for a given $(u_G - u_L)$. The term u_G is the actual gas velocity so that

$$u_{GS} = u_G \frac{h_G}{B} \tag{2.13}$$

for a channel flow and

$$u_{GS} = u_G \frac{A_G}{A_t} \tag{2.14}$$

for a pipe flow, where A_t is the area of the empty pipe. Equations (2.9) and (2.10) for inviscid flows are consistent with observations that transition from a stratified flow to a slug flow at small u_G is predicted to occur at a larger h_L/d_t in a larger diameter pipe for air–water flows. However, comparisons with measurements indicate that the critical h_L/d_t predicted by Taitel and Dukler is larger than what is observed; that is, the system is predicted to be more stable (Wallis & Dobson, 1973).

A physical explanation of this deficiency of (2.9) and (2.10) has been given by Lin & Hanratty (1986) and by Wu *et al.* (1987), who associated it with the neglect of viscous effects. They retained the assumption that the wavelength is long compared with h_G and h_L, but abandoned the assumption of inviscid flow by including the effects of the drag of the gas and the resisting stress of the wall on the liquid. Equations (2.9) and (2.10) still hold provided the velocities in the gas and the liquid can be approximated as uniform flows. The principal difference found in their viscous large-wavelength analysis is that the wave velocity is not given by (2.08) or (2.11). Rather, it is the kinematic wave velocity defined by Lighthill & Whitham (1955).

The first term on the left side of (2.6) represents the destabilizing effects of liquid inertia. For air–water flow at atmospheric conditions, $\rho_G/\rho_L = 1.2 \times 10^{-3}$, so (2.8) simplifies to $c = u_L$ for the range of conditions over which transition is observed. Consequently, the inviscid analysis, equation (2.6), predicts no influence of liquid inertia on stability. It represents a static instability. In contrast, the viscous large-wavelength analysis gives non-zero values of $(c/u_L - 1)$ at the critical condition. As a result, an important destabilizing effect of liquid inertia is considered. The inclusion of viscous stresses has the surprising effect of causing the air–water system to be more unstable, in that the initiation of large-wavelength interfacial disturbances is predicted to occur at lower gas velocities (consistent with experimental observations) than is given by an inviscid analysis.

Viscous large-wavelength theory has been used by Hanratty & Hershman (1961) to describe the initiation of roll waves. Also, as indicated in this subsection, it has been used to predict a critical height for the transition from a stratified air–water flow to a slug/plug flow at $u_{SL} = 10$ m/s in a 9.53 cm pipe and at $u_{LS} = 4$ m/s in a 2.54 cm pipe (Hurlburt & Hanratty, 2002). Experiments with air–water by Soleimani & Hanratty (2003) show that the theory correctly predicts the initiation of slugs/plugs at low gas velocities and the initiation of pseudo-slugs at high gas velocities. It is of interest to note that the observed mechanism for transition to slug/plug flow at low gas velocities (see Section 2.2.6 and Figure 2.4) is quite different from what is conjured by theory, that is, the monotonic growth of very large-wavelength waves.

As is illustrated in the discussions in this subsection, theoretical constructs that represent the transition to slug/plug flow provide a critical h_L/d_t or h_L/B for a given u_{GS}. An important issue in utilizing these theories is the specification of u_{LS}. This requires a proper model for a stratified flow, a topic to be covered in Chapter 5. A discussion about the sensitivity of predicting the critical u_{LS}, to determining interfacial drag, to modeling the liquid flow and to predicting gradients of h_L in the flow direction, is presented by Hurlburt & Hanratty (2002). The effect of an underdeveloped flow is particularly important at low gas velocities, for which changes of h_L in the flow direction can have effects comparable to the pressure gradient.

2.4 Effect of liquid viscosity on flow regimes in a horizontal pipe

The influence of liquid viscosity on flow regimes has been studied by Taitel & Dukler (1987), Andritsos & Hanratty (1987) and Andritsos *et al.* (1989). Results for a 9.53 cm pipe, with $L/d_t = 250$, for air and 1 cP water, and for air and a 100 cP solution are given in Figure 2.6. Results for a 2.54 cm pipe, with $L/d_t = 400$, for a 70 cP solution are shown in a paper by Andritsos & Hanratty (1987). Figure 2.7 presents a graph, from

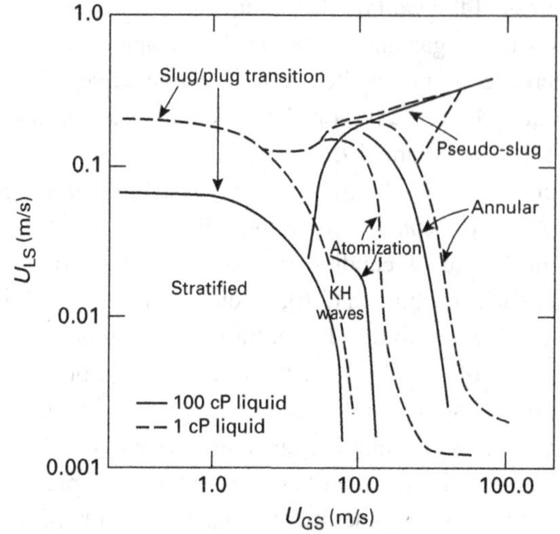

Figure 2.6 Comparison of flow regime maps for air–water and an air–glycerine solution in a horizontal pipe with a diameter of 9.53 cm. Andritsos *et al.*, 1989.

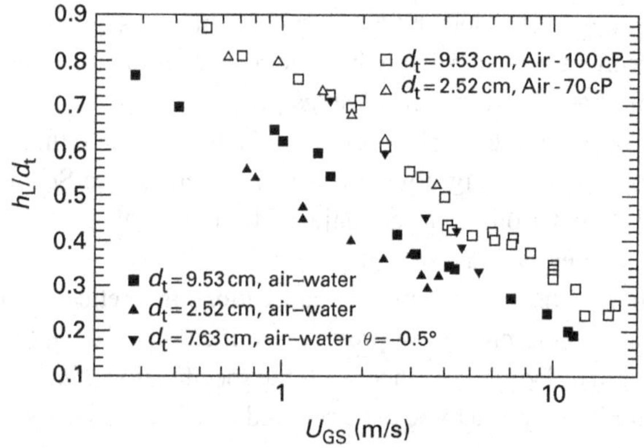

Figure 2.7 Critical h_L/d_t for transition from stratified to slug/plug flow for d_t = 9.53 cm, d_t = 7.63 cm and d_t = 2.52 cm. Hurlburt & Hanratty, 2002.

Hurlburt & Hanratty (2002), which plots the critical h_L/d_t versus superficial gas velocity for the transition from a stratified flow to a slug flow.

Despite the differences between water and high-viscosity glycerine–water solutions, the curves in the flow regime map, in Figure 2.6, representing the initiation of atomization, the formation of annular flows, the appearance of Kelvin–Helmholtz waves and the transition to slug flow at high gas velocities, are similar. The chief differences are associated with the initiation of slugs at low gas velocities. This can be understood by considering Figure 2.7.

Of particular interest is the observation that the critical height is the same for high-viscosity liquids in 9.53 cm and 2.52 cm pipes, and for flow in a 7.63 cm pipe declined

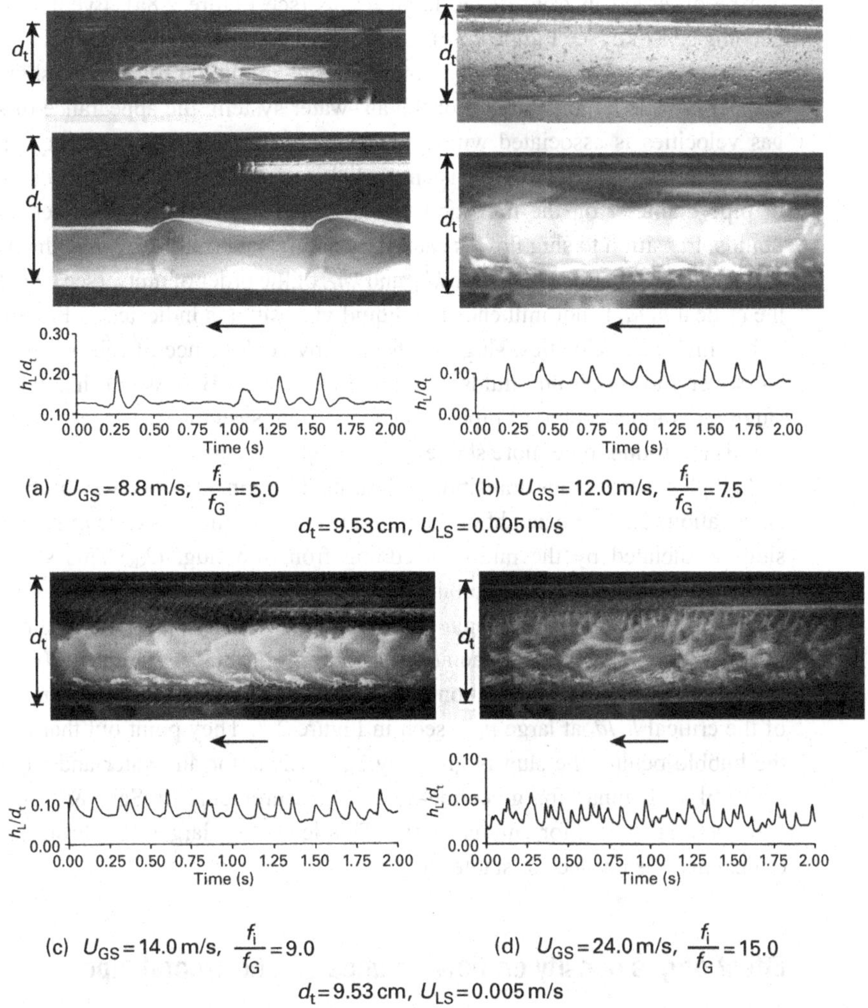

Figure 2.8 Large-amplitude waves on a viscous liquid. Andritsos & Hanratty, 1987.

at an angle of 0.5° to the horizontal (discussed in the chapter on stratified flow). The common feature of all three of these studies is that the interface is smooth at transition. The results for declined flows are interpreted in Section 5.5.3, Figures 5.8 and 5.9 and, in more detail, in a paper by Woods *et al.* (2000). Only effects of liquid viscosity in horizontal pipes are considered in this section.

Consider a stratified flow of a very viscous liquid. The liquid rate is fixed. The interface is smooth. The gas velocity is increased (so h_L/d_t decreases). The first interfacial disturbances that are observed, with increasing gas velocity, in stratified high-viscosity liquids have small wavelengths and small amplitudes. They are regular two-dimensional waves (see Figure 2.3a). With just a slight increase in gas velocity, they give way to a few large-amplitude roll waves with steep fronts, smooth troughs, and spacings that can vary from a few centimeters to a meter. Occasionally, several small two-dimensional waves can be observed in front of the large waves (see Figure 2.8a). Eventually, the waves become cell-like (see Figure 2.8c). Under these conditions the spacing is 1–2 cm.

Slugs can evolve very rapidly from the large-amplitude roll waves, shown in Figure 2.8a, if h_L/d_t is large enough. For the air–water system, the appearance of slugs at low gas velocities is associated with a viscous large-wavelength instability of a stratified flow (as discussed above). Consistent with this interpretation, Figure 2.7 shows an effect of pipe diameter on the transition to slug flow in the air–water system. For viscous liquids, transition to slug flow is caused by an inviscid small-wavelength KH instability, described by equation (2.7), with kh_L and kh_G of the order of unity. (see Chapter 4). Thus, the critical h_L/d_t is not influenced by liquid viscosity (as indicated in Figure 2.7).

For high gas velocities, slugs are formed by coalescence of roll waves, both for the air–water and for air and fluids with large viscosities. Because of this, the transition to slugging is independent of pipe diameter for both systems. However, the more viscous liquids are found to be more stable.

At high gas velocities, transition is described by considering the stability of slugs, that is, equation (2.5). The height of the liquid needed for the existence of stable or growing slugs is dictated by the rate of shedding from the slug, Q_{sh}. This shedding rate is commonly interpreted by assuming that the back of the slug (see Figure 9.1b) may be considered to be a bubble which is moving at a velocity c_B. The shedding rate is then given as $Q_{sh} = (c_B - u_m)A_1$, if aeration is ignored. Hurlburt & Hanratty (2002) provide the following speculative explanation for the influence of liquid viscosity on measurements of the critical h_L/d_t at large u_{GS}, seen in Figure 2.7. They point out that the velocity of the bubble behind the slug is given by $c_B \approx 1.2u_m$ for air–water and suggest that the slug tail is laminar for glycerine–water. Measurements of Sam & Crowley (1986) show that $c_B = 2u_m$ for laminar flows. This leads to a larger Q_{sh}. From (2.5), a larger critical h_L/d_t is needed for stable slugs.

2.5 Effect of gas density on flow regimes in a horizontal pipe

Well-defined results on the effect of gas density on flow regimes are not plentiful. Studies by Crowley *et al.* (1986) in a 17.8 cm pipe and by Wu *et al.* (1987) in a 20.3 cm pipe are discussed by Hurlburt & Hanratty (2002) and by Andritsos *et al.* (1992), who emphasize the importance of considering slug stability in determining transition at a large gas density. The study by Crowley *et al.* involved the use of freon gas and water. The gas density was 32.5 kg/m^3 (compared with a density of 1.2 kg/m^3 for air at atmospheric conditions). The study by Wu *et al.* involved a natural gas/condensate. The gas density was 65 kg/m^3 and the surface tension was 0.0118 N/m (compared with 0.07 N/m for water). The studies of the transition from slug flow to annular flow by Reimann & Seeger (1981) for steam–water and air–water are also of interest.

The effect of gas flow on generating instabilities in stratified flows, in initiating atomization, and on influencing the flow rate in a stratified liquid should be represented roughly through variations of the gas phase kinetic energy. Thus, many investigators have suggested that flow regime maps of the type shown in Figure 2.1 should use $\rho_G^{0.5}u_{GS}$

as the abscissa, rather than u_{GS}. This approach is partially consistent with the researches cited above.

Since instability of a stratified flow varies roughly as $\rho_G^{0.5}u_{GS}$, it is possible that mechanisms which consider the instability of a stratified flow (such as the viscous large-wavelength and the Kelvin–Helmholtz analyses) will predict the appearance of waves at too low a gas velocity for the instability to evolve into a stable slug. Then, transition is controlled by slug stability, rather than the instability of a stratified flow.

The transition from slug/plug flow to annular flow clearly shows an influence of ρ_G whereby the critical u_{GS} decreases with ρ_G^n with $n \approx 0.4$. (Reimann and Seeger 1981).

2.6 Effect of drag-reducing polymers

One of the more intriguing advances in studies of single phase turbulence is the finding that the introduction of small amounts of long-chain polymers into a liquid can cause decreases in the fluid turbulence and the resistance of the wall to the flow (Toms, 1948). This motivated a number of studies of the effects of drag-reducing polymers on gas–lquid flows. The main focus was the change in the pressure drop. A review of work in this area by Manfield *et al.* (1999) suggests that an understanding of the influence of drag-reducing polymers on multiphase flow was not available in 1999. That is, changes in the pressure gradient could not be directly related to the drag reduction observed in single-phase flows.

More recent research has focused on the effect of drag-reducing polymers on the configuration of the phases. Thus, Al-Sarkhi & Hanratty (2001a,b) have studied the influence of long-chain polymers on annular flows. They found that disturbance waves are destroyed and therefore the production of drops is discontinued. The flow changed to a stratified configuration. Drag-reductions of 48% were realized for a 9.53 cm pipe and 63% for a 2.54 cm pipe.

Soleimani *et al.* (2002) studied the effects of drag-reducing polymers on interfacial drag and on the initiation of slugs for air and water flowing in a 2.54 cm pipe over a range of superficial gas velocities of 1–100 m/s. The interfacial drag can be represented as $\tau_i = f_i \rho_G \bar{u}_G^2$. Soleimani *et al.* found that added polymers cause a decrease in interfacial activity, and consequently a greatly reduced interfacial friction factor, f_i. However, for stratified flows, decreases in the interfacial friction factor are accompanied by increases in the liquid height, and therefore increases in the gas velocity that counterbalance the effect of the decrease in interfacial roughness. Soleimani *et al.* showed that the addition of polymers to stratified flows causes large decreases in f_i but modest decreases in the pressure drop.

A more interesting result is the finding that polymers delay the transition from a pseudo-slug to a slug flow regime, observed with increases in the superficial velocity of the liquid. This is consistent with the finding, discussed in Section 2.2.6, that transition from a pseudo-slug regime to slug flow with increasing liquid flow is dictated by a consideration of the stability of a slug. An increase in liquid viscosity (see Section 2.4) dampens turbulence and decreases the stability of slugs by increasing the shedding rate.

Baik & Hanratty (2003b) extended the work of Soleimani *et al.* by studying the effect of drag-reducing polymers on stratified flows of air and water in a 9.53 cm pipe. They also observed a damping of interfacial waves. This led to decreases in the interfacial stresses. Drag-reductions as high as 19% were observed.

A second effect observed by Baik & Hanratty was a modification of the mechanism for the transition from stratified flow to slug flow at very low gas velocities. For air–water flows, this transition results from the growth of Jeffreys waves into slugs by the bifurcation process described in the first paragraph of Section 2.2.6 and illustrated in Figure 2.4. The addition of polymers inhibits this bifurcation (by mechanisms which are not completely understood). Drag reductions as large as 40% were realized.

2.7 Vertical upflows

2.7.1 Flow patterns

Several investigations of patterns for upward flow of air and water have been reported in the literature. Significant differences exist, even for the same system at roughly the same gas and liquid flows. There are a number of reasons for these apparent contradictions: The patterns have been defined in different ways. A number of techniques have been used to provide criteria to define transitions (Dukler & Taitel, 1986; Delhaye, 1981). The presence of surfactants and the method for introducing the gas have an effect (Hewitt & Hall-Taylor, 1970, p. 13).

For a fixed liquid flow, the following regimes typically appear with increasing gas velocity.

At low liquid flows a bubble pattern is observed (at low gas flows) for which the liquid phase is continuous and the gas is broken into bubbles by inertial forces. The bubbles can have a distribution of sizes. The largest, called cap bubbles, have a hemispherical shape. The bubbles are not distributed uniformly over the pipe cross-section so the voidage can vary with location. Clusters can form. Void fraction waves can exist under certain conditions.

At higher gas velocities a slug flow can appear. It consists of a progression of bullet-shaped bubbles (called Taylor bubbles) that fill most of the pipe cross-section. The fronts resemble cap bubbles and the backs are approximately flat. The slugs of liquid between the Taylor bubbles can be aerated. Large pressure fluctuations are observed. The liquid between the gas and the wall moves downward. Thus, to an observer moving with the Taylor bubble, the liquid flows around the bubble.

With further increases in gas velocity, slugs break down and a chaotic oscillatory motion of a highly aerated liquid appears. This pattern is called churn flow. (In some studies, it is described as a foam flow.) It occurs over a range of gas and liquid velocities which increases with increasing pipe diameter. At low gas flows, a churn pattern may be looked upon as a breakup of slugs by an occasional bridging across the pipe by the liquid phase. At high gas flows "it may be considered as a degenerate form of annular flow with the direction of the wall film changing and very large waves being formed at the interface.

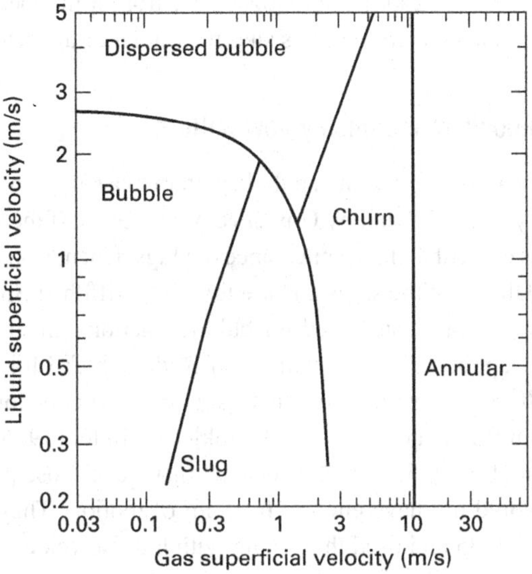

Figure 2.9 Representative map for upward flow of air and water. (constructed by Omebere-Iyari & Azzopardi, 2000)

In this range, the term semi-annular flow has sometimes been used" (Omebere-Iyari & Azzopardi, 2000).

At still larger gas flow ($u_{GS} > \sim 10$ m/s) an annular pattern exists whereby part of the liquid flows upward as a film along the wall and part, as drops entrained in the gas. There is an exchange of liquid between the gas and the flowing film. For air–water flows there is a critical film flow, defined as $\mathrm{Re}_{LF}/4 = W_{LF}/\mu_L \pi d_t$ (Andreussi *et al.*, 1985), below which drops cannot be torn from the film. For a fixed gas velocity, the volume fraction of drops in the gas increases with increasing u_{LS} after the critical liquid flow in the film is reached. At large u_{GS} and large enough u_{LS} the drops reach a concentration such that large structures appear in the gas flow. This pattern, called wispy annular flow, is described by Bennett *et al.* (1965, 1967).

At very large liquid flows and small gas flows, a pattern called a dispersed bubble regime emerges (Taitel *et al.*, 1980). Turbulence in the liquid causes the breakup of the gas into bubbles. This regime is similar to the bubbly pattern observed in horizontal gas–liquid flow. Thus, the critical liquid velocity needed for this type of bubbly flow to appear is approximately the same for these two orientations.

A representative flow pattern map is shown in Figure 2.9 for upward flow of air and tap water in a 29 mm pipe. This was constructed by Omebere-Iyari & Azzopardi (2000) from correlations for the bubble/slug (Taitel *et al.*, 1980), for the bubble/dispersed bubble (Taitel *et al.*, 1980), for the slug/churn (Brauner & Barnea, 1986), and for churn/annular (McQuillan & Whalley, 1985) transitions. It should be noted that the slug pattern was not observed by Cheng *et al.* (1998) in a pipe with a diameter of 150 mm, by Kytömaa & Brennan (1991) in a pipe with a diameter of 102 mm or by a McMaster University research group in a pipe with a diameter of 100 mm (cited by Omebere-Iyari & Azzopardi, 2000).

An article by Omebere-Iyari *et al.* (2008) summarizes data from a number of sources. These show that a churn-turbulent pattern replaces slug flow in large-diameter pipes.

2.7.2 Mechanisms for the breakdown of the bubbly flow pattern

Discussions of early work on the change from a bubble pattern to a slug flow are given in books by Hewitt & Hall-Taylor (1970) and by Govier & Aziz (1972). If the inlet mixing process creates bubbles, it is evident that the appearance of slugs depends on coalescence of the bubbles and on the stability of the slugs that are formed. Griffith & Snyder (1964) suggested that bubbly flow cannot be sustained for bubble fractions greater than 35%. This observation has been supported by calculations of Radovich & Moisis (cited by Hewitt & Hall-Taylor, 1970, p. 12), which show that the number of collisions increases rapidly with void fraction in the range of $\alpha = 0.30$. Dukler & Taitel (1986) argue that collisions cannot be a factor at $\alpha = 0.52$, which is characteristic of a close packed cubic lattice, since the bubbles would not have enough freedom of motion. They assumed a criterion for transition of $\alpha = 0.25$ and used this, along with the rise velocity of bubbles given as

$$U_S = u_G - u_L \tag{2.15}$$

to predict transition. After substituting

$$u_G = \frac{u_{GS}}{\alpha} \tag{2.16}$$

$$u_L = \frac{u_{LS}}{(1 - \alpha)} \tag{2.17}$$

where α is the void fraction, one gets

$$u_{LS} = u_{GS}\frac{(1-\alpha)}{\alpha} - (1-\alpha)U_S \tag{2.18}$$

Dukler & Taitel (1986) used the following equation developed by Harmathy (1960) for single large bubbles in an infinite medium:

$$U_{S\infty} = 1.53\left[\frac{\sigma g(\rho_L - \rho_G)}{\rho_L^2}\right]^{1/4} \tag{2.19}$$

(See Govier & Aziz, 1973, p. 369.) The rise velocity in a swarm of bubbles decreases with increasing α so an equation developed by Zuber & Hench (cited by Dukler & Taitel, 1986) was used to relate U_S to $U_{S\infty}$:

$$U_S = U_{S\infty}(1-\alpha)^{1/2} \tag{2.20}$$

Thus, (2.19) and (2.20), with $\alpha = 0.25$, can be substituted into (2.18). It is noted that these relations give a transition condition which is independent of pipe diameter.

The notion that the transition can be explained solely by considering the gradual coalescence in a uniform swarm of bubbles has been questioned. This classical explanation implies that a transition can always occur if a pipe is long enough. Cheng *et al.* (1998) quote studies which show a small change in bubble properties for an air/tap water flow along a pipe, and no effect of pipe length on the void fraction. These results suggest that transition occurs throughout the column by a stability mechanism, rather than by gradual coalescence.

2.7.3 Mechanism for the slug flow to churn flow transition

The main thrust of theories on the transition from a slug flow to a churn flow is the consideration of the stability of slugs. The interpretation most favored is a breakdown of the film between Taylor bubbles and the wall, whereby it changes from a net downward to a net upward flow. Large (flooding) waves carry liquid upward and the liquid film between the flooding waves carries fluid downward (Jayanti & Brauner, 1994; Hewitt & Hall-Taylor, 1970).

However, Brauner & Barnea (1986) suggest that instability arises from excessive aeration in the liquid slugs between the Taylor bubbles. Jayanti & Brauner (1994) compare these two approaches in their review of research on churn flow. Little, if any, attention has been given to the possibility of associating this transition with the breakdown of the front of a Taylor bubble. Yet, this could be an important consideration in explaining why slug flow is not observed in large diameter pipes. (See Omebere-Iyari *et al.*, 2008.)

2.7.4 Mechanism for the churn/annular transition

Transition from churn flow to annular flow can be explained by considering the conditions needed for the existence of an annular flow, such as the necessity for all of the liquid in the film to be moving upward or for the droplets to be carried upward. Most of the attention has been given to the first of these. A reversal of the direction of film flow is consistent with the existence of a zero velocity gradient (or a zero shear stress) at the wall. A force balance on a small length of film gives the following:

$$0 = \pi d_t \tau_W dz + \pi d_t \tau_i dz - \rho_L g h_L \pi d_t dz \tag{2.21}$$

where the thickness of the film, h_L, is assumed to be small compared with d_t. If $\tau_W = 0$ and $\tau_i = f_i \frac{1}{2}\rho_G u_G^2$,

$$\frac{1}{2} d_t f_i \rho_G u_G^2 = \rho_L g h_L \tag{2.22}$$

at transition. Wallis (1969) gives the following rough approximation for the friction factor in an annular flow

$$f_i = 0.005\left(1 + 300\frac{h_L}{d_t}\right) \tag{2.23}$$

If (2.23) is substituted into (2.22) the following relation is obtained if $u_G \approx u_{GS}$

$$\frac{u_{GS}\rho_G^{1/2}}{g^{1/2} d_t^{1/2} \rho_L^{1/2}} = C \tag{2.24}$$

where C is approximately equal to 0.87. McQuillan & Whalley (1985) used (2.24) with $C = 1$ to correlate data on the transition to annular flow. The agreement is approximate but quite good, considering that the data covered a range of gas densities from 1.29 to 107.0 kg/m^3, tube diameters of 0.01 to 0.105 m, surface tensions of 0.0074 to 0.072 N/m, liquid viscosities of 0.096 to 1.06 N s/m^2 × 10^3.

2.7.5 Mechanism for the transition to dispersed bubbly flow

Dukler & Taitel (1976) and Taitel *et al.* (1980) point out that the breakup of bubbles by turbulence at large liquid velocities has been observed in a number of situations and recommended the use of the following correlation developed by Hinze to calculate the bubble size, d_P, created by turbulence:

$$\frac{d_P \varepsilon^{2/5}}{(\sigma/\rho_L)^{3/5}} = \text{constant} \tag{2.25}$$

where σ is the surface tension and ε is the rate of dissipation of mechanical energy per unit volume in the liquid. Since $\varepsilon = \rho_M F$, where F, the frictional loss per unit mass of liquid, can be calculated with (1.44).

The notion behind (2.25) is that turbulent velocity fluctuations may be viewed as containing a number of scales (or wavelengths). Spatial variations of the pressure fluctuations that have a scale close to the bubble diameter can exert a force on the bubbles. This is balanced by surface tension forces, characterized as σ/d_P. For bubbles in equilibrium with the turbulence field, these forces will be equal so that

$$\rho_L v^2_{\lambda=d_P} = C\sigma/d_P \tag{2.26}$$

where v_λ is the characteristic velocity of fluctuations with a wavelength λ and C is a constant. It is assumed that λ is small enough to be independent of the method by which turbulence is produced and large enough for the fluctuations to be independent of viscosity. Komogoroff has proposed that, in this range of wavelengths, the velocity fluctuations in a homogeneous isotropic field depend only on ε and λ. Thus, from dimensional reasoning,

$$v_\lambda \approx (\varepsilon\lambda)^{1/3} \tag{2.27}$$

provided $\lambda/d_t << 1$ and $\lambda v_\lambda/\upsilon_L >> 1$. From (2.26) and (2.27), the equilibrium bubble size depends on σ/ρ_L, ε, d_P (with d_P being substituted for λ). Equation (2.25) results from dimensional analysis. It is presumed to be valid for all turbulent liquid flows. Taitel *et al.* (1980) argue that, in addition, the dispersed bubbles must keep a spherical shape. They use this criterion and (2.25) to provide a criterion for transition from a dispersed bubble flow to a bubbly pattern or a slug pattern Thus, in Figure 2.9, the dispersed bubble regime represents conditions for which bubbles are created by liquid turbulence, and the bubble regime represents conditions for which bubbles are created by liquid inertia.

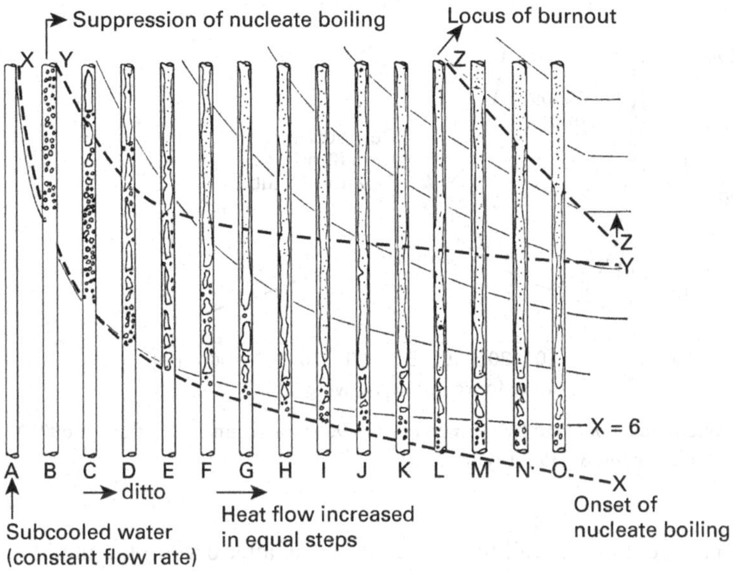

Figure 2.10 Patterns for upward flow of a liquid in a heated tube. Figure 2.11 in Hewitt & Hall-Taylor, 1970.

2.8 Vapor patterns in evaporating and condensing flows / critical heat flux

An important application involving a gas–liquid mixture is the evaporation of an upwardly flowing liquid which is supplied with a constant heat flux at the wall, shown in Figure 2.10 (taken from Hewitt & Hall-Taylor, 1970). The sketches depict situations with a constant mass flow rate and increasing heat flux. Boiling first appears as a bubbly pattern at the downstream end of the pipe. With increasing heat flux, the location at which boiling is initiated moves upstream and more vapor is generated. Thus, the bubbly pattern can evolve into a slug or churn flow. At large enough heating, an annular flow occurs. At still larger heat fluxes, the liquid in the wall film evaporates and a vapor/droplet flow exists for which the pipe wall is in contact with the vapor flow. The light curves represent constant qualities defined as the fraction of the mass flow that is vapor.

Hewitt & Hall-Taylor describe Figure 2.10 along the following lines: The dark curve labeled X in the sketch represents the location at which boiling is initiated. In order for this to occur, the liquid at the wall must be superheated. However, over most of the channel cross-section the liquid is below its boiling point, so that bubbles generated at the wall collapse after they leave the wall. Under these conditions, the average temperature of the flowing liquid is below the boiling point and the evaporation process is known as subcooled boiling. Eventually, phase change occurs not only by boiling at the wall but also by evaporation at the interfaces. The latter process becomes more important as the fluid progresses along the pipe so that the liquid at the wall is not sufficiently superheated for boiling to occur. In the annular flow region, vapor generation occurs primarily by evaporation at the interface of the wall film. Curve Y indicates the loci at which vapor generation at the wall ceases. Curve Z indicates

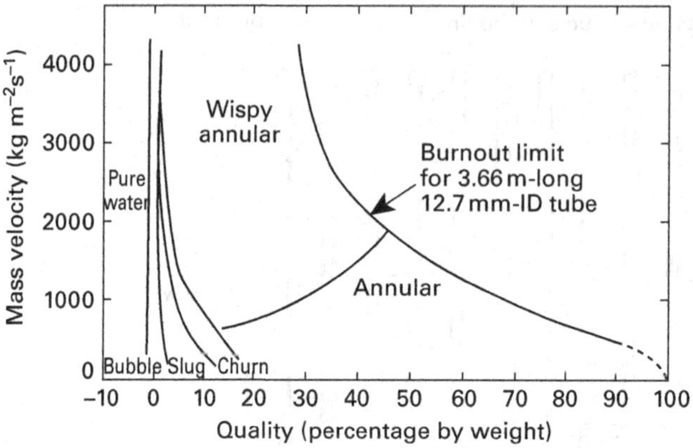

Figure 2.11 Flow pattern data collected by Bennett *et al.* (1965) for steam–water flow at 6.89 MP pressure. Figure 2.6 in Hewitt & Hall-Taylor, 1970.

whether dryout at the wall has occurred. Since, after dryout, the liquid blanket at the wall is no longer available the wall temperature increases rapidly. In situations for which the heat input is constant, this can lead to a deterioration of the integrity of the pipe wall. Operating conditions (for example, heat flux) need to be selected so as to avoid this burnout.

A critical heat flux need not be associated with a dryout of the wall film in annular flow. For example, burnout can also occur at low flows and high fluxes. The number of boiling sites at the wall can then increase to such an extent that a film of gas forms at the wall. This film boiling is not able to remove as much heat at the wall as does nucleate boiling, so the wall temperature increases rapidly. This is similar to the boiling crisis observed for pool boiling.

The analysis of evaporating flows is very difficult. Flow maps, such as shown in Figure 2.11 (Hewitt & Hall-Taylor, 1970), can be useful. These measurements were made for the flow of steam and water up a 3.66 m length of 12.7 mm pipe subjected to a constant heat flux. The mass flow was kept constant and the heat flux was increased. From (1.11) and the assumption that the liquid is uniformly distributed over the pipe cross-section, the following energy balance can be written for a flowing mixture if the contributions to the enthalpy from changes in kinetic energy and work against gravity are negligible compared with the heat flux:

$$(h - h_{\mathrm{i}})GA = \overline{q}_{\mathrm{W}}Pl \tag{2.28}$$

where h_{i} is the enthalpy at the inlet. The term in parentheses in (2.28) represents the enthalpy change over a pipe length l for a given heat flux at the wall and Pl is the inside area, with P equal to the pipe perimeter. Now, if equilibrium is assumed to exist between the phases the quality can be calculated from tables of thermodynamic functions. (Hewitt and Hall-Taylor (1970) call this a thermodynamic quality.) A quality obtained in this manner is plotted as the abscissa in Figure 2.11. The important role of the annular flow region is emphasized.

This method for representing flow patterns is a way to use gas–liquid flow data (with no phase change) to approximate, very roughly, flow regimes in evaporative and condensing flows. This would simply involve plotting $(\rho_G u_{GS})/(\rho_G u_{GS} + \rho_L u_{LS})$ as the abscissa.

2.9 Downflows in vertical pipes

Bubbles in a liquid experience a buoyancy force given by

$$\vec{F}_b = V_P(\rho_L - \rho_G)\vec{g} \tag{2.29}$$

This is the major contributor to the slip velocity of bubbles, $(u_G - u_L)$. In a shear flow, bubbles also experience a lateral force perpendicular to the mean flow direction described by

$$F_{Lr} = -C_L(\overline{u}_G - \overline{u}_L)\rho_L \frac{d\overline{u}_z}{dr} \tag{2.30}$$

where C_L is an empirically determined lift coefficient and V_P is the volume of the bubble (see Section 10.4).

In an upward dispersed flow, the slip velocity is positive and du_L/dr is negative, so bubbles will tend to move toward the wall. However, for downward flow, the gravitational force is such that bubbles move in a direction opposite to the flow. Thus, the slip is negative. The lift force is away from the wall, so bubbles will tend to move toward the center of the pipe.

Another difference between upflows and downflows is that, at zero gas velocity, one can have a falling film along the wall, which resembles an annular flow.

Oshimoto & Charles (1974) report on the coring described above for downflows. They identified the appearance of the following patterns with increasing gas flow: (1) bubble-coring, (2) bubble-slug, (3) falling film, (4) froth, (5) annular flow.

Slug flow is initiated by coalescence of the coring bubbles. The Taylor bubbles appear to be the same as observed in upward flow. However, they are upside-down and move counter to the liquid flow. Thus, the nose of the bubble points upstream.

The falling film identified by Oshimoto & Charles is thicker than the film observed in annular flows. Bubbles occluded in the film escape and bridges of liquid form across the pipe. The froth flow is similar to the froth flow observed in vertical upflows, except that the mixture of gas and liquid is less turbulent. The annular flow that appears at very large gas velocities is the same as observed in upflows.

Sekoguchi *et al.* (1996) and Ishii (2004) studied air and water flowing down 2.58 cm and 5.08 cm pipes. Air was mixed in the flowing liquid by using a centrally located sparger. A specially designed array of needle conductivity probes enabled an observation of the configuration of the air–water interfaces. At small u_{GS}, the gas in the bubble was found to be moving upstream relative to the liquid (a negative slip velocity). Under these circumstances the large bubble is asymmetric in the flow direction. At intermediate u_{GS}, a range of conditions exist for which the slip velocity is close to zero and the front and back of the bubble are the same.

2.10 Microgravity flows

2.10.1 General comments

Microgravity effects on gas–liquid flows are an important consideration in space applications. Moreover, they supply a test for theories developed from 1g data. Situations for which $g = 0$ are considered in this section. Gas–liquid flows change in several ways: (1) The pattern is not affected by the orientation of the conduit or by the direction of flow, as is found when g is not close to zero. (2) The flows tend to be axisymmetric, so that stratified patterns do not exist. (3) The rise velocity of bubbles, U_S, is zero, so that local gas and liquid velocities are equal.

However, the spatially averaged gas, $\langle u_G \rangle$, and liquid, $\langle u_L \rangle$, velocities need not be equal if the gas is not uniformly distributed, even though $U_S = 0$ locally. Zuber & Findlay (1965) suggest that, for air–liquid flow at 1g,

$$\langle u_G \rangle = C u_m + U_S \tag{2.31}$$

where

$$u_m = u_{LS} + u_{GS} \tag{2.32}$$

and C can vary with flow pattern. Nicklin *et al.* (1962) suggest that C=1.2 for vertical slug flows at 1g (see Section 8.5). Colin *et al.* (1991) claim C=1.1 for small bubbles at 1g.

2.10.2 Flow regimes

Measurements of gas–liquid flow under microgravity conditions have been made for a range of conditions by Dukler *et al.* (1988), Colin *et al.* (1991), Bousman *et al.* (1996), Zhao & Rezkallah (1993). These studies were carried out in drop towers and in flights of jet aircraft. By following the parabolic path of a jet, periods of microgravity of 15–25 s were realized. Photographs of the different flow patterns are provided by Bousman *et al.* (1996) and by Colin *et al.* (1991). All of these experiments show that under microgravity conditions the pattern is independent of orientation.

Measurements by Colin *et al.* (1991) in a microgravity environment confirm that $U_S = 0$ and give $C = 1.2$ in (2.31) both for bubbly and slug flow. By definition,

$$u_{LS} = \langle u_L \rangle \langle 1 - \alpha \rangle \tag{2.33}$$

$$u_{GS} = \langle u_G \rangle \langle \alpha \rangle \tag{2.34}$$

Substitution of (2.31) into (2.34), using $U_S = 0$, yields

$$\frac{u_{GS}}{u_{LS} + u_{GS}} = \langle \alpha \rangle C \tag{2.35}$$

Equation (2.35) can be used to calculate $\langle \alpha \rangle$ as a function of u_{LS} and u_{GS}. Values obtained in this way, with $C = 1.2$, agree with measurements for bubbly and slug flows in microgravity.

Three regimes have been defined for microgravity flows. These are described by Colin *et al.* (1991) as follows:

(1) "bubbly flow at low gas rates, where the bubble size is typically smaller than the pipe diameter,"
(2) "slug flow for moderate gas and liquid rates, consisting of Taylor bubbles longer than $1d_t$ and a liquid slug in which smaller bubbles may be dispersed; these bubbles travel axially without significant drift with respect to the Taylor bubbles,"
(3) "annular flow for high gas rates, in which a wavy liquid film exists at the wall, on which roll waves propagate."

The transition from bubbly flow to slug flow with increasing gas rate is more gradual than is observed at $1g$, so a transitional regime is defined for a range of conditions.

The bubbly pattern is similar to what has been observed in downflows at $1g$, in that coring is observed. This could explain why the constant, C, in (2.31) is the same for slug and bubbly flow, in that the gas accumulates in the central region of the pipe for both regimes. Coring also seems to affect bubble coalescence.

Bousman *et al.* (1996) present flow regime maps for water ($\mu = 1$ cP, $\sigma = 72$ dynes/cm), a 50% water–glycerol solution ($\mu = 6$ cP, $\sigma = 63$ dynes/cm) and a surfactant solution ($\mu = 1$ cP, $\sigma = 21$ dynes/cm). Tubes with diameters of 12.7 mm and 25.4 mm were used. The initiation of annular flow for these systems occurred at roughly the same conditions.

The transition from bubbly to slug flow is given by $u_{GS} = u_{LS}$ for all of the experiments in a 12.7 mm tube and for air and water–glycerine solutions in a 25.4 mm tube. Experiments with water and with surfactant solutions in a 25.4 mm tube are reported to occur at a smaller u_{GS} for a given u_{LS}, that is,

$$u_{GS} = \frac{1}{2.5} u_{LS} \tag{2.36}$$

The difference in the results for air and water in a 25.4 mm tube from those in a 12.7 mm tube, cited above, should be noted. Some support for this finding on the effect of pipe diameter is obtained from the study of air–water in a 40 mm tube by Colin *et al.* (1991). Their critical gas velocities are represented by

$$u_{GS} = \frac{1}{3.2} u_{LS} \tag{2.37}$$

which, within experimental error, is the same as (2.36).

The result $u_{GS} = u_{LS}$ for the transition from bubbly to slug flow, obtained with a 12.7 mm tube, corresponds to a critical $\langle\alpha\rangle$ of about 0.40, that is, a closely packed bubble swarm. The results (2.36) and (2.37) obtained for air and water in 25.4 mm and 40 mm tubes correspond to an $\langle\alpha\rangle$ of about 0.20. Bousman *et al.* (1996) examined movie films of the air–water system near transition. In the 25.4 mm tube:

> the gas–liquid bubble interfaces were in a continuous state of fluctuations while those in the 12.7 mm tube under the same flow conditions were more stable. These oscillations in the large tube, which can be attributed to turbulence, give the bubbles a larger effective diameter, thus increasing the probability of contacting nearby bubbles. The result is a transition to slug flow at a lower void

fraction. This effect was not observed in the air–water/glycerine experiments in the large tube due to a reduction in turbulence. . .

2.10.3 Comparison of flow regimes for microgravity and 1*g* flows

The influence of gravity is most noticeable in horizontal flows where it can have a direct effect on the distribution of liquid. The stratified pattern is not observed and the slug pattern is quite different under microgravity conditions. It would seem that comparisons should be made with vertical flows. However, even for these flows, differences can be observed.

In vertical upflow, gravity opposes the flow. Thus, churn flow, which involves intermittent reversals close to the wall due to gravity, does not exist in microgravity flows. In vertical downflows, gravity aids the flow, so that annular flows at $1g$ are more easily formed and can be observed over a wider range of gas velocities. This is not the case for microgravity conditions.

However, since the absence of gravity leads to a zero value of the rise velocity, there are a number of less obvious reasons to expect differences in flow regime boundaries. As already discussed, this can lead to differences in $\langle \alpha \rangle$, in the spatial distribution of bubbles (because of lift forces), in the interaction of bubbles with one another, and in the wake structure of small bubbles in the liquid.

Comparisons of experiments with and without gravity can be complicated because they might not involve identical systems and might involve the use of different methods to mix the liquid and the gas. Thus, the microgravity experiments by Colin *et al.* (1991) in a 40 mm tube are welcome. They show a transition from bubbly to slug flow which is roughly the same for microgravity and for upflow in a $1g$ environment.

Transition to annular flow in microgravity is observed at a lower u_{GS} than is found for upward flow at $1g$. This can be explained because transition at $1g$ involves a change from a churn flow. At $g = 0$, it involves a breakdown of a slug flow. At large u_{GS} (say 10 m/s) the annular flow resembles what is usually found in vertical $1g$ flows. That is, a thin wall film is agitated by disturbance waves. There is an exchange of drops between the gas flow and the wall film. Annular flow in microgravity at $u_{\mathrm{GS}} < 10$ m/s is characterized by a relatively thick wall layer with large-amplitude waves and a gas flow that contains only a small amount of entrained drops. Sometimes these differences are acknowledged by defining a new flow regime.

2.11 Capillaries and microchannels

2.11.1 Capillary tubes

Microgravity flows are situations where surface tension effects are much larger than gravitational effects. The relative importance of these two forces is characterized by the Bond number

$$\mathrm{Bo} = \frac{d_{\mathrm{h}}}{\left[\sigma/(\rho_{\mathrm{L}} - \rho_{\mathrm{G}})g\right]^{1/2}} \tag{2.38}$$

where $\lambda = [\sigma/g(\rho_L - \rho_G)]^{1/2}$ is the Laplace length scale and d_h is the hydraulic diameter of a passage, equal to 4 times the cross-sectional area divided by the wetted perimeter. For a circular tube $d_h = d_t$. Thus, surface tension can dominate gravitational effects for small d_h, as well as for small g. Serizawa & Feng (2000) point out that $\lambda = 6.6$ mm for boiling water and $\lambda = 2.7$ mm for air–water at atmospheric conditions.

Studies by Damianides & Westwater (1988) in horizontal capillaries with $d_t = 1$–5 mm and by Fukano and his co-workers (Fukano & Kariyasaki, 1993) show that stratified flows do not exist for air–water flows in horizontal tubes with $d_t \leq 5$ mm. Furthermore, Fukano & Kariyasaki (1993) show that bubbles have a zero rise velocity in a still liquid for $d_t \leq 5$ mm.

Photographs, as well as flow regime maps of the configurations of the phases, are given by Triplett *et al.* (1999) and by Fukano & Kariyasaki (1993). Three principal patterns bubbly, slug (intermittent) and annular, are identified. The maps are observed to be roughly the same for tubes with diameters of 1–5 mm and to be independent of orientation.

The annular flow pattern and the transition from slug flow to annular flow are similar to what is found for microgravity situations.

However, the range of conditions associated with the bubbly pattern is much smaller for air–water flows in capillary tubes ($d_t = 1 - 5$ mm) than in microgravity flows ($d_t = 12.7$ mm, 25.4 mm, 40 mm). It appears that the small-diameter capillary tubes promote the formation of slugs so that bubbly flows of the type discussed in Section 2.7.5 do not exist. However, at large liquid velocities the dispersed bubbly pattern is observed. Thus, in the air–water studies of Fukano & Kariyasaki (1993), patterns with small bubbles occupy only a small region (roughly, $u_{LS} > 0.8$ m/s, $u_{GS} < 1$ m/s). Since the properties of dispersed bubbles are believed to be governed by liquid phase turbulence, the picture presented here could change in more viscous liquids (which would be laminar).

2.11.2 Microchannels

Tubes with diameters much smaller than 1 mm, say 20–100 microns (μm), are called microchannels. Gas–liquid flows in these tiny conduits are of interest in the design of compact heat exchange devices that involve boiling. Studies in microchannels include air–water flow in 25 micron tubes by Serizawa & Feng (2000, 2001) and in 100 micron tubes by Kawahara *et al.* (2002). The gas and liquid were admitted at the inlet to the test section where inertia and liquid phase turbulence mixes the two phases.

The dispersed bubble pattern defined for capillary tubes is not observed. As pointed out by Serizawa & Feng (2000), the decrease in channel size can result in the existence of a laminar environment. This could explain the absence of the dispersed regime which depends on the interaction of liquid phase turbulence and bubbles. Thus, the main flow regimes are slug flow (elongated bubbles) and a separated regime in which all of the liquid flows along the wall as a smooth or wavy layer, or as a sequence of rings. These rings could result from the breakdown of a slug flow which has very short bridges.

The paper by Serizawa & Feng (2000) suggests the existence of an annular pattern with a thin liquid film and droplets entrained in the gas at large gas flows. They show photographs for steam–water flow in a 50 micron heated silicon tube. Tiny steam bubbles generated at the wall are observed close to the wall at a low heating rate. By increasing the heating rate the following patterns are observed: bubbles of the same size as the tube, short elongated bubbles, long bubbles (slug flow), a liquid ring flow, and annular-dispersed flow. A discussion of forced flow boiling in microchannels is given by Ghiaassiaan (2003).

References

Al-Sarkhi, A. & Hanratty, T.J. 2001a. Effect of drag-reducing polymers on annular gas–liquid flow in a horizontal pipe. *Int. J. Multiphase Flow* 27, 1152–1162.

Al-Sarkhi, A. & Hanratty, T.J. 2001b Effect of pipe diameter on the performance drag-reducing polymers in annular gas–liquid flows. *Trans IChemE*, 79 A, 402–408.

Andreussi, P. & Persen, L.N. 1987 Stratified gas–liquid flow in downwardly inclined pipes. *Int. J. Multiphase Flow* 13, 565–575.

Andreussi, P., Asali, J.C. & Hanratty, T.J. 1985 Initiation of roll waves in gas–liquid flows. *AIChE Jl* 21, 119–126.

Andritsos, N. & Hanratty, T.J. 1987 Interfacial instabilities for horizontal gas–liquid flows in pipelines. *Int. J. Multiphase Flow* 13, 583–603.

Andritsos, N., Williams, L. & Hanratty, T.J. 1989 Effect of liquid viscosity on the stratified-slug transition in horizontal pipe flow. *Int. J. Multiphase Flow* 15, 877–892.

Andritsos, N., Bontozoglou, V. & Hanratty, T.J. 1992 Transition to slug flow in horizontal pipes. *Chem. Eng. Comm.* 118, 361–385.

Baik, S. & Hanratty, T.J. 2003a Concentration profiles of droplets and prediction of the transition from stratified to annular flow in horizontal pipes. *Int. J. Multiphase Flow* 29, 329–338.

Baik, S. & Hanratty, T.J. 2003b Effects of a drag-reducing polymer on stratified gas–liquid flow in a large diameter horizontal pipe. *Int. J. Multiphase Flow* 29, 1749–1757.

Barnea, D., Shoham, O., Taitel, Y. & Dukler, A.E. 1980 Flow pattern transition for gas–liquid flow in horizontal and inclined pipes. *Int. J. Multiphase Flow* 6, 387–397.

Bennett, A.W., Hewitt, G.F., Kearsey, H.A., Keeys, R.K.F. & Lacey, P.M.C. 1965 Flow visualization studies of boiling at high pressure. AERE-R4874.

Bennett, A.W., Hewitt, G.F., Kearsey, H.A. & Keeys, R.K.F. 1967 Heat transfer to steam–water mixtures flowing in uniformly heated tubes in which the critical heat flux has been exceeded. *Proc. Inst. Mech. Eng.* 180 (3C): 1–11. AERE-R15373, presented in Hewitt & Hall-Taylor 1970.

Bousman, W.S., McQuillen, J.B. & Witte, L.C. 1996 Gas–liquid flow patterns in microgravity: effects of tube diameter, liquid viscosity and surface tension. *Int. J. Multiphase Flow* 22, 1035–1053.

Brauner, N. & Barnea, D. 1986 Slug/churn transition in upward gas–liquid flow. *Chem. Eng. Sci.* 40, 159–163.

Cheng, H., Hills, J.H. & Azzopardi, B.J. 1998 A study of the bubble to slug transition in vertical gas–liquid flow in columns of different diameter. *Int. J. Multiphase Flow* 24, 431–452.

Cohen, L.S. & Hanratty, T.J. 1965 Generation of waves in the concurrent flow of air and a liquid. *AIChE Jl* 11, 138–144.

Colin, C., Fabre, J. & Dukler, A.E. 1991 Gas liquid flow at microgravity conditions:1 Dispersed bubble and slug flow. *Int. J. Multiphase Flow* 17, 533–544.

Crowley, C.J., Sem, R.G. & Rothe, P.H. 1986 Investigation of two-phase flow in horizontal and inclined pipes at large pipe size and high gas density. Report TN-399 for the Pipeline Research Committee, American Gas Associates, Hanover, NH.

Damianides, C.A. & Westwater, J.W. 1988. Two-phase flow patterns in a compact heat exchanger and in small tubes. *Second UK National Conference on Heat Transfer*, Glasgow, 14–16 September. London: Mechanical Engineering Publications, pp. 1257–1268. (See also Damianides, C. A. 1987 Horizontal two-phase flow of air–water mixtures in small diameter tubes and compact heat exchangers, Ph.D. thesis, University of Illinois at Urbana-Champaign.)

Delhaye, J.M. 1981 Two-phase flow patterns. In *Two-Phase Flow and Heat Transfer in the Power and Process Industries*, eds A.E. Bergles, J.G. Collier, J.M. Delhaye, G.F. Hewitt, & F. Mayinger, Washington, DC: Hemisphere.

Dukler, A.E. & Hubbard, M.G. 1975 A model for gas–liquid slug flow in horizontal tubes. *Ind. Eng. Chem. Fundamentals* 14, 337–347.

Dukler, A.E. & Taitel, Y. 1986 Flow pattern transitions in gas–liquid systems: Measurement and modelling. *Multiphase Sci. Technol.* 2, 1–94.

Dukler, A.E., Fabre, J.A., McQuillen, J.B. & Vernon, R. 1988 Gas–liquid flow at microgravity conditions: flow patterns and their transitions. *Int. J. Multiphase Flow* 14, 389–400.

Fan, Z., Lusseyran, F. & Hanratty, T.J. 1993 Initiation of slugs in horizontal gas–liquid flows. *AIChE Jl* 39, 1742–1753.

Fukano, T. & Kariyasaki, A. 1993 Characteristics of gas–liquid two-phase flow in a capillary tube. *Nucl. Eng. Des.* 141, 59–68.

Ghiaasiaan, M. 2003 Gas–liquid two-phase flow and boiling in microchannels. *Multiphase Sci. Technol.* 15, 323–333.

Govier, G.W. & Aziz, K. 1972 *The Flow of Complex Mixtures in Pipes*. New York: VanNostrand Reinhold.

Griffith, P. & Snyder, G.A. 1964 The bubbly-slug transition in a high velocity two-phase flow. MIT Report 5003–29 (TID-20947).

Hanratty, T.J. & Engen, J.M. 1957 Interaction between a turbulent air stream and a moving water surface. *AIChE Jl* 3, 294–304.

Hanratty, T.J. & Hershman, A. 1961 Initiation of roll waves *AIChE Jl* 7, 488–497.

Hanratty, T.J., Theofanous, T., Delhaye, J.-M., Eaton, J., McLaughlin, J., Prosperetti, A., Sundaresan, S. & Tryggvason, G. 2003 Workshop on scientific issues in multiphase flow. *Int. J. Multiphase Flow* 29, 1042–1116.

Harmathy, T.Z. 1960 Velocity of large drops and bubbles in media of infinite or restricted extent. *AIChE Jl* 6, 281–288.

Hewitt. G.F. & Hall-Taylor, N.S. 1970 *Annular Two-Phase Flow*. Oxford: Pergamon Press.

Hurlburt, E.T. & Hanratty, T.J. 2002 Prediction of the transition of from stratified to slug and plug flow for long pipes. *Int. J. Multiphase Flow* 28, 707–729.

Ishii, M., Pavanjope, S.S., Kim, S. & Sun, X. 2004 Interfacial structure and interfacial area transport in downward two-phase bubbly flow. *Int. J. Multiphase Flow* 30, 779–801.

Jayanti, S. & Brauner, N. 1994 Churn flow. *Multiphase Sci. & Technol.* 8, 471–521.

Jeffreys, H. 1925 On the formation of water waves by wind. *Proc. R. Soc. Lond.* A107, 189–206.

Kawahara, A., Chung, P.M.-Y., & Kawaji, M. 2002 Investigation of two-phase flow pattern and pressure drop in a microchannel. *Int. J. Multiphase Flow* 28, 1411–1435.

Kocamustafaogullari, G. & Wang, Z. 1991 An experimental study on local interfacial parameters in a horizontal bubbly two-phase flow. *Int. J. Multiphase Flow* 17, 553–572.

Kordyban, E.S. & Ranov, T. 1970 Mechanism of slug formation in horizontal two-phase flow. *J. Basic Eng.* 92, 857–864.

Kytömaa, H.K. & Brennan, C.E. 1991 Small amplitude kinematic wave propagation in two-component media. *Int. J. Multiphase Flow* 17, 13–26.

Lighthill, M.J. & Whitham, G.B. 1955 On kinematic waves. 1. Flood movement in long rivers; 2. Theory of traffic flow on long narrow roads. *Proc. R. Soc. Lond.* A229, 281–345.

Lin, P.Y. & Hanratty, T.J. 1986 Prediction of the initiation of slugs with linear stability theory. *Int. J. Multiphase Flow* 12, 79–98.

Lin, P.Y. & Hanratty, T.J. 1987a Detection of slug flow from pressure measurements. *Int. J. Multiphase Flow* 15, 13–21.

Lin, P.Y. & Hanratty, T.J. 1987b Effect of pipe diameter on flow patterns for air–water flow in horizontal pipes. *Int. J. Multiphase Flow* 13, 549–563.

Mandhane, J.M., Gregory, G.A. & Aziz, K.A. 1974 A flow pattern map for gas–liquid flow in horizontal pipes. *Int. J. Multiphase Flow* 1, 537–551.

Manfield, C.J., Lawrence, C. & Hewitt, G. 1999 Drag-Reduction with additives in multiphase flow. *Multiphase Sci. Technol.* 11, 197–201.

McQuillan, K.W. & Whalley, P.B. 1985 Flow patterns in vertical two-phase flow. *Int. J. Multiphase Flow* 11, 161–176.

Niklin, D.J., Wilkes, J.O. & Davidson, J.F. 1962 Two-phase flow in vertical tubes. *Trans. IChem E* 40, 61–68.

Omebere-lyari, N.K. & Azzopardi, B.J. 2000 Links across flow-patterns in gas–liquid flow in vertical pipes. Paper presented at the 2nd Japanese-European Two-Phase Group Meeting, Tsukuba, 25–29 September.

Omebere-lyari, N.K., Azzopardi, B.J., Lucas, D., Beyer, M. & Prasser, H-M. 2008 The characteristics of gas/liquid flow in large risers at high pressures. *Int. J. Multiphase Flow* 34, 461–476.

Oshimoto, T. & Charles, M.E. 1974 Vertical two-phase flow, Part 1. Flow pattern correlations. *Can. J. Chem. Engng* 52, 25–35.

Reimann, J. & Seeger, J.W. 1981 Transition to annular flow in horizontal air–water and steam–water flow. Report of Kernforschungszentrum, Karlsruhe, KfK 3198.

Ruder, Z. & Hanratty, T.J. 1990 A definition of gas–liquid flow in horizontal pipes. *Int. J. Multiphase Flow* 16, 233–242.

Ruder, Z., Hanratty, P.J. & Hanratty, T.J. 1989 Necessary conditions for the existence of stable slugs. *Int. J. Multiphase Flow* 15, 209–226.

Sam, R.G. & Crowley, C.J. 1986. Investigation of two-phase processes in coal slurry/hydrogen heaters. Creare Report TH-1085 for Department of Energy.

Sekoguchi, K., Mori, K. & Masuo, K. 1996 Interfacial profiles and flow characteristics in vertical downward two-phase plug and foam flows. *Chem. Eng. Comm.* 141, 415–441.

Serizawa, A. & Feng, Z.P. 2000 Reviews of two-phase flow in microchannels. In The US-Japan Seminar on Two-Phase Flow Dynamics, June 4–9, Santa Barbara, CA.

Serizawa, A. & Feng, Z.P. 2001 Two-phase flow in microchannels. In *Proceedings of the 4th International Conference on Multiphase Flow,* May 27–June 1, New Orleans, LA.

Soleimani, A. & Hanratty, T.J. 2003 Critical liquid flows for the transition from pseudo-slug and stratified patterns to slug flow. *Int. J. Multiphase Flow* 29, 51–67.

Soleimani, A., Al-Sarkhi, A. & Hanratty, T.J., 2002. Effect of drag-reducing polymers on pseudo-slugs, interfacial drag and transition to slug flow. *Int. J. Multiphase Flow* 28, 1911–1927.

Taitel, Y. & Dukler, A.E. 1976 A model for predicting flow regime transitions in horizontal and near-horizontal gas–liquid flow. *AIChE Jl* 22, 47–55.

Taitel, Y. & Dukler, A.E. 1987 Effect of pipe length on the transition boundaries for high viscosity liquids. *Int. J. Multiphase Flow* 13, 577–581.

Taitel, Y., Barnea, D. & Dukler, A. 1980 Modelling flow transitions for steady upward gas–liquid flow in vertical tubes. *AIChE Jl* 26 345–354.

Toms, B.A. 1948 Some observations on the flow of linear polymer solutions through straight tubes at large Reynolds numbers. In *Proceedings of the 1st International Conference on Rheology,* Vol. 2, Amsterdam: North Holland, 135–141.

Triplett, K.A., Ghiaasiaan, S.M., Abdel-Khalik, S.I. & Sadowski, D.L. 1999 Gas–liquid two-phase flow in microchannels Part 1: two-phase flow patterns. *Int. J. Multiphase Flow* 25, 377–394.

Wallis, G.B. 1969 *One Dimensional Two-phase Flow.* New York: McGraw-Hill.

Wallis, G.B. & Dobson, J.E. 1973 The onset of slugging in horizontal stratified air–water flow. *Int. J. Multiphase Flow* 1, 173–193.

Woods, B.D. & Hanratty, T.J. 1996 Relation of slug stability to shedding rate. *Int. J. Multiphase Flow* 22, 809–828.

Woods, B.D., Hurlburt, E.T. & Hanratty, T.J. 2000 Mechanism of slug formation in downwardly inclined pipes. *Int. J. Multiphase Flow* 26, 977–998.

Wu, H.L., Pots, B.F.M., Hollenburg, J.F. & Meerhof, R. 1987. Flow pattern transitions in two-phase gas/condensate flow at high pressre in an 8-inch horizontal pipe. In *Proceedings of the 3rd BHRA Conference on Multiphase Flow,* The Hague, The Netherlands, 13–21.

Zhao, L. & Rezkallah, K.S. 1993 Gas–liquid flow patterns at microgravity conditions. *Int. J. Multiphase Flow* 19, 751–763.

Zuber, N. & Findlay, J.A. 1965 Average volumetric concentration in two phase flow systems. *J. Heat Transfer* 87, 453–468.

3 Film flows

3.1 Free-falling layer

3.1.1 Laminar flow

The simplest example of a wall film is the free-fall of a liquid along a wall. Consider the case of a flat wall where y is the distance from the wall and z is the distance in the direction of flow. Liquid moves down the wall under the force of gravity. It is desired to calculate the thickness of the film as a function of the volumetric flow.

Consider the differential element, shown in Figure 3.1, of size $dzdy$, and of length l in the z-direction. The time-rate of change of momentum in the volume is zero so, from the momentum theorem, the net force is zero. That is, the force of gravity is balanced by shear stresses defined by τ and $\tau + d\tau$.

The force of gravity is $+\rho gldzdy$. The convention is used that the shear stress is positive when the forces on the two sides of a plane have the same sign as the normals to the plane. Thus, from Figure 3.1, the force due to the shear stresses equals $-\tau ldz + (\tau + d\tau)ldz$, so

$$0 = \rho gldydz - \tau ldz + (\tau + d\tau)ldz$$

or

$$\frac{d\tau}{dy} = -\rho g \tag{3.1}$$

This can be integrated to give

$$\tau = -\rho gy + c$$

where c is the constant of integration. Since $\tau = 0$ at $y = m$, $c = \rho gm$ and

$$\tau = -\rho gy + \rho gm \tag{3.2}$$

where m is the height of the liquid layer. From (3.2), the stress at the wall is $\tau_W = \rho gm$.

For laminar flow τ is given by Newton's law of viscosity,

$$\tau = \mu_L \frac{du}{dy} \tag{3.3}$$

where μ_L is the viscosity of the liquid. Substituting (3.3) into (3.2) and integrating twice gives

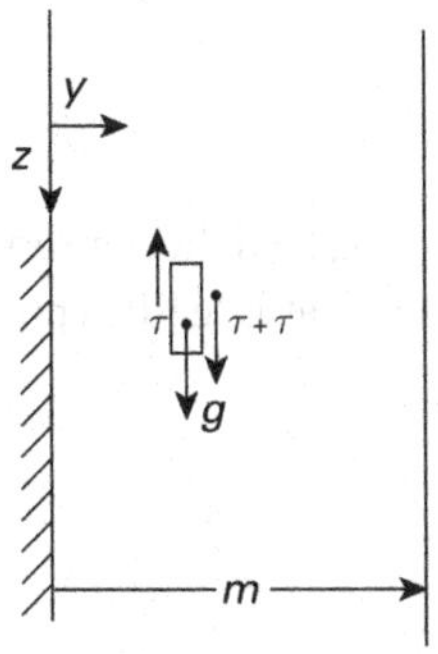

Figure 3.1 Differential volume in a free film of height m on a vertical wall.

$$\Gamma = \int_0^m u dy = \frac{gm^3}{3\upsilon_L} \tag{3.4}$$

where $\upsilon_L = \mu_L/\rho_L$ is the kinematic liquid viscosity, and Γ is the volumetric flow per unit length in the spanwise direction. The hydraulic diameter, defined as 4 times the area divided by the wetted perimeter, is given as 4 m. A film Reynolds number is defined as $\mathrm{Re}_{LF} = 4\Gamma/\upsilon_L$. Since $\tau_W = \rho g m$, (3.4) can be written as

$$\mu_L \Gamma = \frac{m^2}{3}\tau_W \tag{3.5}$$

or as

$$m^+ = 0.866\,\mathrm{Re}_{LF}^{0.5} \tag{3.6}$$

where $m^+ = mv^*/\upsilon_L$ and a friction velocity is defined as $v^* = (\tau_W/\rho)^{1/2}$.

In Section 3.2.2, a characteristic stress, $\tau_c = (2/3)\tau_W + (1/3)\tau_i$, is defined. A friction velocity based on this stress is designated as $v_c^* = (\tau_c/\rho)^{1/2}$. For the case being considered, the interfacial stress, τ_i, is zero, so that an alternative form of equation (3.6), for a free-falling film, is

$$m_c^+ = 0.707\mathrm{Re}_{LF}^{0.5} \tag{3.7}$$

where $m_c^+ = mv_c^*/\upsilon_L$.

For axisymmetric flow, such as would exist for flow down the outside of a tube of radius r_t, the above analysis gives

$$\Gamma = \frac{gm^3}{3\upsilon_L}\left[1 + \frac{m}{r_t} + \frac{3}{20}\left(\frac{m}{r_t}\right)^2 - \frac{1}{40}\left(\frac{m}{r_t}\right)^3 + \frac{1}{140}\left(\frac{m}{r_t}\right)^4 + \cdots\right] \tag{3.8}$$

3.1.2 Turbulent flow

At high Reynolds numbers the flow is turbulent, so momentum can be transferred both by molecular and turbulent motions. Thus, the stress is defined as

$$\frac{\tau}{\rho} = \upsilon \frac{du}{dy} + \upsilon^{t} \frac{du}{dy} \tag{3.9}$$

A number of models for the turbulent kinematic viscosity, υ^{t}, have been proposed. These are discussed in the book by Fulford (1964). Measurements at large Reynolds numbers are fitted by

$$m_{c}^{+} = 0.031\mathrm{Re}_{\mathrm{LF}}^{0.90} \tag{3.10}$$

3.1.3 Interpolation formula

The following interpolation relation for free-falling films, suggested by Henstock and Hanratty (1976), can be used for all Reynolds numbers.

$$m_{c}^{+} = \left[\left(0.707\mathrm{Re}_{\mathrm{LF}}^{0.5}\right)^{5} + \left(0.031\mathrm{Re}_{\mathrm{LF}}^{0.9}\right)^{5}\right]^{1/5} \tag{3.11}$$

Note that it has the correct behavior for $\mathrm{Re}_{\mathrm{LF}} \to 0$ and for $\mathrm{Re}_{\mathrm{LF}} \to \infty$.

3.2 Gas–liquid flows

3.2.1 Horizontal flow in a channel

For gas–liquid flow in a horizontal rectangular channel, the liquid is pulled along by the drag of the gas at the interface, τ_{i}, which is not zero. For a fully developed film flow, the momentum theorem indicates that the net force on a control volume, such as shown in Figure 3.1, is zero. The distance from the wall is given as y and the coordinate in the direction of flow is z. The gas flow has a pressure gradient, dp/dz, and imposes an interfacial stress, τ_{i}. Since the flow is fully developed, there is no mean flow in the y-direction and the pressure gradient in the gas flow is the same as the pressure gradient in the liquid film.

The volume $l\,dzdy$ is acted on by the pressure gradient (which is negative) and shear stresses. Thus a force balance gives

$$(\tau + d\tau)ldz - \tau ldz + pldy - (p + dp)ldy = 0$$

This can be rearranged to give

$$\frac{d\tau}{dy} = \frac{dp}{dz} \tag{3.12}$$

where, as indicated above, dp/dz is a constant for a fully developed flow.

Equation (3.12) can be integrated from y to m to give

$$\tau_{i} - \tau = (m - y)\frac{dp}{dz} \tag{3.13}$$

so that at the wall ($y = 0$)

$$\tau_W = \tau_i - m\frac{dp}{dz} \tag{3.14}$$

Note that $\tau_W > \tau_i$ since dp/dz is negative. (For a horizontal gas–liquid flow $\tau_w \approx \tau_i$ and $\tau_c \approx \tau_W$. This is not the case for vertical flows, to be considered in Section 3.2.2.)

Colburn & Carpenter (1949) assumed that a film behaves the same as single-phase turbulent flow. This implies that the law of the wall relation

$$u^+ = f(y^+) \tag{3.15}$$

with $y^+ = yv^*\rho/\mu$ and $u^+ = u/v^*$, can be used to describe the velocity profile. The friction velocity is defined with the shear stress at the wall, $v^* = (\tau_W/\rho)^{1/2}$. The film Reynolds number is obtained by integrating (3.15) between $y^+ = 0$ and m^+. Since $\mathrm{Re}_{LF} = 4\Gamma/\upsilon$ and $\Gamma = \int_0^m u dy$

$$\frac{\mathrm{Re}_{LF}}{4} = \int_0^{m^+} u^+ dy^+ \tag{3.16}$$

The Colburn–Carpenter hypothesis gives (Henstock & Hanratty, 1976)

$$m_c^+ = \left[\left(0.707\mathrm{Re}_{LF}^{0.5}\right)^{2.5} + \left(0.0379\mathrm{Re}_{LF}^{0.9}\right)^{2.5}\right]^{0.4} \tag{3.17}$$

Equation (3.17) is similar to (3.11), which was developed for free-falling films. It has the same or similar behavior at high and low liquid Reynolds numbers. However, the transition from laminar to turbulent flow is more gradual.

When the interface is smooth, the flow is laminar. When three-dimensional pebbled waves are introduced, the Colburn–Carpenter hypothesis becomes valid and equation (3.17) can be used to calculate m^+. This is illustrated in the paper by Cohen & Hanratty (1966), which shows that m^+ is only a function of Re_{LF}.

Studies of horizontal gas–liquid flows are described in a number of papers (see, for example, Hanratty & Engen, 1957; Cohen & Hanratty, 1966; Andreussi *et al.*, 1985). These consider the case of a liquid layer flowing along the bottom of an enclosed channel and air flowing concurrently with it. For a fixed Re_{LF}, the interface is smooth at low gas velocities. As the gas velocity increases, a critical condition is reached at which two-dimensional waves appear at the interface (see Figure 6.2). At higher gas velocities, a three-dimensional wavy interface is observed. At still higher gas velocities, roll waves are observed. (Of interest is the finding that equation (3.17) also holds when roll waves are present.) With increasing gas velocity the average height of the film decreases and the intermittent flow surges change their appearance in that they are a complicated collection of capillary waves. They have the appearance of patches of turbulence. Hewitt & Hall-Taylor (1970) call them "disturbance waves." At high enough gas velocities, capillary waves are entrained in the gas flow.

3.2.2 Film flow in a vertical pipe

Consider the common case of a film on the wall of a pipe through which a gas is flowing. Equation (3.8) and the paper by Henstock & Hanratty (1976) show that the film may be treated as a two-dimensional flow, similar to what would be found for flow along a flat wall. The y-axis is perpendicular to the wall. The z-axis is in the direction of the mean flow.

A force balance on a differential volume yields the following equations

$$\frac{d\tau}{dy} = \frac{dp}{dz} + \rho g \text{ upflow} \tag{3.18}$$

$$\frac{d\tau}{dy} = \frac{dp}{dz} - \rho g \text{ downflow} \tag{3.19}$$

since the film is acted upon both by the pressure gradient and by the force of gravity. Thus, the negative pressure gradient is augmented in downflow and opposed in upflow by the force of gravity. Because of the gravitational force, there can be a large spatial variation of the shear stress in the film (not seen in horizontal flows), so the choice of a characteristic stress for the liquid film is not obvious. For convenience, define

$$\hat{g} = \frac{1}{\rho}\left|\frac{dp}{dz}\right| + g \tag{3.20}$$

for downflow and

$$\hat{g} = \frac{1}{\rho}\left|\frac{dp}{dz}\right| - g \tag{3.21}$$

for upflow, where $|dp/dz|$ signifies the absolute value of the pressure gradient.

The integration of (3.18), (3.19) gives

$$\tau - \tau_W = \rho\hat{g}y \tag{3.22}$$

The stress at the interface is

$$\tau_i - \tau_W = \rho\hat{g}m \tag{3.23}$$

Therefore, from (3.22)

$$\mu\frac{du}{dy} = \tau_W + \rho\hat{g}y \tag{3.24}$$

for a laminar flow. Integration of (3.24) gives

$$\mu_L u = \tau_W y + \rho\hat{g}\frac{y^2}{2} \tag{3.25}$$

The volumetric flow is obtained by integrating (3.25) and by using (3.23) to eliminate $\rho\tilde{g}m$

$$\mu_L \Gamma = \frac{m^2}{2}\left(\frac{2}{3}\tau_W + \frac{1}{3}\tau_i\right) \tag{3.26}$$

Note that equation (3.5) is recovered for $\tau_i = 0$. Thus, the use of a characteristic stress of

$$\tau_c = (2/3)\tau_W + (1/3)\tau_i \tag{3.27}$$

represents both a free-falling film and a film in the presence of a gas flow for laminar films.

The velocity field and the volumetric flow can be calculated for a turbulent field by substituting (3.9) into (3.22) and integrating. The Colburn–Carpenter hypothesis can be employed by assuming that the spatial variation of the kinematic turbulent viscosity is the same as for single-phase flows. The calculated velocity profiles can be given as $u_c^+ = u/v_c^*$ versus $y_c^+ = yv_c^*/\upsilon$. Only one additional parameter is needed to characterize the profiles, if m/r_t is small enough. This is

$$am^+ = -\frac{\hat{g}m}{v_c^{*2}} \tag{3.28}$$

where $a = -\hat{g}\upsilon/v_c^{*3}$. For typical annular flows, the friction velocity is large so am^+ is close to zero.

For upward flows, am^+ is a positive quantity. For increasing am^+, the profile is distorted by the gravitational force and at $am^+ = 3$ the velocity gradient at the wall is zero. For am^+ larger than 3, the calculated average velocity shows negative values close to the wall. Downward flowing liquid films in an upwardly flowing gas have negative values of am^+. For free-falling layers, $\tau_i = 0$ and $am^+ = -3/2$. For downward film flows the interfacial stress imposed by the countercurrent gas velocity retards the flow so that the maximum velocity in the film is not at the outer edge. For still larger gas flows the velocity at the interface can assume a zero value. Thus, between $am^+ = -3/2$ and $am^+ = 0$ the flow changes from a free-falling downward flow to an upwardly flowing film. The critical condition for which the flow starts to reverse its direction is reached at an intermediate negative value of am^+.

Henstock & Hanratty (1976) have used the equation for the eddy viscosity developed by van Driest to integrate (3.22) and (3.9). The calculated relation between the volumetric flow and the height of the film is plotted as m_c^+ versus Re_{LF} with am^+ as a parameter in Figure 3.2. Because of the use of τ_c, rather than τ_W, as a characteristic stress, the plot is independent of am^+ at small Re_{LF} where the film is laminar. For larger Re_{LF}, the parameter am^+ has a smaller effect than would be expected, considering its large influence on the shape of the velocity profile. Measurements available to Henstock & Hanratty did not include conditions with extreme values of am^+, so they recommended equation (3.17), which agrees approximately with their calculation for $am^+ = 0$.

Thus, equation (3.17) provides the value of $m_c^+ = mv_c^*/\upsilon$ for a given flow rate of the film. Since $v_c^* = \left(\tau_c^*/\rho\right)^{1/2}$ and τ_c are given by equation (3.27), one needs a relation for the drag on the interface, τ_i, to calculate m for a given film Reynolds number. This matter is addressed later in Section 3.3.

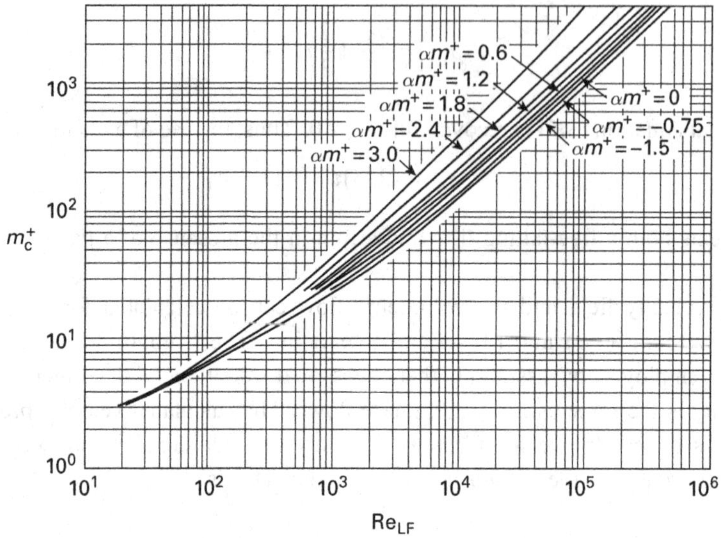

Figure 3.2 Plot of m_c^+ versus Re_{LF}, calculated using a van Driest model for the mixing length with $A = 26$. Parameter am^+ is defined by equation (3.28). Henstock & Hanratty, 1976.

3.2.3 Flow reversal and flooding

Hewitt (1982), Delhaye (1981) and others have commented on flow reversal and flooding as examples of situations for which gravity has a strong effect on the behavior of a vertically flowing film.

Consider the case of an upward flow of a gas and a liquid. Equation (3.18) presents a force balance for the liquid film. Integration of this equation yields

$$\tau_i - \tau = \frac{dp}{dz}(m - y) + \rho g(m - y) \tag{3.29}$$

or

$$\tau = \tau_i - \frac{dp}{dz}(m - y) - \rho g(m - y) \tag{3.30}$$

Since dp/dz is negative, the gas phase interfacial stress and pressure gradient supply the motive forces pulling the film upward. These are opposed by gravity. As the gas velocity decreases, the interfacial stress decreases and it is no longer able to keep the entire film moving upward. From (3.30), the shear stress at the wall is

$$\tau_W = \tau_i - m\frac{dp}{dz} - \rho g m \tag{3.31}$$

where $\tau_W = \mu_L(du/dy)_W$.

The gas velocity at which $\tau_W = 0$ represents a situation for which the average velocity gradient at the wall is zero. A plot of pressure gradient versus gas flow rate presented in

Hewitt (1982) shows a minimum at $\tau_W = 0$. For gas velocities below this critical condition a net negative flow would exist in the region close to the wall. Figure 2.1.20 in the paper by Hewitt (1982) shows that beyond this critical condition the pressure drop increases with decreasing flow rate, the flow is haphazard and alternating its direction. At a small enough gas flow the film completely changes its direction. Section 2.4.4 of the Hewitt paper uses the criterion of $\tau_w = 0$ to correlate data on the transition from annular flow to churn flow.

Flooding is observed when the fluid in a downward flow changes direction. It is suggested in Section 3.2.2 that this is initiated when the liquid velocity in the film at the interface changes direction. Measurements indicate that at gas velocities larger than this critical value an intermittent pattern exists, for which large-amplitude upwardly moving waves alternate with a downwardly flowing film.

As indicated by Hewitt (1982) and by Delhaye (1981), a knowledge of when this occurs is important in many applications. Flooding can limit the operation of reflux condensers and mass transfer operations involving falling film absorbers or packed columns. In a loss-of-coolant accident in a nuclear reactor, flooding can limit the rate at which emergency cooling water can enter the reactor. Slug flow in vertical tubes involves the passage of elongated bubbles which have a thin downwardly flowing film at the wall. The breakdown of this pattern as the gas velocity increases has been explained as resulting from breakdown of the film by a process similar to flooding.

The analysis of Henstock & Hanratty (1976) can explain changes in the shape of the velocity profiles in a film and the dependence of flow reversal and flooding on system variables.

3.3 Interfacial stress for vertical flows

3.3.1 Waves in annular flows

This section focuses on annular flows in vertical pipes for gas velocities that are high enough for flow reversal or flooding not to be considered. The behavior is similar for vertical and for horizontal films. At very large gas flows and very small liquid flows, the film is covered with long crested ripples that have steep fronts. Between these ripples the interface is smooth and the flow appears laminar.

At sufficiently large liquid flows, disturbance waves (Section 3.2.1) appear on the film. They travel at much larger velocities than the ripples in the base film and are coherent in that they keep their identity over a long length of pipe (see Figure 3.3). The disturbance waves are several times thicker than the base film over which they move.

At high gas velocities, this transition depends mainly on the liquid Reynolds number; that is, it is only weakly dependent on the gas velocity. For flow in a horizontal channel (Andreussi *et al.*, 1985), this critical liquid Reynolds number (defined as $4W_{LF}/\mu_L P$ with P signifying the pipe perimeter) is equal to 280. The same critical Reynolds number is found for flow in vertical pipes (Andreussi *et al.*, 1985).

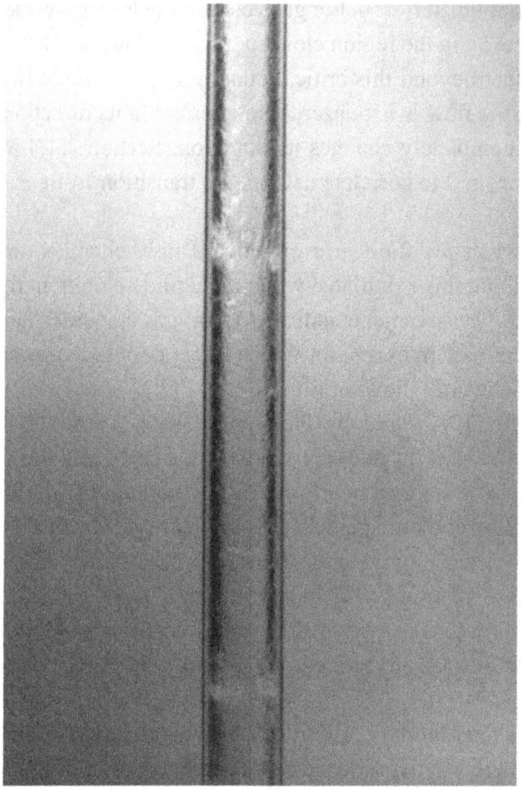

Figure 3.3 Photograph of annular flow in a vertical pipe that shows the intermittent disturbance waves.

3.3.2 Scaling

The interfacial stress, τ_i, represents the drag on the liquid film. Negative τ_i indicates the resistance of the film to the gas flow. It is desired to predict the dependence of τ_i on the flows of the gas and of the liquid film. The solution of this problem is central to predicting the behavior of gas–liquid flows. The interfacial stress is determined by measuring the gas-phase pressure gradient and the height of the liquid layer under conditions of a fully developed symmetric flow.

Consider a control volume with a length, l, extending from the center of the pipe to the edge of the wall layer, which has a height m. Liquid can be injected into the gas flow from disturbance waves at a velocity, U_A, at a rate per unit area of R_A. Drops deposit at the wall with a velocity, U_D, at a rate per unit area of R_D. Under fully developed conditions, $R_A = R_D$. These processes can lead to a net transfer of momentum equal to $(R_D U_D - R_A U_A) 2\pi((d_t/2) - m)l$.

Applying the momentum theorem to the gas-phase control volume defined above yields

$$2\pi l\left(\frac{d_t}{2} - m\right)(R_D U_D - R_A U_A) = -\pi\left(\frac{d_t}{2} - m\right)^2 \frac{dp}{dz} l - 2\pi\left(\frac{d_t}{2} - m\right)\tau_i l \quad (3.32)$$

The term $-\tau_i$ represents the resisting stress of the film to the gas flow. The first term on the right side is the force on the control volume due to the pressure gradient.

The following equation is obtained from (3.32):

$$(\tau_i + \tau_P) = -\left(\frac{d_t}{2} - m\right)\frac{1}{2}\frac{dp}{dz} \tag{3.33}$$

where

$$\tau_P = (R_D U_D - R_A U_A) \tag{3.34}$$

represents an effective stress due to deposition and atomization. This term has been considered by several investigators (see, for example, Moeck & Stachiewicz, 1972). A difficulty in using (3.34) is the evaluation of U_D and U_A. A number of suggestions have been made. For example, U_A has been set equal to the velocity of the disturbance waves and U_D has been represented as the average axial velocity of the drops in the gas space.

The interfacial stress is usually described in terms of a friction factor, f_i, so that

$$\tau_i = \frac{1}{2} f_i u_G^2 \tag{3.35}$$

Calculations by Fore & Dukler (1995) and by Schadel & Hanratty (1989) suggest that τ_P is of the order of $0.1 - 0.2\tau_i$. Because of the uncertainties in evaluating τ_P, it is not treated separately in this chapter. Its influence is included in the specification of f_i.

If the interface is smooth, f_i is the same as is found for flow over smooth solid surfaces, designated as f_S. However, the presence of waves at the interface can lead to very large values of f_i/f_S, which are related to the amplitudes of the waves. Two approaches have been taken to represent this effect. One argues that the wave amplitude varies as the thickness of the wall layer, m, and that this, in turn, varies as the dimension of the conduit, d_t.

$$\frac{f_i}{f_S} \propto \frac{m}{d_t} \tag{3.36}$$

The other uses the viscous length in the gas phase, υ_G/v^*, where υ_G is the kinematic viscosity of the gas and $v^* = (\tau_i/\rho_G)^{1/2}$. Thus

$$\frac{f_i}{f_S} \propto \frac{mv^*}{\upsilon_G} \tag{3.37}$$

3.3.3 Correlation for f_i using $d_t/2$ as a scaling factor

For a smooth surface,

$$f_S = 0.046\ \mathrm{Re}_G^{-0.20} \tag{3.38}$$

where $Re_G = d_t W_G / A_t \mu_G$, W_G is the mass flow of the gas, A_t is the tube area, μ_G is the viscosity of the gas. A difficulty in using (3.36) is that the fundamental equation for film height (3.17) contains the unknown τ_i. Henstock & Hanratty (1976) proposed a method that avoids a trial and error solution.

Equation (3.17) can be written in the form

$$m_c^+ = \gamma(Re_{LF}) \tag{3.39}$$

where $\gamma(Re_{LF})$ is defined by the right side of (3.17). This can be rearranged to

$$\frac{m(\tau_c/\tau_i)^{1/2}(\tau_i/\rho_L)^{1/2}}{\upsilon_L} = \gamma(Re_{LF}) \tag{3.40}$$

After some algebra Henstock & Hanratty obtained

$$\frac{m}{d_t} = \frac{6.59F\sqrt{\tau_i/\tau_c}}{\sqrt{f_i/f_S}} \tag{3.41}$$

where

$$F = \frac{\gamma(Re_{LF})}{Re_G^{0.9}} \frac{\upsilon_L}{\upsilon_G} \sqrt{\frac{\rho_L}{\rho_G}} \tag{3.42}$$

From (3.36) and (3.41), they suggested that f_i/f_S is strongly dependent on F and weakly dependent on am^+

$$\frac{f_i}{f_S} = g(F, am^+) \tag{3.43}$$

The advantage of using (3.43) is that F can be calculated directly if Re_{LF} and Re_G are known.

Henstock & Hanratty chose to correlate measurements of f_i/f_S by using (3.43). They obtained the following relation for upflows and downflows at high gas velocities:

$$\frac{f_i}{f_S} - 1 = 1400F \tag{3.44}$$

For downflows with moderate and small gas velocities, the effect of am^+ should be considered. They suggested the following approximate equation

$$\frac{f_i}{f_S} = 1 + 1400F\left[1 - \exp\left(-\left|\frac{\tau_i}{\rho_L g m}\right|\right)\right] \tag{3.45}$$

For both upflows and downflows, the following equation was obtained for m:

$$\frac{m}{d_t} = \frac{6.59F}{(1 + 1400F)^{1/2}} \tag{3.46}$$

3.3.4 Correlation using the viscous length for the gas phase

Asali *et al.* (1985) used additional measurements of interfacial stress for upward vertical annular flows and chose a scaling based on the viscous length for the gas phase. Particular attention was given to low liquid rates and high gas velocities for which the film can have thicknesses of 30 to 200 μm. The following relation for the friction factor was obtained at $u_G > 25$ m/s

$$\frac{f_i}{f_S} - 1 = 0.45\left(m_G^+ - 4\right) \tag{3.47}$$

where $m_G^+ = mv_G^*/\upsilon_G$ and $v_G^* = (\tau_i/\rho_G)^{1/2}$. Surprisingly, the same equation represents measurements under conditions that roll waves are present. Measurements for downward flow by Andreussi & Zanelli (1978), Charvonia (1959), Chien & Ibele (1964) and Webb (1970) for pipe diameters ranging from 2.4 cm to 6.35 cm are represented by the relation

$$\frac{f_i}{f_S} - 1 = 0.45\left(m_G^+ - 5.9\right) \tag{3.48}$$

If the wave height varies as the thickness of the film, m, equations (3.47) and (3.48) have a kinship to what has been found for sand-roughened solid boundaries. That is, the wavy interface is similar to a sand-roughened surface in that it may be considered to be hydraulically smooth if the dimensionless roughness, k_S^+, is small enough.

The evaluation of f_i/f_S from either (3.47) or (3.48) requires the specification of m_G^+. Asali *et al.* (1985) suggest approximations for m_G^+ both in the ripple and disturbance wave regimes.

References

Andreussi, P. & Zanelli, S. 1978 Downward annular and annular-mist flow of air–water mixtures. Paper presented at International Seminar, Dubrovnik, Yugoslavia, September 4–9.

Andreussi, P., Asali, J.C. & Hanratty, T.J. 1985 Initiation of roll waves in gas–liquid flows. *AIChE Jl* 31, 119–126.

Asali, J.C., Hanratty, T.J. & Andreussi, P. 1985 Interfacial drag and film height for vertical annular flow. *AICh Jl* 31, 895–902.

Charvonia, D.A. 1959 A study of the mean thickness of the liquid film and the characteristics of the interfacial surface in annular two-phase flow in a vertical pipe. Jet Propulsion Center Report No. I-59–1, Purdue University and Purdue Research Foudation, Lafayette, IN.

Chien, S.F. & Ibele, W. 1964 Pressure drop and liquid film thickness in annular mist flows. *J. Heat Transfer* 86, 164.

Cohen, L.S. & Hanratty, T.J. 1966 Height of a liquid film in a horizontal concurrent gas–liquid flow. *AIChE Jl* 12, 290–292.

Colburn, A.P. & Carpenter, F.G. 1949 Heat transfer lecture. *Nucl. Energy USAEC* 2, 105.

Delhaye, J.M. 1981 Two-phase flow patterns. In *Two-phase Flow and Heat Transfer in Power and Process Industries,* Section 1–8, eds A.E. Bergles, J.G. Collier, J.M. Delhaye, G.F. Hewitt, and F. Mayinger. Washington, DC: Hemisphere.

Fore, L.B. & Dukler, A.E. 1995 The distribution of drop size and velocity in gas–liquid annular flow. *Int. J. Multiphase Flow* 21, 137–149.

Fulford, G.D. 1964 The flow of liquids in thin films. In *Advances in Chemical Engineering*, Vol. 5. New York: Academic Press, 151–236.

Hanratty, T.J. & Engen, J.M. 1957 Interaction between a turbulent air stream and a moving water surface. *AIChE Jl* 3, 299–304.

Henstock, W.H. & Hanratty, T.J. 1976 The interfacial drag and the height of the wall layer in annular flows. *AIChE Jl* 22, 990–1000.

Hewitt, G.F. 1982 Flow regimes, In *Handbook of Multiphase Systems*, ed. G. Hetsroni. Washington, DC: Hemisphere, pp. 2–25.

Hewitt, G.F. & Hall-Taylor, N.S. 1970 *Annular Two-Phase Flow.* Oxford: Pergamon Press.

Moeck, E.O. & Stachiewicz, J.W. 1972 A droplet interchange for annular-dispersed, two-phase flow. *Int. J. Heat Mass Transfer* 15, 673–683.

Schadel, S.A. & Hanratty, T.J. 1989 Interpretation of atomization rates of the liquid film in gas–liquid annular flow. *Int. J. Multiphase Flow* 15, 893–900.

Webb, D. 1970 Studies of the characteristics of downward annular two-phase flow. AERE Report R6426.

4 Inviscid waves

4.1 Inviscid, incompressible, irrotational flow

A Reynolds number characterizing waves at an interface is $\lambda c/\nu$, where λ is the wavelength, c is the wave velocity and ν is the kinematic viscosity. If this has a large magnitude the assumption of an inviscid irrotational flow field is a useful approximation.

An irrotational flow is one for which fluid particles are not rotating. It is defined by

$$\operatorname{curl} \vec{u} = 0 \tag{4.1}$$

where $\vec{u}$ is the velocity given by

$$\vec{u} = \vec{i}_x u_x + \vec{i}_y u_y + \vec{i}_z u_z \tag{4.2}$$

and $\vec{i}_x, \vec{i}_y, \vec{i}_z$ are unit vectors along the x, y, z coordinates. The definition of the curl is

$$\operatorname{curl} \vec{u} = -\vec{i}_x\left(\frac{\partial u_z}{\partial y} - \frac{\partial u_y}{\partial z}\right) - \vec{i}_y\left(\frac{\partial u_x}{\partial z} - \frac{\partial u_z}{\partial x}\right) - \vec{i}_z\left(\frac{\partial u_y}{\partial x} - \frac{\partial u_x}{\partial y}\right) \tag{4.3}$$

Equation (4.1) is satisfied if the three amplitudes are respectively zero.

$$\frac{\partial u_z}{\partial y} = \frac{\partial u_y}{\partial z} \qquad \frac{\partial u_x}{\partial z} = \frac{\partial u_z}{\partial x} \qquad \frac{\partial u_y}{\partial x} = \frac{\partial u_x}{\partial y} \tag{4.4}$$

These equations can be satisfied if $\vec{u}$ is given by a potential function, $\phi(x, y, z, t)$, whereby

$$\vec{u} = -\operatorname{grad}\phi = -\vec{i}_x\frac{\partial \phi}{\partial x} - \vec{i}_y\frac{\partial \phi}{\partial y} - \vec{i}_z\frac{\partial \phi}{\partial z} \tag{4.5}$$

so that

$$u_x = -\frac{\partial \phi}{\partial x} \qquad u_y = -\frac{\partial \phi}{\partial y} \qquad u_z = -\frac{\partial \phi}{\partial z} \tag{4.6}$$

This can be seen simply by substituting (4.6) into (4.4).

The equation of conservation of mass is

$$-\frac{\partial \rho}{\partial t} = \frac{\partial(\rho u_x)}{\partial x} + \frac{\partial(\rho u_y)}{\partial y} + \frac{\partial(\rho u_z)}{\partial z} \tag{4.7}$$

This equates the time change of mass in a fixed volume to the net flow of mass out of the volume. For an incompressible flow, ρ = constant, so equation (4.7) simplifies to

$$0 = \frac{\partial u_x}{\partial x} + \frac{\partial u_y}{\partial y} + \frac{\partial u_z}{\partial z} \tag{4.8}$$

The substitution of (4.6) into (4.8) gives

$$0 = \frac{\partial^2 \phi}{\partial x^2} + \frac{\partial^2 \phi}{\partial y^2} + \frac{\partial^2 \phi}{\partial z^2} \tag{4.9}$$

This equation describes the velocity field for an incompressible, irrotational flow.

Equation (4.9) cannot satisfy the conditions that at a solid boundary both the normal and tangential velocity components are equal to the velocity of the boundary. The usual practice is to satisfy only the condition on the normal component. Thus, the tangential velocity of the fluid can be different from that of the boundary; that is, there can be slippage.

The equation of conservation of momentum needs to be considered in order to calculate the pressure. For an inviscid flow

$$\rho\left(\frac{\partial \vec{u}}{\partial t} + u_x\frac{\partial \vec{u}}{\partial x} + u_y\frac{\partial \vec{u}}{\partial y} + u_z\frac{\partial \vec{u}}{\partial z}\right) = -\operatorname{grad} p + \rho\vec{g} \tag{4.10}$$

where $\vec{g}$ is the acceleration of gravity

$$\vec{g} = \vec{i}_x g_x + \vec{i}_y g_y + \vec{i}_z g_z \tag{4.11}$$

$$\operatorname{grad} p = \vec{i}_x\frac{\partial p}{\partial x} + \vec{i}_y\frac{\partial p}{\partial y} + \vec{i}_z\frac{\partial p}{\partial z} \tag{4.12}$$

The left side of (4.10) represents the time rate of change of momentum in a fixed differential volume in space plus the net flow of momentum out of that volume. The right side gives the net force on the differential volume due to pressure and gravity.

The operator on the left side of (4.10)

$$\frac{D}{Dt} = \frac{\partial}{\partial t} + u_x\frac{\partial}{\partial x} + u_y\frac{\partial}{\partial y} + u_z\frac{\partial}{\partial z} \tag{4.13}$$

is the substantial derivative. It gives the change of a property of a field with time, as seen by an observer moving with a fluid particle. This can be understood by defining $f(x, y, z, t)$ as a property of the field. Then

$$df = \frac{\partial f}{\partial t}dt + \frac{\partial f}{\partial x}dx + \frac{\partial f}{\partial y}dy + \frac{\partial f}{\partial z}dz \tag{4.14}$$

Divide by dt to get the change of f over any arbitrary path.

$$\frac{df}{dt} = \frac{\partial f}{\partial t} + \frac{\partial f}{\partial x}\frac{dx}{dt} + \frac{\partial f}{\partial y}\frac{dy}{dt} + \frac{\partial f}{\partial z}\frac{dz}{dt} \tag{4.15}$$

Suppose the path is that of a fluid particle. Then

$$u_x = \frac{dx}{dt} \qquad u_y = \frac{dy}{dt} \qquad u_z = \frac{dz}{dt} \tag{4.16}$$

and (4.14) is the same as (4.13). Thus (4.10) can be written as

$$\rho \frac{D\vec{u}}{Dt} = -\text{grad}\, p + \rho \vec{g} \tag{4.17}$$

That is, the terms on the left side of (4.10) that are enclosed in parentheses represent the acceleration of a fluid particle.

Equation (4.10) has x-, y-, z-components which can be written as

$$\rho\left(\frac{\partial u_x}{\partial t} + u_x \frac{\partial u_x}{\partial x} + u_y \frac{\partial u_x}{\partial y} + u_z \frac{\partial u_x}{\partial z}\right) = -\frac{\partial p}{\partial x} + \rho g_x \tag{4.18a}$$

$$\rho\left(\frac{\partial u_y}{\partial t} + u_x \frac{\partial u_y}{\partial x} + u_y \frac{\partial u_y}{\partial y} + u_z \frac{\partial u_y}{\partial z}\right) = -\frac{\partial p}{\partial y} + \rho g_y \tag{4.18b}$$

$$\rho\left(\frac{\partial u_z}{\partial t} + u_x \frac{\partial u_z}{\partial x} + u_y \frac{\partial u_z}{\partial y} + u_z \frac{\partial u_z}{\partial z}\right) = -\frac{\partial p}{\partial z} + \rho g_z \tag{4.18c}$$

where

$$g_x = -g\frac{\partial h}{\partial x} \qquad g_y = -g\frac{\partial h}{\partial y} \qquad g_z = -g\frac{\partial h}{\partial z} \tag{4.19}$$

and h is the height above some datum plane. From (4.4), $\partial u_y/\partial x$, $\partial u_z/\partial y$ can be substituted for $\partial u_x/\partial y$, $\partial u_y/\partial z$ in (4.18a) to give

$$\rho\left(-\frac{\partial^2 \phi}{\partial x \partial t} + \frac{1}{2}\frac{\partial q^2}{\partial x}\right) = -\frac{\partial p}{\partial x} - \rho g \frac{\partial h}{\partial x} \tag{4.20}$$

where (4.6) is substituted for u_x and

$$q^2 = u_x^2 + u_y^2 + u_z^2 \tag{4.21}$$

Equation (4.20) can be integrated with respect to x, keeping y, z and t constant to give

$$-\frac{\partial \phi}{\partial t} + \frac{q^2}{2} + \frac{p}{\rho} + gh = F(y,z,t) \tag{4.22}$$

In the same way, (4.18b,c) can be integrated with respect to y and z, keeping x, z, t and x, y, t constant. These produce (4.22) with the right sides replaced by $F(x, z, t)$ and $F(x, y, t)$. This indicates that

$$F(y,z,t) = F(x,z,t) = F(x,y,t) = F(t) \tag{4.23}$$

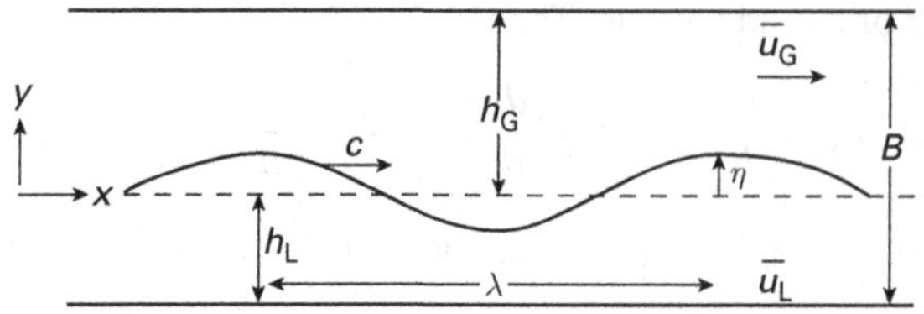

Figure 4.1 Sketch of a propagating wave.

Thus

$$-\frac{\partial \phi}{\partial t}+\frac{q^2}{2}+\frac{p}{\rho}+gh=F(t) \tag{4.24}$$

This is the Bernoulli equation for an incompressible fluid. It neglects viscous effects and assumes an irrotational flow.

4.2 Propagation of small-amplitude two-dimensional waves

Consider a two-dimensional wave propagating on a liquid layer in a channel, as depicted in Figure 4.1. The heights of the channel, the liquid space and the gas space for the undisturbed flow are B, h_L, h_G. Plug flows with velocities $\bar{u}_L$ and $\bar{u}_G$ are also assumed. The coordinate y is perpendicular to the wall and $y_0 = h_L$ is the average location of the interface. The displacement from y_0 is designated by η. It is convenient to represent η by the real part of

$$\eta = a \exp ik\,(x - ct) \tag{4.25}$$

The amplitude a and the wave number, $k=2\pi/\lambda$ are real. The wave velocity is complex

$$c = c_R + ic_I \tag{4.26}$$

where c_R is the velocity with which the wave is propagating downstream.

If (4.26) is substituted into (4.25)

$$\eta = [a \exp ik(x - c_R t)][\exp kc_I t] \tag{4.27}$$

It is noted that if c_I is positive the disturbance is growing in time. Now

$$\exp ik(x - c_R t) = \cos k(x - c_R t) + i \sin k(x - c_R t) \tag{4.28}$$

Therefore, the real part of (4.27) is

$$\eta = [a \cos k\,(x - c_R t)] \exp(kc_I t) \tag{4.29}$$

This represents a wavy disturbance with wave number k and a propagation velocity c_R that is growing or decaying in time.

We are interested in calculating the velocity field introduced by the waves. The assumption of an inviscid, incompressible, irrotational flow is made. The velocities in the gas and liquid are represented by

$$\vec{u}_{\mathrm{L}} = \bar{u}_{\mathrm{L}} + \vec{u}'_{\mathrm{L}} \tag{4.30}$$

$$\vec{u}_{\mathrm{G}} = \bar{u}_{\mathrm{G}} + \vec{u}'_{\mathrm{G}} \tag{4.31}$$

where $\bar{u}_{\mathrm{L}}, \bar{u}_{\mathrm{G}}$ are the averages and $\vec{u}'_{\mathrm{L}}, \vec{u}'_{\mathrm{G}}$ are the disturbances. Equation (4.9) can be used to calculate the two-dimensional velocity so that

$$\phi = \bar{\phi} + \phi' \tag{4.32}$$

and

$$\frac{\partial^2 \phi'_{\mathrm{L}}}{\partial x^2} + \frac{\partial \phi'_{\mathrm{L}}}{\partial y^2} = 0 \tag{4.33}$$

$$\frac{\partial^2 \phi'_{\mathrm{G}}}{\partial x^2} + \frac{\partial^2 \phi'_{\mathrm{G}}}{\partial y^2} = 0 \tag{4.34}$$

Two of the boundary conditions are that the velocities of the gas and liquid normal to the walls are zero:

$$\frac{\partial \phi'_{\mathrm{G}}}{\partial y} = 0 \ \text{at}\, y = h_{\mathrm{G}} \tag{4.35}$$

$$\frac{\partial \phi'_{\mathrm{L}}}{\partial y} = 0 \quad \text{at}\, y = -h_{\mathrm{L}} \tag{4.36}$$

Two other conditions at the interface are needed.

One of these is that the normal velocities of the fluids on both sides of the interface are equal to the normal velocity of the interface. The substantial derivative follows the change of a property of a fluid particle with time. The time change of the height of a particle at the interface is the normal velocity of that particle. Thus

$$\frac{D\eta}{Dt} = \left(u'_{\mathrm{L}y}\right)_{\mathrm{i}} = -\left(\frac{\partial \phi'_{\mathrm{L}}}{\partial y}\right)_{\mathrm{i}} \tag{4.37}$$

where, by definition, η is a function only of x and t. Thus, from (4.13)

$$\frac{\partial \eta}{\partial t} + \bar{u}_{\mathrm{L}} \frac{\partial \eta}{\partial x} + u'_{\mathrm{L}} \frac{\partial \eta}{\partial x} = -\left(\frac{\partial \phi'_{\mathrm{L}}}{\partial y}\right)_{\mathrm{i}} \tag{4.38}$$

This is known as the kinematic condition at the interface. The assumption of small-amplitude waves is made so that $\eta \propto a$, as do ϕ' and $u'_{\mathrm{L}x}$. Thus, all the terms in (4.38) vary with a, except the third term on the left side, which varies as a^2. Since a is assumed to be a small number (4.38) simplifies to

$$\frac{\partial \eta}{\partial t} + \bar{u}_{\mathrm{L}} \frac{\partial \eta}{\partial x} = -\left(\frac{\partial \phi'_{\mathrm{L}}}{\partial y}\right)_{\mathrm{i}} \tag{4.39}$$

Since the normal velocity of the gas at the interface equals the normal velocity of the liquid,

$$\frac{\partial \eta}{\partial t} + \bar{u}_{\mathrm{G}} \frac{\partial \eta}{\partial x} = -\left(\frac{\partial \phi'_{\mathrm{G}}}{\partial y}\right)_{\mathrm{i}} \tag{4.40}$$

One would like to apply (4.39) and (4.40) at a fixed location, $y = 0$, rather than at the interface. A Taylor series expansion yields

$$\left(\frac{\partial \phi'_{\mathrm{L}}}{\partial y}\right)_{\mathrm{i}} = \left(\frac{\partial \phi'_{\mathrm{L}}}{\partial y}\right)_{y=0} + \left(\frac{\partial^2 \phi'}{\partial y^2}\right)_{y=0} \eta + \ldots \tag{4.41}$$

If η is a small number the second term on the right side of (4.41) can be neglected. The linearized form of the kinematic conditions

$$\frac{\partial \eta}{\partial t} + \bar{u}_{\mathrm{L}} \frac{\partial \eta}{\partial x} = -\left(\frac{\partial \phi'_{\mathrm{L}}}{\partial y}\right)_0 \tag{4.42}$$

$$\frac{\partial \eta}{\partial t} + \bar{u}_{\mathrm{G}} \frac{\partial \eta}{\partial x} = -\left(\frac{\partial \phi'_{\mathrm{G}}}{\partial y}\right)_0 \tag{4.43}$$

specifies the derivatives of ϕ'_{L} and ϕ'_{G} at $y = 0$. Thus the disturbed velocity fields in the liquid and in the gas can be calculated with equations (4.33) and (4.34) using boundary conditions (4.42), (4.43), (4.35), (4.36).

Assume a solution of (4.33) of the form

$$\phi'_{\mathrm{L}} = \hat{\phi}_{\mathrm{L}}(y) \exp ik(x - ct) \tag{4.44}$$

Note that $\hat{\phi}_{\mathrm{L}}$ needs to be linearly dependent on amplitude a in order to satisfy (4.42). After substituting into (4.33), an equation for $\hat{\phi}_{\mathrm{L}}(y)$ is obtained.

$$\frac{d^2 \hat{\phi}_{\mathrm{L}}}{dy^2} - k^2 \hat{\phi}_{\mathrm{L}} = 0 \tag{4.45}$$

If (4.25) is substituted into the kinematic boundary conditions, (4.42) and (4.43)

$$iak(c - \bar{u}_{\mathrm{L}}) = -\left(\frac{\partial \hat{\phi}_{\mathrm{L}}}{\partial y}\right)_0 \tag{4.46}$$

$$iak(c - \bar{u}_{\mathrm{G}}) = -\left(\frac{\partial \hat{\phi}_{\mathrm{G}}}{\partial y}\right)_0 \tag{4.47}$$

Two fundamental solutions of (4.45) are $\exp(ky)$ and $\exp(-ky)$. These are used if an infinite field is considered. For a finite field, it is more convenient to use hyperbolic functions

$$\sinh(ky) = \frac{1}{2}[\exp(ky) - \exp(-ky)] \tag{4.48}$$

which equals 0 at $y = 0$ and

$$\cosh(ky) = \frac{1}{2}[\exp(ky) + \exp(-ky)] \tag{4.49}$$

which equals 1 at $y = 0$. Thus the solution for the flow field in the liquid is

$$\hat{\phi}_L = [A]\sinh(ky) + [B]\cosh(ky) \tag{4.50}$$

In order to satisfy the boundary conditions (4.46) and (4.36)

$$[A] = -i(\overline{u}_L - c)a \tag{4.51}$$

$$[B] = -i(\overline{u}_L - c)\coth(kh_L)a \tag{4.52}$$

Following the same procedure, equation (4.34) can be solved to give

$$\phi'_G = \hat{\phi}_G \exp ik(x - ct) \tag{4.53}$$

$$\hat{\phi}_G = [D]\sinh(ky) + [E]\cosh(ky) \tag{4.54}$$

where

$$[D] = -i(\overline{u}_G - c)a \tag{4.55}$$

$$[E] = -i(\overline{u}_G - c)\coth(kh_G)a \tag{4.56}$$

4.3 Dispersion relation for propagating waves

An important use of the results in Section 4.2 is the derivation of the relation between the wave velocity, c, and the wave number, k. A starting point is a consideration of the Laplace formula, which specifies the pressure difference across an interface separating two media (Landau & Lifshitz, 1959).

Take, for example, a gas–liquid interface. Define $R_1$and R_2 as the principal radii of curvature of the surface, where R is positive for a convex liquid surface and negative for a concave liquid surface. According to the Laplace formula the pressure change across the interface is

$$p_{Li} - p_{Gi} = \sigma\left(\frac{1}{R_1} + \frac{1}{R_2}\right) \tag{4.57}$$

where σ is the surface tension. For example, a bubble with radius r_P has radii of curvature that are negative and equal to $-r_P$ so that

$$p_{Gi} - p_{Li} = \frac{2\sigma}{r_P} \tag{4.58}$$

Thus, the pressure in the gas is greater than the pressure in the liquid. This can be understood by considering the sketch of one half of a bubble shown in Figure 4.2. The force due to the surface tension by the half which is not pictured is $\sigma 2\pi r_P$. This force is

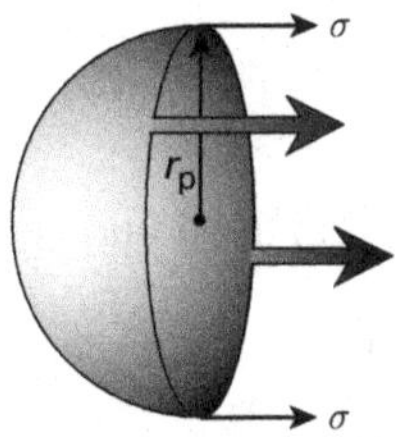

Figure 4.2 Sketch of one half of a bubble.

balanced by the force due to the pressure difference between the gas and the liquid, that is, the product of Δp and the projected area, $\Delta p \pi r_P^2$. Equating these two forces gives (4.58).

For a wavy interface, such as shown in Figure 4.1, radii of curvature can be approximated for small wave amplitudes as indicted below (Landau & Lifshitz, 1959):

$$p_{Li} - p_{Gi} = \sigma\left(-\frac{\partial^2\eta}{\partial x^2} - \frac{\partial^2\eta}{\partial z^2}\right) \tag{4.59}$$

For a two-dimensional wave this simplifies to

$$p_{Li} - p_{Gi} = -\sigma\frac{\partial^2\eta}{\partial x^2} \tag{4.60}$$

The pressures p_{Li} and p_{Gi} can be calculated by using the Bernoulli equation (4.24).

Let the reference level y_0 be equal to zero. At $t = 0$, $\partial\phi\ \partial t = 0$ and the fluid is at rest. Define $p_0 = p$ at $y = 0$ and $t = 0$. Then, from the liquid and gas phase Bernoulli equations,

$$F_L(t) = \frac{p_0}{\rho_L} + \frac{\overline{u}_L^2}{2} \tag{4.61}$$

$$F_G(t) = \frac{p_0}{\rho_G} + \frac{\overline{u}_G^2}{2} \tag{4.62}$$

The Bernoulli equation (4.24) for the liquid can then be written as

$$0 = -\frac{\partial\phi}{\partial t} + \frac{1}{2}u_{Lx}^2 + \frac{1}{2}u_{Ly}^2 + gy + \left(\frac{p_L - p_0}{\rho_L}\right) - \frac{\overline{u}_L^2}{2} \tag{4.63}$$

$$\frac{1}{2}u_{Lx}^2 = \frac{1}{2}\left(\overline{u}_L + u'_{Lx}\right)^2 = \frac{\overline{u}_L^2}{2} + \overline{u}_L u'_{Lx} + \frac{u'^2_{Lx}}{2} \tag{4.64}$$

$$u_{Ly}^2 = u'^2_{Ly} \tag{4.65}$$

Because of the assumption of small-amplitude waves, u'^2_{Lx} and u'^2_{Ly} are negligible so (4.63) can be rewritten as

$$0 = -\frac{\partial\phi'_L}{\partial t} + \overline{u}_L u'_{Lx} + gy + \frac{p'_L - p_0}{\rho_L} \tag{4.66}$$

Similarly, for the gas

$$0 = -\frac{\partial \phi'_{\mathrm{G}}}{\partial t} + \bar{u}_{\mathrm{G}} u'_{\mathrm{G}x} + gy + \frac{p'_{\mathrm{G}} - p_0}{\rho_{\mathrm{G}}} \tag{4.67}$$

where, from (4.6)

$$u'_{\mathrm{L}x} = -\frac{\partial \phi'_{\mathrm{L}}}{\partial x} \qquad u'_{\mathrm{G}x} = -\frac{\partial \phi'_{\mathrm{G}}}{\partial x} \tag{4.68}$$

The interface is located at $y = \eta$. The substitution of (4.66) and (4.67) into (4.60) yields

$$-\sigma \frac{\partial^2 \eta}{\partial x^2} = \rho_{\mathrm{L}} \left(\frac{\partial \phi'_{\mathrm{L}}}{\partial t} \right)_{\mathrm{i}} - \rho_{\mathrm{G}} \left(\frac{\partial \phi'_{\mathrm{G}}}{\partial t} \right)_{\mathrm{i}} + \rho_{\mathrm{L}} \bar{u}_{\mathrm{L}} \left(\frac{\partial \phi'_{\mathrm{L}}}{\partial x} \right)_{\mathrm{i}} - \rho_{\mathrm{G}} \bar{u}_{\mathrm{G}} \left(\frac{\partial \phi'_{\mathrm{G}}}{\partial x} \right)_{\mathrm{i}} - (\rho_{\mathrm{L}} - \rho_{\mathrm{G}}) g_y \eta \tag{4.69}$$

Because of the assumption of small-amplitude waves

$$\left(\frac{\partial \phi'_{\mathrm{L}}}{\partial t} \right)_{y=\eta} = \left(\frac{\partial \phi'_{\mathrm{L}}}{\partial t} \right)_{y=0} \qquad \left(\frac{\partial \phi'_{\mathrm{G}}}{\partial t} \right)_{y=\eta} = \left(\frac{\partial \phi'_{\mathrm{G}}}{\partial t} \right)_{y=0} \tag{4.70}$$

$$\left(\frac{\partial \phi'_{\mathrm{L}}}{\partial x} \right)_{y=\eta} = \left(\frac{\partial \phi'_{\mathrm{L}}}{\partial x} \right)_{y=0} \qquad \left(\frac{\partial \phi'_{\mathrm{G}}}{\partial x} \right)_{y=\eta} = \left(\frac{\partial \phi'_{\mathrm{G}}}{\partial x} \right)_{y=0} \tag{4.71}$$

Equations (4.25), (4.44) and (4.53), with $\hat{\phi}_{\mathrm{L}}$ and $\hat{\phi}_{\mathrm{G}}$ given by (4.50) and (4.54), are substituted into (4.69). Using the approximations given by (4.70) and (4.71), the following relation is obtained:

$$\sigma k^2 a = -\rho_{\mathrm{L}} ikc[B] + \rho_{\mathrm{G}} ikc[E] + \rho_{\mathrm{L}} \bar{u}_{\mathrm{L}} i[B] - \rho_{\mathrm{G}} \bar{u}_{\mathrm{G}} ik[E] - (\rho_{\mathrm{L}} - \rho_{\mathrm{G}}) g_y a \tag{4.72}$$

Since constants $[B]$ and $[E]$ are given by (4.52) and (4.56) the following equation, relating c to k, is obtained:

$$\sigma k^2 + (\rho_{\mathrm{L}} - \rho_{\mathrm{G}}) g_y = \rho_{\mathrm{G}} (\bar{u}_{\mathrm{G}} - c)^2 k \coth(kh_{\mathrm{G}}) + \rho_{\mathrm{L}} (\bar{u}_{\mathrm{L}} - c)^2 k \coth(kh_{\mathrm{L}}) \tag{4.73}$$

which is equation (2.6) in Chapter 2.

4.4 Propagation of waves at the interface of horizontal stationary fluids

Consider (4.73) for the case of $\bar{u}_{\mathrm{L}} = 0$, $\bar{u}_{\mathrm{G}} = 0$ and ρ_{G} much smaller than ρ_{L}.

$$c^2 = \left(\frac{\sigma k}{\rho_{\mathrm{L}}} + \frac{g}{k} \right) \tanh(kh_{\mathrm{L}}) \tag{4.74}$$

where the subscript y on the gravity term is dropped since a horizontal channel is being considered. Equation (4.74) is the classical relation between wave velocity and the wavelength. It describes a situation for which the stabilizing effects of gravity and surface tension balance destabilizing effects of inertia.

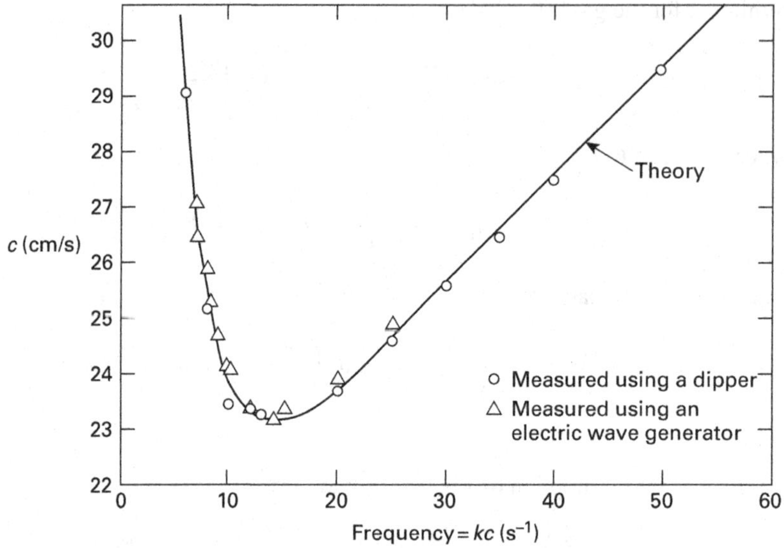

Figure 4.3 Dispersion relation for distilled water at 23.8 °C.

For $kh_L \geq 2, \tanh(kh_L) = 1$, so

$$c^2 = \left(\frac{k\sigma}{\rho_L} + \frac{g}{k}\right) \tag{4.75}$$

This is called the deep liquid assumption, even though $kh_L = 2$ corresponds to $\lambda/h_L = \pi$. Equation (4.75) is plotted in Figure 4.3 for air–water at atmospheric pressure. Agreement between theory and the experiments of Kim (1968) is noted. Small wavelengths for which $k\sigma/\rho_L$ is much greater than g/k are called capillary waves. The wave velocity increases with decreasing wavelength or increasing frequency. Gravity waves, for which the wave velocity increases with increasing wavelength, or decreasing frequency, are observed for g/k is much greater than $k\sigma/\rho_L$. Thus, there is a wavelength at which a minimum in the wave velocity occurs. As seen in Figure 4.3, this occurs at $\lambda = 1.78$ cm and $c = 23.5$ cm/s for an air–water interface. The Reynolds number for this system at the minimum is $\lambda c/\upsilon_L = 4200$, which supports the assumption that viscous effects can be ignored.

For small kh_L and $\tanh(kh_L) = 1$

$$c^2 = \left(\frac{k^2 h_L \sigma}{\rho_L} + g h_L\right) \tag{4.76}$$

These waves are known as shallow liquid waves. If effects of surface tension can be ignored,

$$c^2 = g h_L \tag{4.77}$$

The velocity of shallow gravity waves increases with liquid height and is independent of wavelength.

4.5 Kelvin–Helmholtz instability

4.5.1 General conditions for a KH instability

The wave velocity, calculated above, for small-amplitude waves at a gas velocity of zero is real. That is, the system is neutrally stable. This is not the case for high gas velocities, for which c can be complex. This situation is called a Kelvin–Helmholtz instability, which can have a number of outcomes. It has been used widely to explain phenomena in multiphase systems.

Define c_0 as the velocity when $\overline{u}_L = 0$, $\overline{u}_G = 0$.

$$c_0^2 = \frac{\sigma k^2 + (\rho_L - \rho_G)g}{k(\rho_L^* + \rho_G^*)} \tag{4.78}$$

$$\rho_L^* = \rho_L \coth(hk_L) \qquad \rho_G^* = \rho_G \coth(kh_G) \tag{4.79}$$

It is noted that (4.78) differs from (4.75) because the upper fluid need not be a gas. Thus, the density ρ_G is assumed to be smaller than ρ_L but not negligible.

The general relation (4.73) can be written as follows:

$$(\rho_G^* + \rho_L^*)c_0^2 = \rho_G^*(\overline{u}_G^2 - 2\overline{u}_G c + c^2) + \rho_L^*(\overline{u}_L^2 - 2\overline{u}_L c + c^2) \tag{4.80}$$

This can be solved for c to get

$$c = \frac{\rho_G^* \overline{u}_G + \rho_L^* \overline{u}_L}{\rho_G^* + \rho_L^*} \pm \left[c_0^2 - \frac{\rho_G^* \rho_L^* (\overline{u}_G - \overline{u}_L)^2}{(\rho_G^* + \rho_L^*)^2}\right]^{\frac{1}{2}} \tag{4.81}$$

From (4.78), c_0^2 is a positive number so long as $\rho_L > \rho_G$. From (4.81), it is seen that c is a complex number if

$$c_0^2 < \frac{\rho_G^* \rho_L^* (\overline{u}_G - \overline{u}_L)^2}{(\rho_G^* + \rho_L^*)^2} \tag{4.82}$$

Equation (4.82) defines the $(\overline{u}_G - \overline{u}_L)$ needed to initiate a Kelvin–Helmholtz instability. The time constant characterizing the growth rate of the wave, defined by (4.27), is

$$kc_I = k\left[\frac{\rho_G^* \rho_L^* (\overline{u}_G - \overline{u}_L)^2}{(\rho_G^* + \rho_L^*)^2} - c_0^2\right]^{\frac{1}{2}} \tag{4.83}$$

The velocity of unstable waves, defined by the real part of (4.81), is

$$c_R = \frac{\rho_G^* \overline{u}_G + \rho_L^* \overline{u}_L}{\rho_G^* + \rho_L^*} \tag{4.84}$$

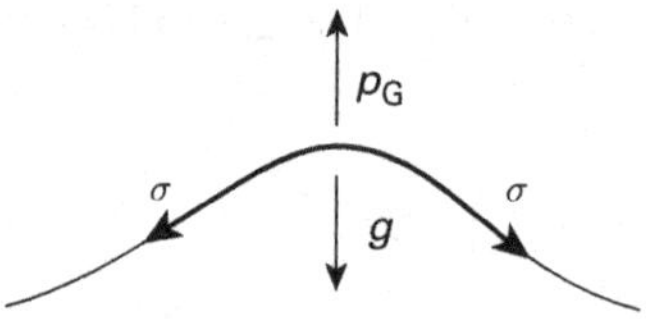

Figure 4.4 Forces acting at a wave crest.

4.5.2 Gas–liquid flows with deep gas and liquid layers

A consideration of a Kelvin–Helmholtz instability of waves at an interface dividing deep gas and liquid layers with ρ_G much smaller than ρ_L provides a good example. For this case, $\rho_G^* = \rho_G$ and $\rho_L^* = \rho_L$. Equation (4.81) simplifies to

$$c = \frac{\rho_G \overline{u}_G + \rho_L \overline{u}_L}{\rho_L} \pm \left[\frac{\sigma k}{\rho_L} + \frac{g}{k} - \frac{\rho_G}{\rho_L} (\overline{u}_G - \overline{u}_L)^2 \right]^{\frac{1}{2}} \tag{4.85}$$

At small $(\overline{u}_G - \overline{u}_L)^2$ the wave velocity is real; it decreases with increasing $(\overline{u}_G - \overline{u}_L)$. Eventually a critical condition is reached where

$$\frac{\rho_G}{\rho_L} (\overline{u}_G - \overline{u}_L)^2 = \frac{\sigma k}{\rho_L} + \frac{g}{k} \tag{4.86}$$

The wave velocity, c, will be complex for large $(\overline{u}_G - \overline{u}_L)$.

This can be interpreted as follows: The streamlines representing the gas flow are squeezed at the crest. This leads to an increase in the gas velocity and, from a consideration of the Bernoulli equation, a decrease in the gas pressure. As depicted in Figure 4.4, the left side of (4.86) represents the magnitude of the destabilizing effect of this suction. The right side represents the stabilizing effects of gravity and surface tension. The imbalance of these forces leads to an instability and a rapidly growing wave. At neutral stability and for all super-critical conditions

$$c_R = \frac{\rho_G \overline{u}_G + \rho_L \overline{u}_L}{\rho_L} \tag{4.87}$$

$$k^2 c_I^2 = k^2 \frac{\rho_G}{\rho_L} (\overline{u}_G - \overline{u}_L)^2 - \frac{\sigma k^3}{\rho_L} - gk \tag{4.88}$$

Figure 4.5 sketches the above results for a deep liquid in a plot of wavelength, λ, versus $(\overline{u}_G - \overline{u}_L)$. The solid curve indicates neutral stability. The domain inside this curve represents conditions for which $c_I > 0$. The critical $(\overline{u}_G - \overline{u}_L)$ below which $c_I = 0$ for all wavelengths is indicated by the solid vertical line. This can be obtained by taking the derivative of (4.86) and setting $d(\overline{u}_G - \overline{u}_L)/dk = 0$ to find the extremum

$$k_{\text{critical}} = \left(\frac{\rho_L g}{\sigma} \right)^{\frac{1}{2}} \tag{4.89}$$

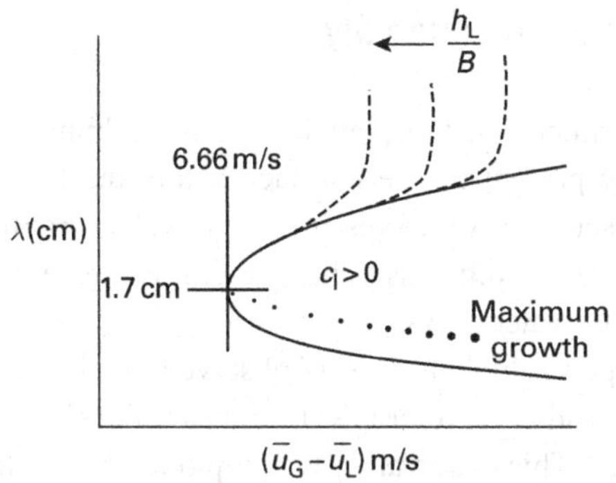

Figure 4.5 Sketch of the neutral stability curve for air flowing over water.

The critical $(\overline{u}_G - \overline{u}_L)$ corresponding to k_{critical} is obtained by solving (4.86).

For air–water at atmospheric conditions

$$\lambda_{\text{critical}} = 1.7\,\text{cm} \tag{4.90}$$

$$(\overline{u}_G - \overline{u}_L)_{\text{critical}} = 6.66\,\text{m/s} \tag{4.91}$$

As can be seen from (4.27), the growth rate is represented by kc_I. From (4.88), the wave number at which the growth rate is a maximum for a given $(\overline{u}_G - \overline{u}_L)$ is

$$\frac{\rho_G}{\rho_L}(\overline{u}_G - \overline{u}_L)^2 = \frac{3\sigma k_M}{2\rho_L} + \frac{g}{2k_M} \tag{4.92}$$

The loci of maximum growth are sketched in Figure 4.5 by the dotted curve. The argument is usually given that for $(\overline{u}_G - \overline{u}_L)$ larger than the critical the waves that appear are the most rapidly growing ones given by (4.92).

4.5.3 Effect of h_G

For very long wavelengths (coth $kh_L = 1\,kh_L$), the deep liquid assumption is not valid. The critical condition predicted by KH theory simplifies to

$$(\overline{u}_G - \overline{u}_L)^2_{\text{critical}} = \frac{\rho_L}{\rho_G} g h_L \tag{4.93}$$

Thus, the neutral stability curve shown in Figure 4.5 for coth $kh_L = 1$ needs to be modified by including the dashed curves which are sketched on the figure, where h_L/B is a parameter and σ/ρ_L, g, ρ_G/ρ_L, B are specified. For small λ, the plot is independent of h_L/B. However, for large λ the neutral stability changes so that (4.93) is satisfied.

As discussed in Section 2.3, equation (4.93) has been used to predict the initiation of slugging. The predicted critical gas velocity is found to be too large. Better agreement with observation is obtained if viscous effects are considered for very large wavelengths (viscous large-wavelength theory). The dashed curves then move to the left (see Andritsos *et al.*, 1989).

4.6 Experimental verification of the KH instability

There are not many clear-cut experiments that display the Kelvin–Helmholtz instability. This is because waves are usually present on the interface before the KH critical gas velocity is attained. Thus, the discussion of waves in an air–water stratified flow in Section 2.2.2 characterizes the critical condition as a change of the character of the waves from "regular" to "irregular" for air–water flows.

However, as pointed out in Chapter 2 "regular waves" observed for air–water flow are not observed for high-viscosity liquids. Therefore, KH critical conditions are realized when the liquid interface is smooth. This observation was exploited by Andritsos *et al.* (1989), who used horizontal pipelines with diameters of 2.52 cm and 9.53 cm to study stratified flows of air and water–glycerine solutions. For a 70 cP liquid, the interface was smooth at low gas velocities. At a critical gas velocity, sinusoidal disturbances with a wavelength of 1–2 cm appeared. This is to be compared with KH theory which predicts $\lambda_{\text{critical}} = 1.7$ cm at the critical gas velocity of 6.66 m/s for air–water. Depending on the height of the stratified liquid layer, the KH instability leads to tumbling large-amplitude waves or to slugs.

Figure 4.6 is taken from the paper by Andritsos *et al.* (1989). The data indicate either the transition to slugs (large h_L/d_t) or the initiation of large-amplitude waves (small h_L/d_t). Good agreement is found between the KH analysis and experiment. It is noted that both experiment and theory indicate that transition is independent of pipe diameter (as already shown in Figure 2.7).

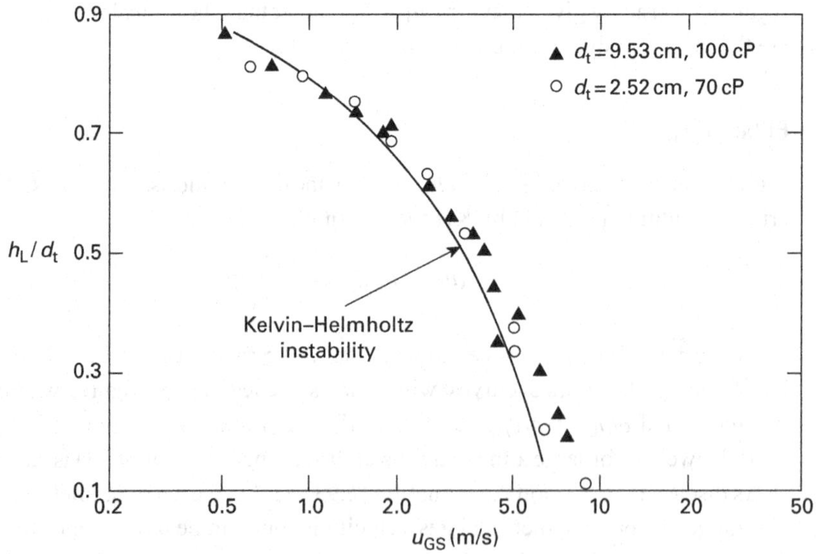

Figure 4.6 Comparison of Kelvin–Helmholtz stability calculations with observed initiation of slugs or large-amplitude waves on 70 cP (d_t = 2.52 cm) and 100 cP (d_t = 9.53 cm) liquids. Andritsos & Hanratty, 1989.

Earlier experiments by Francis (1954, 1956) used an oil with a viscosity of 2.2 Poise and a surface tension of 34 dynes/cm. Air was blown over the oil surface. Both the air and oil layers were deep. For this liquid, KH theory predicts a critical air velocity of 5.16 m/s, a critical wavelength of 1.27 cm and a critical wave velocity of 0.67 cm/s. A transition from a smooth interface to one with small-amplitude waves with $\lambda \approx 1$ cm and $c_R = 1$ cm/s was observed when the air velocity was 9.67 m/s at 8 cm above the interface. This critical velocity is higher than predicted by KH theory.

However, the very large liquid viscosity used in the experiments of Francis would greatly decrease the growth rates of waves. Therefore, a much longer channel might be needed to observe the critical condition. (It should be pointed out that Francis first observed waves at the end of his channel.)

4.7 Group velocity

4.7.1 Physical interpretation of group velocity

Large energies can be associated with wave motion at gas–liquid interfaces. The group velocity, c_g, is related to the speed with which this energy is transported downstream. It can be different from the wave velocity and is important in understanding many phenomena. Two-dimensional gravity waves, in a situation for which there is no mean flow and the waves have a small amplitude, will be discussed. A unit width in the spanwise direction is considered. Waves can have kinetic energy, T, associated with fluctuations in the velocity and a potential energy, P, associated with the displacement of the interface in a gravitational field.

The wave motion being considered is depicted in Figure 4.1, with $\bar{u}_L = 0$ and $\bar{u}_G = 0$. The displacement of the interface from its average location, h_L, is given as

$$\eta = a\cos(kx - \omega t) \tag{4.94}$$

where $\omega = kc$ is the circular frequency. Consider a liquid element, $dxdy$, in the region where η is positive. The net force due to gravity and buoyancy is $-g(\rho_L - \rho_G)dxdy$ for a unit width in the z-direction. The work done to lift this element to a height y above the average location of the interface is $g(\rho_L - \rho_G)ydxdy$. Integration with respect to y in the region where the interface has a positive displacement gives $g(\rho_L - \rho_G)(\eta^2/2)dx$.

Similarly, in regions where the interface has negative displacements the net force on a gaseous element due to gravity and buoyancy is given by $g(\rho_L - \rho_G)dxdy$. Since y will have negative values, the work needed to move a gaseous particle to a location y is $-gy(\rho_L - \rho_G)dydx$. Integration between $y = \eta$ and $y = 0$ gives $g(\rho_L - \rho_G)(\eta^2/2)dx$. Therefore, the potential energy associated with one wavelength is

$$P = \int_0^\lambda g(\rho_L - \rho_G)\frac{\eta^2}{2}dx = \frac{1}{4}a^2 g\rho_L\lambda \tag{4.95}$$

where ρ_G has been neglected and η is a function of x and t, given by (4.94).

The kinetic energy in one wavelength is given by

$$T = \int_0^h \int_0^\lambda \frac{\rho_L q^2}{2} dy dx = \frac{1}{4} a^2 g \rho_L \lambda \tag{4.96}$$

where

$$q^2 = u_x^2 + u_y^2 \tag{4.97}$$

and u_x, u_y are defined with (4.6), (4.44), (4.50). The total energy is

$$E = P + T = \frac{1}{2} a^2 g \rho_L \lambda \tag{4.98}$$

This result is independent of h_L so it is valid both for shallow and deep layers.

Now consider the rate at which fluid to the left of a *y–z* plane at a given *x* is doing work on the fluid to the right side of the plane. Again, a unit slice in the *z*-direction is considered. The force on a small area perpendicular to the direction of propagation is *pdy*. The rate of work on this differential area is $pu_x dy$. The rate of work over the plane at a given time can be obtained by integrating the force over the area

$$\text{Rate of work} = \int_0^{h_L + \eta} p u_x dy \tag{4.99}$$

Integrate over one period to obtain a time-averaged rate of work. The Bernoulli equation (4.24) gives $p = \partial \phi_L / \partial t$ if terms of second order are neglected. Since u_x is given by (4.6), the average rate of work over one period is given by

$$\text{Average rate} = \frac{\omega}{2\pi} \int_0^{2\pi/\omega} \int_0^{y_L} -\frac{\partial \phi_L}{\partial t} \frac{\partial \phi_L}{\partial x} dy dt \tag{4.100}$$

Since

$$\phi_L = \hat{\phi}_L \cos(kx - \omega t) \tag{4.101}$$

equation (4.100) can be written as

$$\text{Average rate} = \frac{\omega}{2\pi} \int_0^{2\pi/\omega} \int_0^{y_L} \{\sin^2(kx - \omega t)\} k \omega \hat{\phi}_L^2(y) dy dt \tag{4.102}$$

where $\hat{\phi}_L(y)$ is given by (4.50). The following result is obtained:

$$\text{Average rate} = \frac{1}{4} g \rho_L a^2 c \left(1 + \frac{2kh_L}{\sinh 2kh_L}\right) \tag{4.103}$$

where the relation

$$2\cosh kh_L \sinh kh_L = \sinh 2kh_L \tag{4.104}$$

has been used (see Lamb, 1953, p. 383, equation 11).

The energy per length in the spanwise direction is $\left(\frac{1}{2}a^2 g\rho_L\right)$. Define a group velocity as the velocity with which this energy propagates. Thus

$$c_g\left(\frac{1}{2}a^2 g\rho_L\right) = \frac{1}{4}g\rho_L a^2 c\left(1 + \frac{2kh_L}{\sinh 2kh_L}\right) \tag{4.105}$$

Solve for c_g.

$$c_g = \frac{c}{2}\left(1 + \frac{2kh_L}{\sinh 2kh_L}\right) \tag{4.106}$$

Thus, for gravity waves on a deep liquid

$$c_g = \frac{c}{2} \tag{4.107}$$

For a shallow liquid (small kh_L)

$$c_g = c \tag{4.108}$$

Thus, the group velocity is less than the phase velocity for gravity waves on a deep liquid.

Consider the packet of gravity waves on a deep liquid shown in Figure 4.7. The packet would move downstream at a velocity c_g. The waves in the packet have a velocity c, which is greater than c_g, so the waves are formed at the front of the packet and disappear at the back. If the packet contains more than one wavelength (or phase velocity), the group of waves could break up to form two or more packets. Experiments with wave packets have been reported by Kim (1970) and by Kim & Hanratty(1971). Waves were generated at the center of a long tank. A brass rod with a diameter of 3/16 inch that extended over the whole width of the tank was placed 5 mm above the water surface. A constant 9 kV voltage with a superposed AC signal between the electrode and the water was applied. By modulating the AC voltage, a wave packet with a maximum steepness of about 0.03 could be formed.

A general relation for group velocity is given in Lamb (1953, p. 381) as

$$c_g = \frac{d\omega}{dk} = c + k\frac{dc}{dk} \tag{4.109}$$

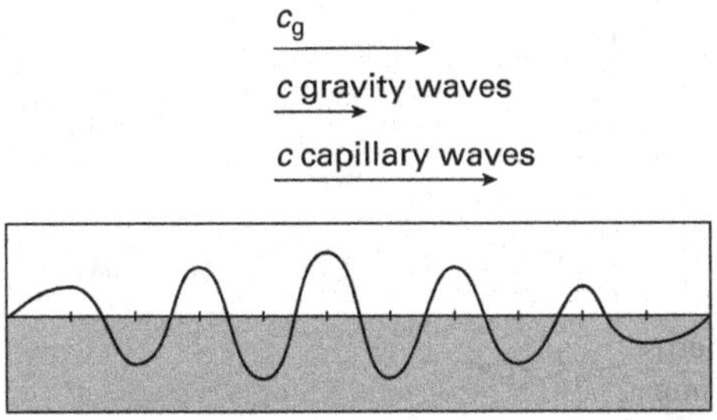

Figure 4.7 Motion of a packet of waves at the interface of a deep liquid.

since $\omega = kc$. This is valid for all types of waves. Sound waves are non-dispersive, that is, c is independent of k, so

$$c_g = c \tag{4.110}$$

For capillary waves on deep water

$$c^2 = \frac{k\sigma}{\rho_L} \tag{4.111}$$

and

$$c_g = \frac{3}{2}c \tag{4.112}$$

Thus, for deep capillary waves, the group velocity is larger than the phase velocity. Capillary waves in a packet would be formed at the back of the packet and disappear at the front.

4.7.2 Wave resistance

A boat moving through a liquid with velocity c forms gravity waves in its rear. The consequence is that more power is needed. That is, the waves may be considered to provide an additional resistance.

A sketch of a very simplified example is shown in Figure 4.8. The energy of the waves per unit length is given by (4.98). The length of the wave train produced per unit time is the velocity of the object, if the train is attached. Thus, the energy per unit time needed to create this wake is

$$\text{energy/time} = c\left(\frac{1}{2}\rho_L g a^2\right) \tag{4.113}$$

The energy flowing through the dashed reference plane shown in Figure 4.8 is

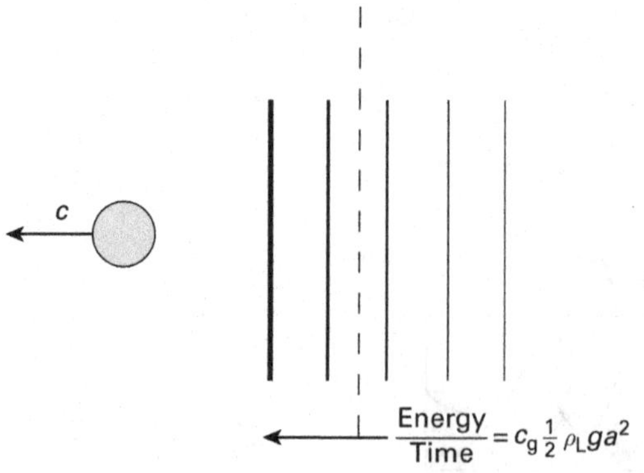

Figure 4.8 Wave train behind a partially submerged body moving with a velocity c.

$$\text{energy/time} = c_{\mathrm{g}}\left(\frac{1}{2}\rho_{\mathrm{L}} g a^2\right) \tag{4.114}$$

The work needed to produce the waves can be written as the product of the velocity of the object and a wave resistance, R.

$$c\frac{1}{2}\rho_{\mathrm{L}} g a^2 = c_{\mathrm{g}}\frac{1}{2}\rho_{\mathrm{L}} g a^2 + Rc \tag{4.115}$$

Thus, the wave resistance is given by

$$R = \frac{c - c_{\mathrm{g}}}{c}\frac{1}{2}\rho_{\mathrm{L}} g a^2 \tag{4.116}$$

For gravity waves,

$$R = \frac{1}{4}\rho_{\mathrm{L}} g a^2\left(1 - \frac{2kh_{\mathrm{L}}}{\sinh(kh_{\mathrm{L}})}\right) \tag{4.117}$$

If the body moves faster than the maximum wave velocity, $(gh_{\mathrm{L}})^{1/2}$, it cannot have an attached train. Wave resistance vanishes.

4.7.3 Waves which decay in space, rather than time

The wave height has been given as

$$\eta = a \exp ik(x - ct) \tag{4.118}$$

$$\eta = a \exp i(kx - \omega t) \tag{4.119}$$

where ω is the circular frequency and is equal to kc. The analysis in this chapter represents the interfacial displacement given by (4.118) with k real and c complex. As shown in (4.29) this represents a wave pattern growing or decaying in time.

The more usual observation is for waves growing or decaying as they propagate. This is represented by (4.118), (4.119) with k complex and c or ω real. The analysis with k complex is more difficult than the one with c complex. Fortunately, work by Huerre & Monkewitz (1985) and by Gaster (1962) show that, for small growth rates,

$$k_{\mathrm{I}} = \frac{k_{\mathrm{R}} c_{\mathrm{I}}}{c_{\mathrm{g}}} \tag{4.120}$$

where c_{g} is the group velocity. Thus, the analysis for temporal changes (c complex) can be used to calculate spatial changes (k complex). This result will be used in Chapter 6.

4.8 Uses of the notion of a KH instability

4.8.1 Initiation of slugging

The usual explanation for the initiation of slugging is the growth of an unstable wave. However, the final stage in the closing of the air space to form a slug occurs when the gas velocity above the crest of the wave reaches the critical value for a KH instability. This is particularly evident for transitions in downwardly inclined pipes, Section 5.5.3, for which the instability is initiated in a stratified configuration at gas flows less than the critical KH velocity. It also determines the critical height at the inlet of an upwardly inclined pipe. For this situation, reversed flow causes an intermittent buildup at the upstream end of the pipe (Section 5.5.2).

4.8.2 Rate of atomization

Taylor (1940) argued that air flowing over a liquid surface can atomize it by removing small ripples through a KH instability. This has been supported by photograpic studies of Woodmansee & Hanratty (1969). See Chapter 11.

Taylor suggested that the volumetric rate of removal varies as the wavelength cubed and the time constant characterizing the rate of growth of the waves

$$Q_A \propto \frac{\lambda^3}{t_c} \propto \frac{1}{k^3 t_c} \tag{4.121}$$

The mass rate per unit area then satisfies the following:

$$R_A \propto \frac{\rho_L}{k^3 t_c \lambda^2} \propto \frac{\rho_L}{k t_c} \tag{4.122}$$

The time-scale is defined as

$$t_c^{-1} \propto k c_I \tag{4.123}$$

where, for capillary ripples,

$$c_I^2 = \frac{\rho_G}{\rho_L}(\overline{u}_G - \overline{u}_L)^2 - \frac{k\sigma}{\rho_L} \tag{4.124}$$

(See equation (4.88).) Consider small $\overline{u}_L$. Then

$$R_A \propto \frac{\rho_L}{k} k \left(\frac{\rho_G}{\rho_L}\right)^{\frac{1}{2}} \overline{u}_G \left(1 - \frac{k}{k_0}\right) \tag{4.125}$$

or

$$R_A \propto \rho_L^{1/2} \rho_G^{1/2} \overline{u}_G f\left(\frac{k}{k_0}\right) \tag{4.126}$$

where

$$k_0 = \frac{\rho_G \overline{u}_G^2}{\sigma} \tag{4.127}$$

Taylor included effects of liquid viscosity, which would dampen the wave growth, so that

$$\frac{R_A}{\rho_L^{1/2} \rho_G^{1/2} \overline{u}_G} = f\left(\frac{k}{k_0}, \theta\right) \tag{4.128}$$

where

$$\theta = \frac{\rho_L}{\rho_G} \frac{\sigma^2}{\mu_L^2 \overline{u}_G^2} \tag{4.129}$$

Increasing θ causes a decrease in R_A. For $\theta > 10$ the influence of viscosity can be neglected since f in (4.128) varies approximately as $\left(1 + \theta^{0.5}\right)^{-1}$. Then

$$\frac{R_A}{\rho_L^{1/2} \rho_G^{1/2} \overline{u}_G} = f\left(\frac{k\sigma}{\rho_G \overline{u}_G^2}\right) \tag{4.130}$$

For deep water, one can argue that k should represent the fastest growing wave. From (4.92)

$$k_M \propto \frac{\rho_G \overline{u}_G^2}{\sigma} \tag{4.131}$$

so $k\sigma/\rho_G \overline{u}_G^2$ of the atomizing capillary waves is constant and, for $\theta < 10$,

$$\frac{R_A}{\rho_G^{1/2} \rho_L^{1/2} \overline{u}_G} = \text{constant} \tag{4.132}$$

This result can be applied to annular flows where the wall film is characterized by the intermittent appearance of disturbance waves. The photographic studies by Woodmansee & Hanratty (1969) show that atomization occurs by the removal of waves in these disturbances. The wavelengths of the atomizing waves are less than λ_{max} calculated for thick films since viscous effects associated with the presence of the wall limits the maximum possible wavelength that can exist. Therefore, measured R_A are less than predicted by (4.132). Tatterson (1975) and Schadel & Hanratty (1989) argued that the unstable waves have lengths that scale with the film height so that

$$\frac{R_A}{\rho_L^{1/2} \rho_G^{1/2} \overline{u}_G} = If\left(\frac{\rho_G \overline{u}_G^2 h_L}{\sigma}\right) \tag{4.133}$$

where I is an intermittency factor which represents the fraction of the time that disturbance waves are present. This equation is complicated to use and the assumptions made to apply Taylor's theory to annular flows have not been substantiated. Therefore, it has not been successfully used for annular flows.

The empirical equation below (Pan & Hanratty, 2002) has been employed in flows for which the rate of atomization varies linearly with the mass flow rate of the wall layer

$$\frac{R_A}{\rho_G^{1/2}\rho_L^{1/2}\overline{u}_G} = k_A \frac{(W_{LF} - W_{LFC})}{P\sigma} \tag{4.134}$$

Equation (4.134) has similarities to (4.133) in that the left sides are the same and R_A varies inversely with surface tension. The representation of the effect of σ as given in (4.134) has been motivated by the papers of Lopez de Bertodano *et al.* (1997) and Assad *et al.* (1998).

4.8.3 Interfacial stress in horizontal stratified flows

The drag for stratified flows can be larger than what is found for a smooth surface if waves are present. This is particularly evident at velocities larger than required for a Kelvin-Helmholtz instability (u_G = 6–7 m/s for air–water flow at atmospheric conditions). This is because large amplitude irregular waves evolve from this instability (see Chapter 5).

References

Andritsos, N., Williams, L. & Hanratty, T. J. 1989 Effect of liquid viscosity on the stratified-slug transition in horizontal pipe flow. *Int. J. Multiphase Flow* 15, 877–892.

Assad, A., Jan, C. S., Lopez de Bertodano, M. & Beus, S. 1998 Scaled entrainment measurements in ripple-annular flow in a small tube. *Nucl. Eng. Des.* 184, 437–447.

Francis, J. R. D. 1954 Wave motions and the aerodynamic drag on a free oil surface. *Phil. Mag.*, Ser. 7, 45: 366, 695–702.

Francis, J. R. D. 1956 Wave motions on a free oil surface. *Phil. Mag.*, Ser. 8, 47: 7, 685–688.

Gaster, M. 1962 A note on the relation between temporally-increasing and spatially-increasing disturbances in hydrodynamic instabililty. *J. Fluid Mech.* 14, 222–224.

Huerre, P. & Monkewitz, P. A. 1985 Absolute and convective instabilities in free shear layers. *J. Fluid Mech.* 159, 151–168.

Kim, Y. Y. 1968 *Wave motion on viscoelastic fluids*. M.Sc. thesis, University of Illinois.

Kim, Y. Y. 1970 *Non-linear effects in wave motion*. Ph.D. thesis, University of Illinois.

Kim, Y. Y. & Hanratty, T. J. 1971 Weak quadratic interactions of two-dimensional waves. *J. Fluid Mech.* 50, 107–132.

Lamb, H. 1953 *Hydrodynamics*. 6th edn, Cambridge: Cambridge University Press.

Landau, L. B. & Lifshitz, E. M. 1959 *Fluid Mechanics*. Reading, MA: Addison-Wesley, p. 231.

Lopez de Bertodano, M. A., Jan, C. S. & Beuss, G. G. 1997 Annular flow entrainment rate experiment in a small vertical pipe. *Nucl. Eng. Des.* 178, 61–70.

Pan, L. & Hanratty, T. J. 2002 Correlation of entrainment for annular flow in vertical pipes. *Int. J. Multiphase Flow* 28, 363–384.

Schadel, S. A. & Hanratty, T. J. 1989 Interpretation of atomization rates of the liquid film in gas–liquid annular flow. *Int. J. Multiphase Flow* 15, 893–900.

Tatterson, D. F. 1975 Rates of atomization and drop size in annular two-phase flow. Ph.D. thesis, University of Illinois.

Taylor, G. I. 1940 Generation of ripples by wind blowing over a viscous fluid. Reprinted in *The Scientific Papers of Sir Geoffrey Ingram Taylor*, Vol. 3, ed. G. K. Batchelor. Cambridge: Cambridge University Press (1963), pp. 244–252.

Woodmansee, D. E. & Hanratty, T. J. 1969 Mechanism for the removal of droplets from a liquid surface by a parallel air flow. *Chem. Eng. Sci.* 24, 299–307.

5 Stratified flow

5.1 Scope

This chapter describes the stratified pattern observed in gas–liquid flows, for which liquid flows along the bottom of a conduit and gas flows along the top. The gas exerts a shear stress on the surface of the liquid. It is desired to calculate the average height of the liquid layer and the pressure gradient for given liquid and gas flow rates. The flow is considered to be fully developed so that the height of the liquid is not changing in the flow direction and the pressure gradient is the same in the gas and liquid flows.

In order to consider stratified flow in circular pipes, the simplified model of the flow pattern, presented by Govier & Aziz (1972), is exploited. The interface is pictured to be flat. At large gas velocities, some of the liquid can be entrained in the gas. This pattern is considered in Section 12.5 entitled "the pool model" for horizontal annular flow.

The approach taken in utilizing the simplified model is to consider the interfacial stress as given and, then, to calculate the behavior of the liquid phase. It will turn out that the prediction of the interfacial stress is the principal problem to be addressed.

At high gas velocities, the liquid layer thins out and climbs up the wall of the pipe. The interface is not flat, so the simplified model could be a poor one. Under these circumstances, the flow could display a behavior intermediate between a stratified and an annular flow. Thus, the treatment of the liquid distribution in the wall layer of a horizontal annular flow (discussed in Chapter 12) could be a good starting point for considering stratified flows with curved interfaces.

Kelvin–Helmholtz waves can cause the interfacial stress to be much larger than would be realized if the interface were smooth. Experiments show that this increase in stress is related to the steepness of the waves. Results from inviscid flow theory are used to develop an equation for the wave height – and, therefore, to determine the parameters affecting interfacial stress.

Small inclinations of a pipeline can have a strong effect on the behavior of a stratified flow. Thus, separate subsections on inclined and declined pipes are presented.

5.2 Stratified flow in a rectangular channel

The main focus will be on circular pipes. However, because of its simplicity, the rectangular channel will first be treated. Materials from Chapter 3 on film flows are used.

The channel is assumed to be wide enough for a two-dimensional flow to be considered. As shown in Chapter 3, the variation of shear stress in the liquid layer is given by

$$\frac{d\tau}{dy} = \frac{dp}{dz} - \rho_L g \sin\theta \tag{5.1}$$

where θ is the inclination angle of the channel to the horizontal and z is the coordinate in the direction of flow (Figure 5.5), ρ_L is the density of the bottom fluid and y is the distance from the wall. This can be integrated to give the stress distribution

$$\tau = \tau_i - (h_L - y)\left(\frac{dp}{dz} - \rho_L g \sin\theta\right) \tag{5.2}$$

By using a Boussinesq approximation,

$$\tau = \mu_L \frac{du_L}{dy} + \mu^t \frac{du_L}{dy} \tag{5.3}$$

where μ_L is the molecular viscosity and μ^t is the turbulent viscosity. Substituting (5.3) into (5.2),

$$(\mu_L + \mu^t)\frac{du_L}{dy} = \tau_i - (h_L - y)\left(\frac{dp}{dz} - \rho_L g \sin\theta\right) \tag{5.4}$$

Integrating from the wall where $y = 0$

$$u_L = \int_0^y \frac{\tau_i - (h_L - y)\left(\frac{dp}{dz} - \rho_L g \sin\theta\right)}{(\mu_L + \mu^t)} dy \tag{5.5}$$

The volumetric flow per unit width can be obtained by integrating (5.5) from $y = 0$ to h_L.

The interfacial stress is obtained from empirical relations or by considering a momentum balance for the upper fluid for given flows of the gas and the liquid:

$$(\mu_G + \mu^t)\frac{du_G}{dy} = \tau_{W2} - (B - y)\left(\frac{dp}{dz} - \rho_G g \sin\theta\right) \tag{5.6}$$

where B is the height of the channel, τ_{W2} is the stress on the top wall and y is the distance from the bottom wall. Equations (5.5) and (5.6) are solved using the boundary conditions and that velocities u_G and u_L are zero at the two walls, and that $u_G = u_L$ and $\mu_G(du_G/dy) = \mu_L(du_L/dy)$ at the interface.

For the case of laminar flows in the upper and lower phases, $\mu^t = 0$. The solution is given in Bird, *et al.* (1960, p. 54).

5.3 Stratified flow of gas and liquid in a pipe

5.3.1 Geometric relations

For flow in a pipe, the simplifying assumption is usually made that the interface is flat. Geometric parameters A_G, A_L are the areas of the pipe occupied by the gas and the liquid,

S_i is the length of the interface, P_G is the length of the pipe wall in contact with the gas phase, P_L is the length of the pipe wall in contact with the liquid phase and h_L is the distance from the center of the interface to the wall.

The following relations are presented by Govier & Aziz (1972, pp. 563–565), where γ is the angle sustended by two radii to the ends of S_i

$$\gamma = 2\cos^{-1}\left(1 - \frac{2h_\mathrm{L}}{d_\mathrm{t}}\right) \tag{5.7}$$

$$\frac{S_\mathrm{i}}{d_\mathrm{t}} = 2\left[\frac{h_\mathrm{L}}{d_\mathrm{t}} - \left(\frac{h_\mathrm{L}}{d_\mathrm{t}}\right)^2\right]^{1/2} \tag{5.8}$$

$$\frac{P_\mathrm{L}}{\pi d_\mathrm{t}} = \frac{\gamma}{2\pi} \tag{5.9}$$

$$P_\mathrm{G} = \pi d_\mathrm{t} - P_\mathrm{L} \tag{5.10}$$

$$\frac{A_\mathrm{L}}{A_\mathrm{t}} = \frac{1}{2\pi}(\gamma - \sin\gamma) \tag{5.11}$$

$$A_\mathrm{G} = A_\mathrm{t} - A_\mathrm{L} \tag{5.12}$$

5.3.2 Laminar flow

The situation of laminar–laminar flow in a horizontal pipe is usually of interest for two immiscible liquids. For a fully developed pipe flow, the assumption is made that secondary flows can be ignored. Then, the streamwise velocities vary in two directions because the geometries of the cross-sections require a variation of the velocity in two dimensions.

Thus, for a fully developed flow, the velocity fields in the upper and lower phases are described by

$$\frac{\partial^2 u_\mathrm{L}}{\partial y^2} + \frac{\partial^2 u_\mathrm{L}}{\partial x^2} = -\frac{1}{\mu_\mathrm{L}}\frac{dp}{dz} \tag{5.13}$$

$$\frac{\partial^2 u_\mathrm{G}}{\partial y^2} + \frac{\partial^2 u_\mathrm{G}}{\partial x^2} = -\frac{1}{\mu_\mathrm{G}}\frac{dp}{dz} \tag{5.14}$$

These equations are usually solved numerically. The interface is smooth and flat. The boundary conditions are that the fluid velocities are zero at the walls and that the shear stresses and velocities are equal at the interface, $u_\mathrm{L} = u_\mathrm{G}$, $\mu_\mathrm{L}(\partial u_\mathrm{L}/\partial y) = \mu_\mathrm{G}(\partial u_\mathrm{G}/\partial y)$. Examples are given in Govier & Aziz (1972, pp. 562–569).

5.3.3 Laminar liquid–turbulent gas flow in a pipe

The case of a laminar liquid and a turbulent gas in a pipe has been considered by Russel *et al.* (1974). The simplified geometry described in Section 5.3.1 was used. In addition, the stress was assumed to be constant along the interface (consistent with ignoring secondary flows).

5.3.4 Model of Taitel & Dukler

The widely used correlation for frictional pressure drop and holdup developed by Lockhart & Martinelli, discussed in Chapter 1, can overpredict pressure drops by as much as 100% (Baker, 1954; Hoogendorn, 1959). Direct calculations of stratified flow in a pipe are confronted with a number of problems, in addition to the modeling of turbulent transport. The interface is not flat. Attempts to deal with this problem have been described by Chen *et al.* (1997), Hart *et al.* (1989), Grolman & Fortuin (1997), Vlachos *et al.* (1999). The interfacial stress is not constant along the interface. Suzanne (1985) has shown that secondary flows can exist in the liquid. For cases in which the liquid contains sediment, the secondary flow is evident by the appearance of a streak of deposited particles along the bottom of the pipe.

Consequently, the simplified approach formulated by Taitel & Dukler (1976a,b) is commonly used in design calculations for stratified flows. The geometric configuration described in Section 5.3.1 is employed and the stress along the interface is assumed to be constant. For fully developed flow in a pipe, one-dimensional momentum balance equations are used to represent the pressure gradients in the gas and the liquid.

$$-A_{\mathrm{G}}\left(\frac{dp}{dz}\right)_{\mathrm{G}} - \tau_{\mathrm{WG}}S_{\mathrm{G}} - \tau_{\mathrm{i}}S_{\mathrm{i}} - \rho_{\mathrm{G}}A_{\mathrm{G}}g\sin\theta = 0 \tag{5.15}$$

$$-A_{\mathrm{L}}\left(\frac{dp}{dz}\right)_{\mathrm{L}} - \tau_{\mathrm{WL}}S_{\mathrm{L}} - \tau_{\mathrm{i}}S_{\mathrm{i}} - \rho_{\mathrm{L}}A_{\mathrm{L}}g\sin\theta = 0 \tag{5.16}$$

where the four terms represent forces in the z-direction due to the pressure gradient (which is a negative quantity), the wall stress, the interfacial stress and gravity. Since the flow is fully developed, $(dp/dz)_{\mathrm{G}} = (dp/dz)_{\mathrm{L}} = dp/dz$. The elimination of dp/dz between (5.15) and (5.16) gives

$$\frac{\tau_{\mathrm{WG}}S_{\mathrm{G}}}{A_{\mathrm{G}}} - \frac{\tau_{\mathrm{WL}}S_{\mathrm{L}}}{A_{\mathrm{L}}} + \tau_{\mathrm{i}}S_{\mathrm{i}}\left(\frac{1}{A_{\mathrm{L}}} + \frac{1}{A_{\mathrm{G}}}\right) - (\rho_{\mathrm{L}} - \rho_{\mathrm{G}})g\sin\theta = 0 \tag{5.17}$$

$$(A_{\mathrm{L}} + A_{\mathrm{G}}) = A \tag{5.18}$$

Equations (5.17) and (5.18) define A_{L} and A_{G}, that is, the holdup

$$\varepsilon_{\mathrm{L}} = A_{\mathrm{L}}/(A_{\mathrm{L}} + A_{\mathrm{G}}) \tag{5.19}$$

which is related to the void fraction, in that $\alpha = 1 - \varepsilon_{\mathrm{L}}$. The dimensionless height, $h_{\mathrm{L}} / d_{\mathrm{t}}$, can also be calculated if A_{L} is known. (See equations (5.7) to (5.12).) Either (5.15) or (5.16) gives the pressure gradient.

Taitel & Dukler represented the stresses with friction factor relations

$$\tau_{\mathrm{WG}} = f_{\mathrm{G}}\frac{\rho_{\mathrm{G}}u_{\mathrm{G}}^2}{2} \tag{5.20}$$

$$\tau_{\mathrm{WL}} = f_{\mathrm{L}}\frac{\rho_{\mathrm{L}}u_{\mathrm{L}}^2}{2} \tag{5.21}$$

$$\tau_{\mathrm{i}} = \frac{f_{\mathrm{i}}\rho_{\mathrm{G}}(u_{\mathrm{G}} - u_{\mathrm{L}})^2}{2} \tag{5.22}$$

Turbulent flows through non-circular passages are treated by using data for single-phase flow through a circular pipe to evaluate the friction factor. This is done by defining a hydraulic diameter as four times the area divided by the wetted perimeter. Thus, for the gas phase,

$$d_{\mathrm{h}} = \frac{4A_{\mathrm{G}}}{P_{\mathrm{G}} + S_{\mathrm{i}}} \tag{5.23}$$

(This relation was tested by Andritsos *et al.* (1989) by placing inserts on the bottom of a pipe to create passages with the same shape as described in equations (5.7) to (5.11).) At the interface, the velocity gradient in the liquid is much smaller than the velocity gradient in the gas. Thus, the flow in the liquid could be considered to be the same as for an open channel. Taitel & Dukler suggested that

$$d_{\mathrm{h}} = \frac{4A_{\mathrm{L}}}{S_{\mathrm{L}}} \tag{5.24}$$

Figure 5.1 presents calculations of $h_{\mathrm{L}} / d_{\mathrm{t}}$ using (5.16) and (5.17), for different values of $f_{\mathrm{i}} / f_{\mathrm{S}}$, where f_{S} is the friction factor for a smooth interface. The abscissa is the Martinelli factor (Section 1.5.3)

$$X = \left[\frac{(dp/dz)_{\mathrm{LS}}}{(dp/dz)_{\mathrm{GS}}}\right]^{0.5} \tag{5.25}$$

where $(dp/dz)_{\mathrm{LS}}$ is the pressure gradient if the liquid were flowing alone in a smooth pipe with diameter d_{t} and $(dp/dz)_{\mathrm{GS}}$ is the pressure drop if the gas were flowing alone in a smooth pipe with diameter d_{t}. Parameter X increases with an increasing ratio of liquid to gas velocity. The top half of Figure 5.1 is for a laminar liquid and a turbulent gas. (This might be realized for a very viscous liquid.)

The bottom half of Figure 5.1 presents calculations for a turbulent gas and a turbulent liquid, for which (5.25) is equation (1.95) in Chapter 1. The friction factors ($f_{\mathrm{G}}, f_{\mathrm{L}}, f_{\mathrm{S}}$) are given as

$$f = 0.046\mathrm{Re}^{-0.2} \tag{5.26}$$

Note that the calculations in Figure 5.1 are sensitive to changes in $f_{\mathrm{i}} / f_{\mathrm{S}}$, the ratio of the interfacial friction factor, f_{i}, to what would be realized for a smooth surface. Taitel & Dukler assumed that $f_{\mathrm{i}} = f_{\mathrm{S}}$. However, if waves are present, $f_{\mathrm{i}}/f_{\mathrm{S}} > 1$.

Figure 5.2 is a plot of the pressure gradient against the Martinelli factor. Here, $\phi_{\mathrm{G}}^2 = (dp/dz)/(dp/dz)_{\mathrm{GS}}$ is the ratio of the actual pressure gradient to the pressure gradient which would exist if the gas were flowing alone in the pipe (see Section 1.5.3). Again, a sensitivity to $f_{\mathrm{i}} / f_{\mathrm{S}}$ is noted.

A limitation of the calculations presented in Figures 5.1 and 5.2 is the calculation of τ_{WL} with (5.21). Because of the sheared interface, the liquid flow can have a kinship with the flows described in Section 3.2.1, rather than with free flow in a conduit. Furthermore, as mentioned earlier in this section, the flow could contain secondary patterns. These issues have not been addressed adequately.

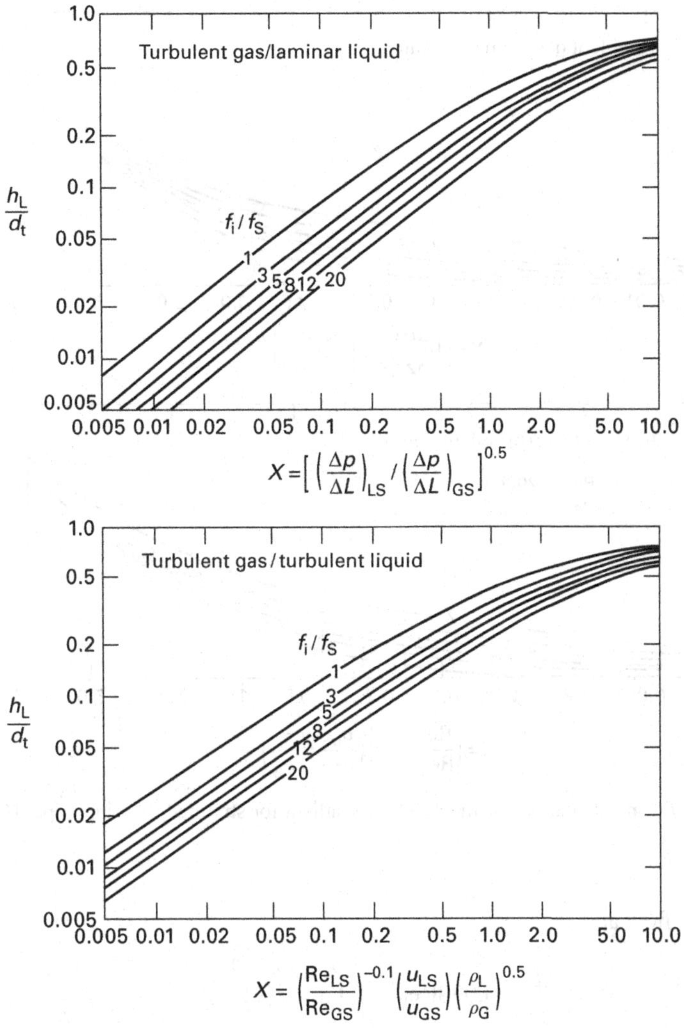

Figure 5.1 Effect of f_i / f_S on calculations of the layer height, h_L / d_t, for flow in a pipe. Henstock & Hanratty, 1976.

Thus, Andritsos & Hanratty (1987) adopted the approach taken by Cheremisinoff & Davis (1979), who developed a correlation between $h_L^+ = h_L(\tau_c\rho)^{1/2}/\upsilon_L$ and $Re_L = d_h\rho_L u_L/\mu_L$, where τ_c is a characteristic stress in the liquid. Andritsos & Hanratty defined

$$\tau_c = \frac{2}{3}\tau_{WL}\left(1 - \frac{h_L}{d_t}\right) + \frac{1}{3}\tau_i \tag{5.27}$$

for the liquid layer. They found that

$$h_L^+ = 1.082 Re_L^{0.5} \tag{5.28}$$

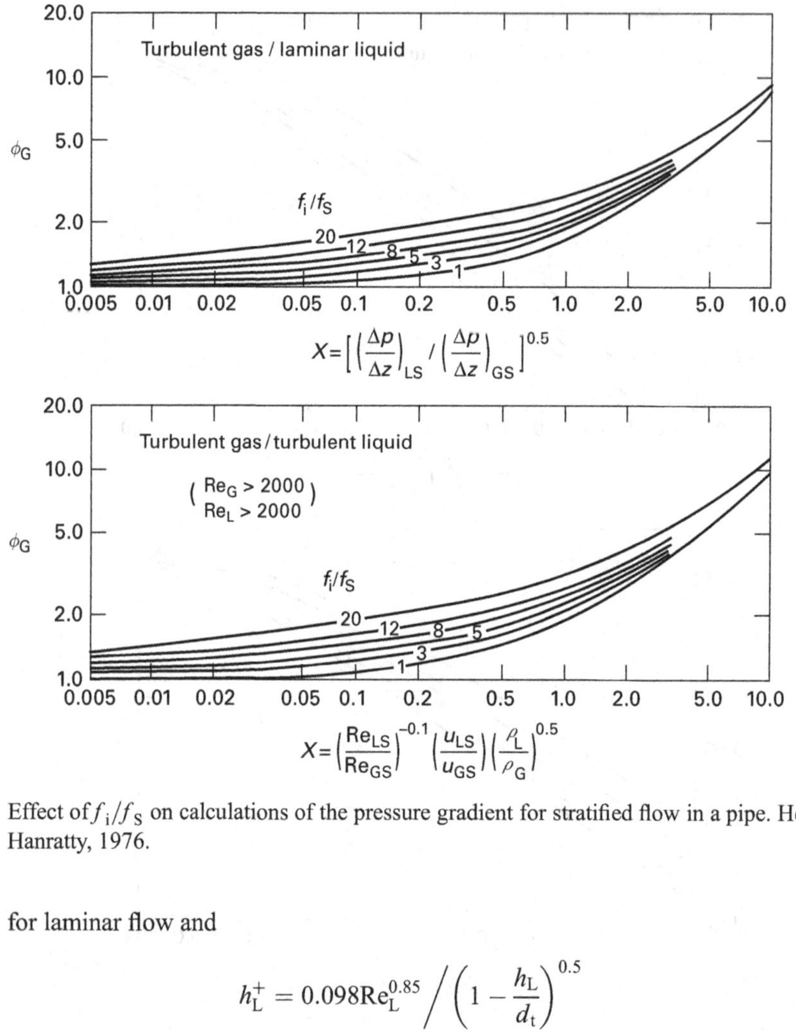

Figure 5.2 Effect of f_i/f_S on calculations of the pressure gradient for stratified flow in a pipe. Henstock & Hanratty, 1976.

for laminar flow and

$$h_L^+ = 0.098\mathrm{Re}_L^{0.85} \Big/ \left(1 - \frac{h_L}{d_t}\right)^{0.5} \tag{5.29}$$

for turbulent flow. The following interpolation formula provides a representation of both laminar and turbulent flow:

$$h_L^+ = \left\{ \left(1.082\mathrm{Re}_L^{0.5}\right)^5 + \left[0.098\mathrm{Re}_L{}^{0.85} \Big/ \left(1 - \frac{h_L}{d_t}\right)^{0.5}\right]^5 \right\}^{0.2} \tag{5.30}$$

5.4 Determination of the interfacial friction factor

5.4.1 Measurements of f_i/f_S

The critical issue in analyzing stratified flows is the determination of the interfacial stress, τ_i, or the interfacial friction factor, f_i, defined by (5.22). The gas flow creates waves and

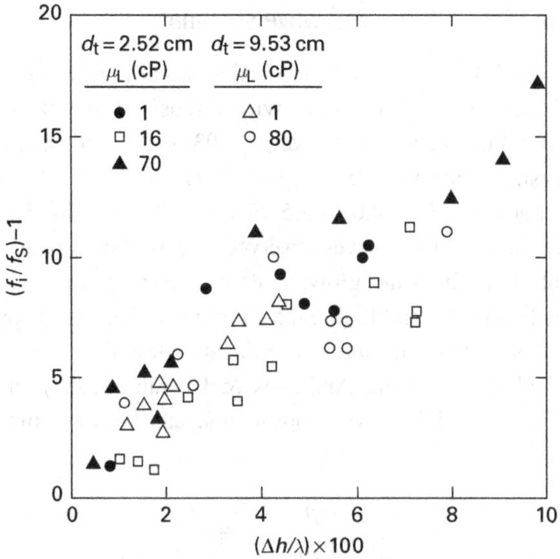

Figure 5.3 Measurements of the effect of wave amplitude on the dimensionless friction factor f_i/f_S. Andritsos & Hanratty, 1987.

these, in turn, can present a roughened surface to the flow. Thus, f_i is related to the structure of the waves.

Measurements of the liquid height provide an average, $\overline{h}_L$, and a deviation from the average, $h_L - \overline{h}_L$, which can be characterized by a standard deviation, $\sigma^2 = \overline{\left(h_L - \overline{h}_L\right)^2}$. A characteristic height of the waves is defined by assuming a sinusoidal wave pattern, that is, $\Delta h = 2\sqrt{2}\sigma$. A characteristic wavelength is defined as the product of the wave velocity and the wave frequency.

Values of f_i (defined by equation (5.22)) are determined by measuring the pressure drop and the height of the liquid layer, $\overline{h}_L$, and calculating τ_i with (5.15). Terms A_G, S_G, S_i are calculated from the knowledge of $\overline{h}_L$. (Thus, the measurements of τ_i are tied into the use of the simple model of a stratified flow defined by equations (5.7)–(5.12).) Figure 5.3, from Andritsos & Hanratty (1987), presents measured values of f_i/f_S, where f_S is what would be obtained if the interface were smooth. It is noted that f_i/f_S can be much larger than unity. The abscissa in Figure 5.3 is the ratio of the wave height, Δh, to the wavelength. It is seen that the interfacial friction factor correlates approximately with the steepness of the waves and that values of f_i/f_S as large as 16 can be realized.

A common way to scale wave heights in engineering practice is to assume that they are proportional to the liquid height. This notion has been applied with reasonable success to very thin films that exist in annular gas–liquid flows (see Chapter 3). However, it does not work for stratified flows. The data of Andritsos & Hanratty (Figure 5.3), for example, show values of $\Delta h/\overline{h}_L$ ranging from 0.09 to 1.7.

5.4.2 Equations for interfacial friction factor and wave steepness

The data shown in Figure 5.3 are part of an extensive study that involved pipelines with diameters of 2.52 cm and 9.53 cm, liquids with viscosities of 1 cP, 4.5 cP, 12 cP, 16 cP, 70 cP, 80 cP, h_L/d_t taking values in the range 0.03–0.22. Andritsos & Hanratty (1987) present the measurements in three graphs, where values of f_i/f_S are plotted against u_{GS} for height ranges of $h_L/d_t = 0.03$–0.05, $h_L/d_t = 0.08$–0.12, $h_L/d_t = 0.17$–0.22. The characteristic wave numbers of the waves observed in these experiments are roughly equal to the wave number at which the growth rate of Kelvin–Helmholtz waves is a maximum, k_M. (See equation (4.92) and Figure 4.5.) For air–water, this is approximately 3.7 cm^{-1}. Under the conditions of the experiments described in Figure 5.3, $(k_M/g)^{1/2}(\rho_G/\rho_L)^{1/2}$ is almost constant. Andritsos & Hanratty (1987) show that the minimum gas velocity at which KH waves appear, $\bar{u}_{Gt}$, can be approximated by the equation

$$(\bar{u}_G - \bar{u}_L)_t \left(\frac{k_M}{g}\frac{\rho_G}{\rho_L}\right)^{0.5} = 1 \tag{5.31}$$

Figure 5.3 shows that f_i/f_S is approximately equal to 1 at low u_{GS}, where the interface is smooth or has regular Jeffreys waves. For $\bar{u}_G/u_{Gt} > {\sim}1$, the measurements increase rapidly with increases in u_{GS}. On the basis of these observations, Andritsos & Hanratty recommended the following approximate equations:

$$f_i/f_S = 1 \quad \text{for } u_{GS} \le u_{GSt} \tag{5.32}$$

$$\frac{f_i}{f_S} = 1 + 15\left(\frac{h_L}{d_t}\right)^{0.5}\left(\frac{u_{GS} - u_{GSt}}{u_{GSt}}\right) \quad \text{for } u_{GS} \ge u_{GSt} \tag{5.33}$$

where u_{GSt} is the critical superficial gas velocity at which KH waves appear (see Section 4.5).

This problem has been revisited by Bontozoglou & Hanratty (1989), who were motivated by Figure 5.3 to use inviscid flow theory to develop an expression for $\Delta h/\lambda$. (A solution for infinitesimal waves is presented in Chapter 4.) The following equations are derived for the appearance of KH waves:

$$\frac{\rho_G}{\rho_L}(\bar{u}_G - \bar{u}_L)^2 = \frac{\sigma k}{\rho_L} + \frac{g}{k} \tag{5.34}$$

$$c_R = \frac{\rho_G \bar{u}_G + \rho_L \bar{u}_L}{\rho_L} \tag{5.35}$$

where c_R is the wave velocity, k is the wave number and σ is the surface tension.

Saffman & Yuen (1982) extended this linear analysis so as to include finite-amplitude waves. They obtained the following relation for gravity waves:

$$\frac{k}{g}c_R^2 + \frac{k\rho_G}{g\rho_L}(\bar{u}_G - c_R)^2 - 1 = \frac{(2\pi)^2}{8}\left(\frac{\Delta h}{\lambda}\right)^2\left[\left(2\frac{k}{g}c_R^2 - 1\right)^2 + 1\right] \tag{5.36}$$

In order to establish a criterion for defining wave height, Saffman & Yuen (1982) and Bontozoglou & Hanratty (1988) assumed that it corresponds to the geometric limit, which occurs when the calculated wave slope becomes unphysical. A criterion which is a measure of closeness to breaking is the ratio, q/c, of the horizontal velocity of the fastest-moving particles on the interface, q, to the wave velocity, c. There is evidence that this ratio is related to the geometric limit. Free-surface gravity waves, for example, are limited in height by the formation of a sharp peak at the crest when $\varepsilon = q/c = 1$ (Stokes, 1847). The value of ε increases with an increase in wave height for waves in the presence of an air flow and $\varepsilon = 1$ is associated with a sharp peak at the crest or with an infinite slope elsewhere along the profile (Holyer, 1979). Because of numerical difficulties, the wave height corresponding to a given value ε, rather than $\varepsilon = 1$, was calculated.

Bongtozoglou & Hanratty (1989) used numerical methods developed by Saffman & Yuen to calculate the change of gravity waves in the presence of an air flow. They chose $\varepsilon = 0.22$ so that the calculated wave steepness is in the range represented by Figure 5.3. Geometrically limited free (no gas flow) surface waves on liquids of arbitrary height have been calculated by Cokelet (1977). For a freely flowing liquid layer, these show that $\alpha_W = \Delta h/\Delta h_{deep}$ (where Δh_{deep} is the height of waves on a deep liquid) increases with increasing $k\overline{h}_L$ and reaches a maximum value of unity at $k\overline{h}_L \approx 2.5$. The calculations of Bontozoglou & Hanratty show that the same functionality is obtained for waves generated by a gas flow (see Figure 5.4). They also show that the dependence on ρ_G/ρ_L is weak. The following result was obtained:

$$\frac{\Delta h}{\lambda} = 0.079(\alpha_W)\left(\frac{\overline{u}_G - \overline{u}_L}{u_{Gt}} - 1\right) \tag{5.37}$$

where α_W is the function given by the solid curve in Figure 5.4 and u_{Gt} is the Kelvin–Helmholtz critical velocity predicted by linear theory. Thus, from Figure 5.3, the following relation for the friction factor is obtained:

$$\frac{f_i}{f_S} - 1 \approx 13\left[(\alpha_W)\left(\frac{\overline{u}_G - \overline{u}_L}{u_{Gt}} - 1\right)\right] \tag{5.38}$$

A comparison of (5.38) with (5.33) shows agreement between the two proposals. The chief difference is the representation of the influence of kh_L with an analytical (rather than an empirical) expression.

5.4.3 Design relations for a horizontal pipe

The simplest method for calculating the liquid holdup and the frictional pressure gradient is to use Figures 5.1 and 5.2. The value of f_i/f_S could be approximated either with (5.33) or (5.38) with $\alpha_W = \Delta h_L/\Delta h_{deep}$ defined by Figure 5.4. An iterative approach is needed. A value of f_i/f_S is assumed. Figure 5.1 is used to calculate $\overline{h}_L/d_t$. Equation (5.33) or (5.38) is used to calculate f_i/f_S. The procedure is repeated until the calculated friction factor agrees with the assumed friction factor.

Figures 5.1 and 5.2 indirectly use the assumption that τ_{WL} can be calculated using (5.21). An alternative approach is to use (5.30) instead of Figures 5.1 and 5.2 to model

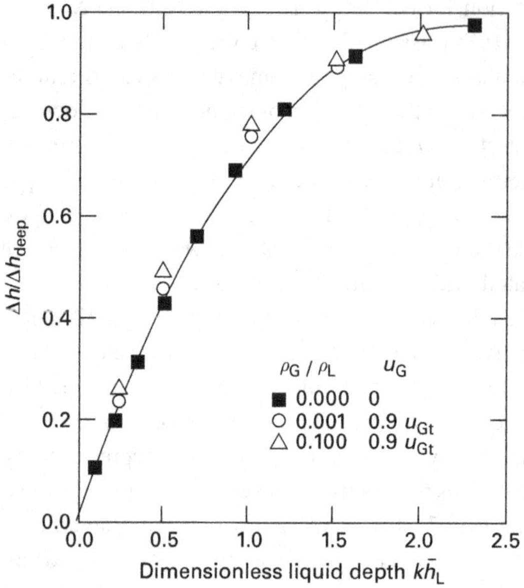

Figure 5.4 Comparison of the calculations of Cokelet (1976) for free-surface gravity flows (the solid curve) with calculations by Bontozoglou for gas–liquid flows, where u_{Gt} is the critical gas velocity at which a Kelvin–Helmholtz instability occurs. Bontozoglou & Hanratty, 1989.

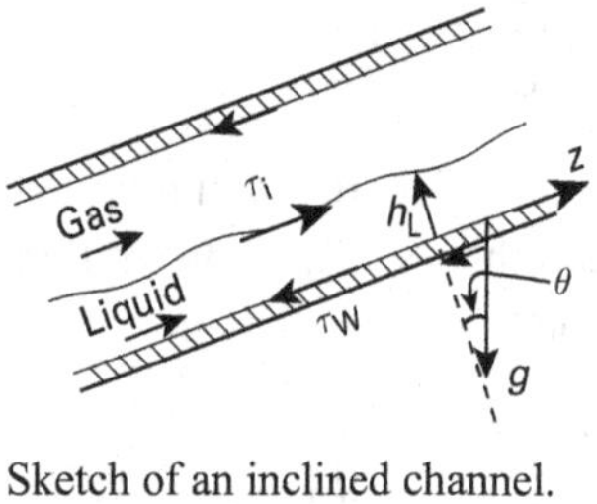

Figure 5.5 Sketch of an inclined channel.

flow in the liquid. Again, this would involve an iterative procedure, which is initiated by assuming $\bar{h}_L/d_t$ and checking this against (5.30).

5.5 Inclined pipes

5.5.1 General comments about the effect of inclination

The inclination of a pipeline to the horizontal at small angles can have a strong effect on the flow patterns. This is particularly evident for upward flows where slugging can be induced. The design of such pipelines needs to be carefully executed so as to avoid slugging. This effect can be understood by considering the momentum balance for the liquid phase.

Assume a fully developed flow in a rectangular channel, as indicated in Figure 5.5. The average height of the liquid layer is designated as $\bar{h}_L$, the drag stress at the interface as τ_i, the resisting stress at the wall as τ_W, and the acceleration of gravity as g. The channel is inclined with an angle θ to the horizontal so the component of gravity in the direction of flow is $-g \sin\theta$; that is, the gravitational force opposes the flow. For the case of a declined flow the component of gravity in the direction of flow is $g \sin\theta$, so the gravitational force augments the flow.

A force balance on the liquid layer is given as follows for a fully developed liquid flow:

$$0 = -\bar{h}_L \frac{dp}{dz} + \tau_i - \bar{h}_L \rho_L g \sin\theta - \tau_W \tag{5.39}$$

The first term represents the force of the pressure gradient; τ_i is the drag force of the gas; the third term is the force of gravity; τ_W is the resisting stress at the wall.

Since τ_i varies as $\rho_G u_{GS}^2$ the influence of gravity can be large if $u_{GS}^2 \rho_G / \bar{h}_L g | \sin\theta | \rho_L$ is small. For upflows, this leads to a decrease in $\bar{u}_L$ for a given $\bar{h}_L$. At small enough u_{GS}, a reversal of flow can occur. For downflows, gravity causes an increase in $\bar{u}_L$ for a given $\bar{h}_L$.

5.5.2 Upflows

Simmons & Hanratty (2001) observed upflows of air and water at inclination angles of 0°, 0.05°, 0.2°, 0.4°, and 1.2° in a pipe with a diameter of 7.63 cm. They found that reversed flows occur at conditions that are close to those for which $\tau_W = 0$. Figure 5.6 presents measurements of the critical conditions for a transition from a steady stratified flow to a slug flow. The thick lines at the bottom of the graph indicate the critical gas velocity below which a stratified flow cannot exist for a given inclination.

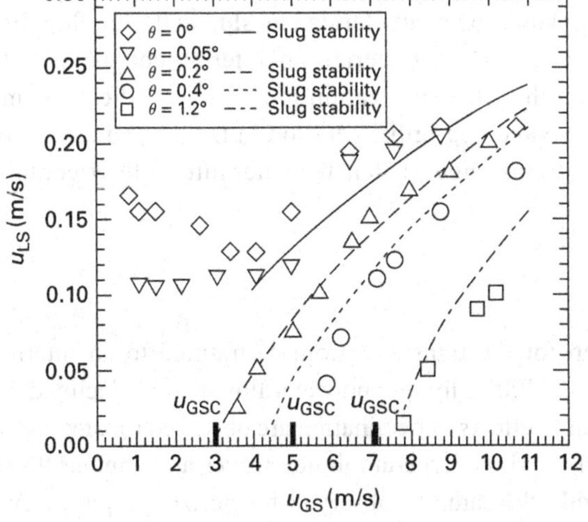

Figure 5.6 Effect of inclination on the transition from stratified flow to slug flow, where u_{GSC} is the critical superficial gas velocity below which slugs always appear. Simmons & Hanratty, 2001.

For very small u_{GS} and u_{LS}, the stratified liquid finds its own level in a pipeline when gas drag is unable to counterbalance gravitational effects. If the angle is very small, pooling could extend over the whole pipeline. The hydraulic gradient causes the liquid to flow out of the pipeline. This behavior will be avoided if the pipeline is long enough for the outlet to be completely above the pipe inlet. This can be realized for

$$\sin\theta < d_t/L \tag{5.40}$$

where L is the length of the pipe. An angle of 0.19° would be required for the system studied by Simmons & Hanratty (2001). Below this angle, measurements would not correspond to what occurs in a very long pipe. Thus, the results for 0.05°, presented in Figure 5.6 for small u_{GS}, do not represent what would be observed in a longer pipe.

Another effect of positive inclination is the existence of reversed flows that occur when the influence of gravity is large compared with the drag of the gas on the liquid. The thick lines on the bottom of Figure 5.6 are the critical superficial gas velocities, u_{GSC}, calculated for the realization of $\tau_W = 0$ at a given inclination. The region to the left of the dashed curves represents conditions for which a calculated steady stratified flow would show reversed flow close to the wall. Observations indicate that an intermittent flow exists in these regions.

The intermittent pattern that occurs in a horizontal pipe is initiated downstream of the entry by growth or coalescence of waves. For the reversed flow regions associated with upflows, transition occurs close to the entry, where slugs or plugs are formed intermittently. These slugs propagate downstream and leave a depleted layer at the inlet. This depleted layer is built up by the inflow and eventually reaches a level at which a Kelvin–Helmholtz instability leads to the formation of slugs or plugs. These carry all of the liquid out of the pipe. The stratified layer that exists between or behind the slugs moves backward.

Simmons & Hanratty (2001) present measurements of the time-varying liquid holdup for two experiments at an inclination of 0.4° ($u_{GS} = 1.54$ m/s, $u_{LS} = 0.064$ m/s; $u_{GS} = 0.54$ m/s, $u_{LS} = 0.078$ m/s), which would be characterized as slug and plug flow by investigators, who use gas occlusion to characterize an intermittent pattern. The holdup measurements agree with this characterization in that they indicate a small amount of aeration in the first case and negligible aeration in the second case. However, the shapes of the holdup patterns suggest that the intermittent flow contains slugs in both cases.

5.5.3 Downward inclinations

Figure 5.7 shows a map for the transition from a stratified to an intermittent flow, developed by Woods *et al.* (2000), for air and water flowing in a declined 7.63 cm pipe with a length of 23 m. Large effects of declination are observed for small gas velocities. However, for $u_{GS} > 8\,\mathrm{m/s}$ (where large-amplitude waves and annular flows exist) the declined flows are roughly the same as observed for horizontal pipes. Two dramatic changes from the flow patterns, shown in Figure 2.1 for air–water flow in a horizontal pipe are observed: (1) At declination angles of 0.2° and greater, regular Jeffreys waves

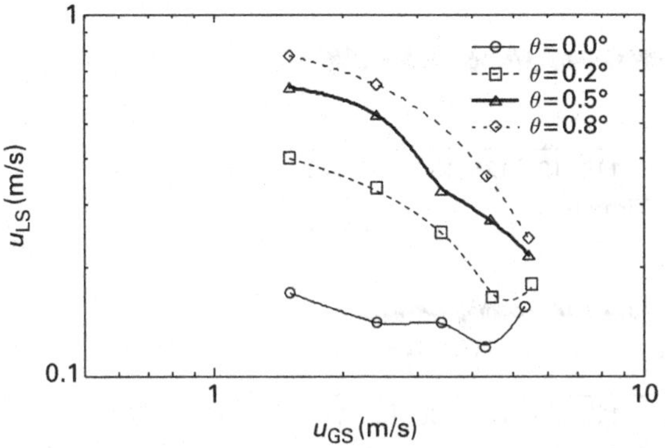

Figure 5.7 Effect of declination on the transition from a stratified flow to a slug flow. Woods *et al.*, 2000.

are not observed in the stratified flows (see Andreussi & Persen, 1987). (2) The transition to slug flow at low gas velocities occurs at much larger superficial liquid velocities than for a horizontal pipe and, from Figure 2.6, at larger h_L/d_t. Theoretical discussions of these effects are given by Woods *et al.* (2000) and by Hurlburt & Hanratty (2002).

As mentioned in Section 5.5.1, the liquid velocities can be much larger for a declined flow, at a given liquid height, if $u_{GS}^2\rho_G/h_L g\rho_L|\sin\theta|$ is small. The damping of Jeffreys waves in declined flows can be associated with this increase in the liquid velocity, which would result in an increase in the wave velocity. Theoretical studies of Jeffreys waves by Miles (1957) and Benjamin (1959) indicate that the transmission of energy from the gas to the waves can depend on the location of the critical layer where the wave velocity equals the gas velocity. This could account for the influence of declination on this process, but this is not an established explanation.

Woods *et al.* (2000) show that the transition to slug/plug flow in declined pipes is predicted by a viscous large-wavelength analysis, as had been found for horizontal air–water flows. However, the mechanism is different from what is described in Section 2.2.6, since Jeffreys waves are not present at the interface on transition.

Figure 5.8 provides an example of the generation of slugs for air–water flow in a pipe which is declined at an angle of 0.5° (Woods *et al.*, 2000). The superficial gas velocity was 2.4 m/s, the superficial liquid velocity was 0.59 m/s and the pipe diameter was 0.0763 m. The interface is disturbed by waves in the early part of the pipe. These are damped; they are not observed at $L/d_t = 106$. A large-wavelength disturbance with a wavelength of ~12 m (approximately one-half of the pipe length) and a period of ~ 8 s appears in the early part of the pipe. It grows with distance downstream and is visible at $L/d_t = 60$. At $L/d_t = 106$ (approximately midway in the pipe) disturbances, which are precursors of slugs, appear. At $L/d_t = 190$, well formed growing slugs are observed.

The eight photographs in Figure 5.9 (flow is from right to left) were obtained sequentially at 1/30 s intervals under the same conditions represented by Figure 5.8. The photographs shown cover a much shorter length of the pipe than the wavelength, so

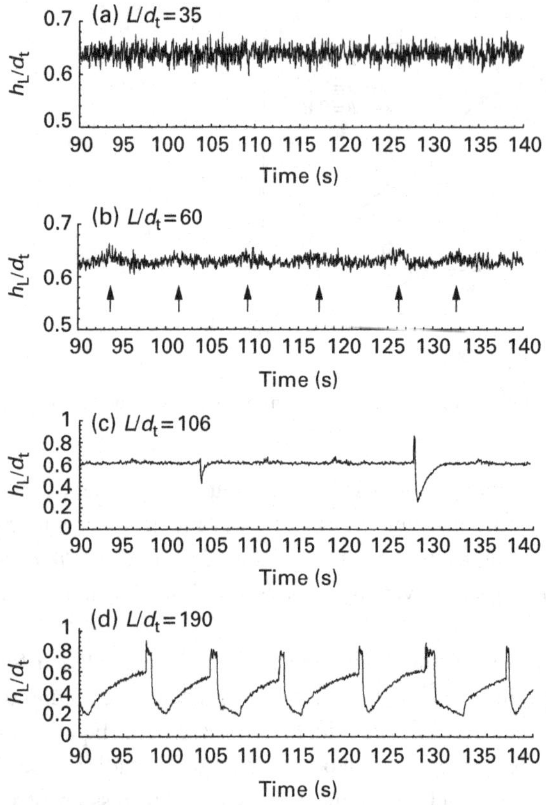

Figure 5.8 Generation of slugs for air–water flow in a pipe declined at 0.5°. Woods *et al.*, 2000.

the interface appears to be flat. The level of this segment of the interface increases as a wave passes (Figure 5.9c and d). The interface in Figure 5.9a is approximately smooth. A small instability, observed at the far right of the second frame is seen to grow in amplitude in Figure 5.9c, d and e. This disturbance eventually touches the top wall to form a slug in Figure 5.9f, g and h.

Slugs are thus seen to be the result of a local KH instability that develops at the crest of a large-wavelength disturbance. This type of behavior has also been observed by Kordyban (1985) and by Kordyban & Ranov (1970), who photographed the initiation of slugs triggered by mechanically generated waves.

Woods *et al.* (2000) show that viscous large-wavelength theory correctly predicts the $u_{SL} - u_{SG}$ curve that describes the initiation of slugs in a 0.0763 m pipe declined at an angle of 0.05° and measurements of Shoham (1982) in 2.52 cm and 5.1 cm pipes declined at 1.0°.

The results of studies of flow regimes in pipes declined at angles of 0° to 80° have been reported by Barnea *et al.* (1982). These show an increasing importance of the stratified regime and a decreased importance of the slug regime with increasing angle.

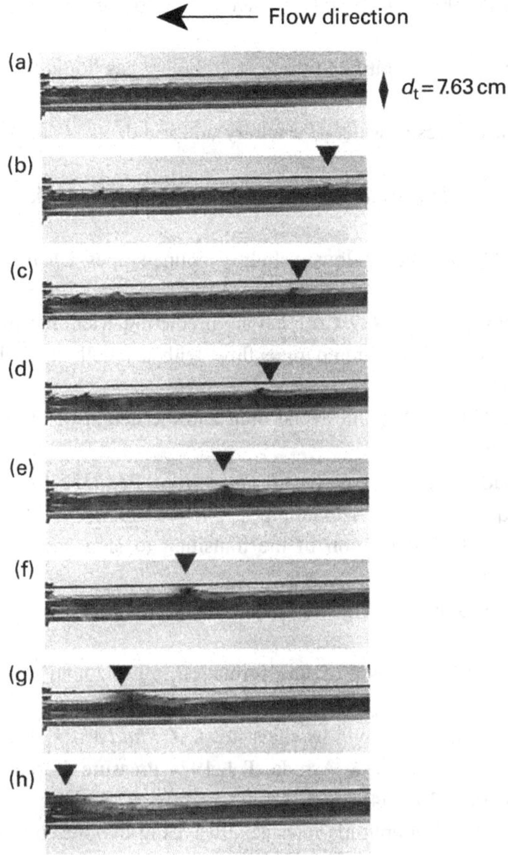

Figure 5.9 Photographs of slug formation for the conditions in Figure 5.8. Woods *et al.*, 2000.

References

Andreussi, P. & Persen, L. N. 1987 Stratified gas–liquid flow in downwardly inclined pipes. *Int. J. Multiphase Flow* 13, 565–575.

Andritsos, N. & Hanratty, T. J. 1987 Influence of interfacial waves in stratified gas–liquid flows. *AIChE Jl* 33, 444–454.

Andritsos, N., Williams, L. & Hanratty, T. J. 1989 Effect of liquid viscosity on the stratified-slug transition in horizontal pipe flow. *Int. J. Multiphase Flow* 15, 877–892.

Baker, O. 1954 Simultaneous flow of oil and gas. *Oil Gas J.* 53, 185.

Barnea, D., Shoham, O. & Taitel, Y. 1982 Flow pattern transition for downward inclined two-phase flow: horizontal to vertical. *Chem. Eng.* 37, 735–740.

Benjamin, T. B. 1959 Shearing flow over a wavy boundary. *J. Fluid Mech.*, 37, 161–205.

Bird, R. B., Stewart, W. E. & Lightfoot, E. N. 1960 *Transport Phenomena*. New York: John Wiley.

Bontozoglou, V. & Hanratty, T. J. 1988 Effects of finite depth and current velocity on large amplitude Kelvin–Helmholtz waves. *J. Fluid Mech.* 196, 187.

Bontozoglou, V. & Hanratty, T. J. 1989 Wave height estimation in stratified gas–liquid flows. *AIChE Jl* 35, 1346–1350.

Chen, T., Cai, X. D. & Brill, J. P. 1997 Gas–liquid stratified wavy flow in horizontal pipelines. *J. Energy Resources* 119, 209–216.

Cheremisinoff, N. P. & Davis, E. J. 1979 Stratified turbulent–turbulent gas liquid flows. *AIChE Jl* 25, 48.

Cokelet, E. D. 1977 Steep gravity waves in water of arbitrary uniform depth. *Phil. Trans. R. Soc. Lond.* A286, 183–230, 183.

Govier, G. W. & Aziz, K. 1972 *The Flow of Complex Mixtures in Pipes.* New York: Van Nostrand Reinhold.

Grolman, E. & Fortuin, M. H. 1997 Gas–liquid flow in slightly inclined pipes. *Chem. Eng. Science* 52, 4461–4471.

Hart, J., Hamersa, P. J. & Fortuin, M. H. 1989 Correlations predicting frictional pressure drop and liquid holdup during horizontal gas–liquid pipe flow with a small liquid holdup. *Int. J. Multiphase Flow* 15, 947–943.

Henstock, W. H. & Hanratty, T. J. 1976 The interfacial drag and the height of the wall layer in annular flows. *AIChE JL* 22, 990–1000.

Holyer, J. Y. 1979 Large amplitude progressive waves. *J. Fluid Mech.* 93, 433.

Hoogendorn, C. J. 1959 Gas–liquid flow in horizontal pipes. *Chem. Eng. Sci.* 9, 205.

Hurlburt, E. T. & Hanratty, T. J. 2002 Prediction of the transition to slug and plug flow. *Int. J. Multiphase Flow* 28, 707–729.

Kordyban, E. S. 1985 Some details of developing slugs in horizontal two-phase flow. *AIChE Jl* 31, 802–806.

Kordyban, E. S. & Ranov, T. 1970 Mechanism of slug formation in horizontal two-phase-flow. *J. Basic Eng.* 192, 857–864.

Miles, J. W. 1957 On generation of surface waves by shear flows. *J. Fluid Mech.* 3, 185–204.

Russel, T. W. F., Etchels, A. W., Jensen, R. H. & Arruda, T. J. 1974 Pressure drop and holdup in stratified gas–liquid flow. *AIChE Jl* 20, 664.

Saffman, O. & Yuen, H. C. 1982 Finite-amplitude interfacial waves in the presence of a current. *J. Fluid Mech.* 123, 459–469.

Shoham, O. 1982 Flow pattern transitions and characteristics in horizontal pipes. Ph.D. thesis, Tel-Aviv University.

Simmons, M. J. H. & Hanratty, T. J. 2001 Transition from stratified to intermittent flows in small angle upflows. *Int. J. Multiphase Flow* 27, 599–616.

Stokes, G. G. 1847 On the theory of oscillatory waves. *Trans. Camb. Phil. Soc.* 8, 441.

Suzanne, C. 1985 Structure de l' ecoulemont stratifie de gaz et de liquide en canal rectangulaire. These d' Etat, Mecanique de Fluides, L'Institut National Polytechnique de Toulouse.

Taitel, Y. & Dukler, A.E. 1976a A theoretical approach to the Lockhart–Martinelli correlation for stratified flow. *Int. J. Multiphase Flow* 2, 591.

Taitel, Y. & Dukler, A. E. 1976b A model for predicting flow regime horizontal and near horizontal gas–liquid flows. *AIChE Jl* 22, 47–55.

Vlachos, N. A., Paras, S. V. & Karabelas, A. J. 1999 Prediction of holdup, axial pressure gradient and wall shear stress in wavy/stratified and stratified/atomization gas–liquid flow. *Int. J. Multiphase Flow* 25, 365–376.

Woods, B. D., Hurlburt, E. T. & Hanratty, T. J. 2000 Mechanism of slug formation in downwardly inclined pipes. *Int. J. Multiphase Flow* 26, 977–998.

6 Influence of viscosity on large Reynolds number interfacial waves; effect of spatially and temporally induced oscillations on a turbulent flow

6.1 Introductory comments

Equation (6.19), developed in this chapter, considers the flow field associated with wave propagation on a viscous liquid. If the field is represented by a length scale equal to the height of a liquid layer, h_L, and a velocity scale equal to $c - u_a$, the following group represents the ratio of the inertia terms to the viscous terms: $[h_L\rho_L(c - u_a)/\mu_L][kh_L]$. Interfacial waves with small or moderate values of (kh_L) are usually characterized by a large value of this dimensionless group. This leads to an expectation that the disturbed velocity over all or most of the field can be described by inviscid equations.

However, if viscous effects are confined to thin layers in the liquid, of thickness δ, a second length scale is needed. These layers are characterized by $[\delta\rho_L(c - u_a)/\mu_L][kh_L]$, so that viscous effects cannot be ignored in these layers. The relation between the wave velocity and the wavelength developed in Chapter 4 remains valid. (This is not the case for large-wavelength waves considered in Chapter 7.) There are a number of situations where these viscous effects need to be considered. One involves the damping of waves downstream of a wavemaker. Another is the generation of waves by a gas flow, discussed in Section 2.2.1. In the second example, energy is fed to the liquid flow through pressure and shear forces exerted by the gas on interfacial waves. This energy is dissipated in viscous layers close to the interface and close to the wall. A stable situation can be realized if the energy fed to the waves is balanced by this dissipation.

The evaluation of the interfacial stresses is a central problem in analyzing gas–liquid flows. This has been attacked by solving for the disturbed velocity field in the gas caused by flow over a small-amplitude wavy interface. Initial efforts have involved the use of a quasi-laminar theory whereby turbulence is taken into account by using a mean velocity profile which is characteristic of a turbulent flow. This approach has been improved by recognizing that turbulent flow over a wavy boundary experiences spatial disturbances of the turbulence, as well as of the mean flow. A turbulence model developed by Loyd *et al.* (1970) has been tested with measurements of the wall shear stress variation along solid wavy surfaces over which a turbulent liquid is flowing, and by studies of the initiation of waves at a gas–liquid interface.

The behavior of a turbulent flow subjected to an imposed temporal oscillation is similar to flow over a wavy boundary, where spatial oscillations are introduced. Thus, analytical methods developed to describe flow over solid waves are applicable to this problem.

Techniques used to analyze flow over small-amplitude waves can also be used to analyze dissolution patterns found on the underside of river ice and in caves. Here, the interest is to predict the wavelength characterizing the pattern.

Small values of $(h_{\mathrm{L}}\rho c/\mu_{\mathrm{L}})(kh_{\mathrm{L}})$ can be realized for very small values of (kh_{L}), that is, very large wavelengths. The flow can be either turbulent or laminar. Viscous and turbulent stresses introduced by the wave motion, then, can extend over the entire liquid layer. These type flows are considered in Chapter 7.

6.2 Equation for small disturbances in a viscous fluid

As in Chapter 4, the assumption is made that the waves are of small enough amplitude, a, that a linearization of the defining equations is justified. A first step is the development of an equation for small disturbances in the liquid velocity field that includes the effect of viscosity. The assumption of incompressibility is made.

A consideration of viscous effects in the momentum balance requires the use of the Navier–Stokes equations (Schlichting, 1968):

$$\rho\left(\frac{\partial u_i}{\partial t} + u_j \frac{\partial u_i}{\partial x_j}\right) = -\frac{\partial \tau_{ij}}{\partial x_j} + \rho g_i \tag{6.1}$$

where i is the component of the velocity vector being considered and the Einstein convention of summation over repeated indices is used. Thus (6.1) represents three equations for which $i = x$, y or z. The term τ_{ij} represents the i component of the stress on a face perpendicular to the j-axis. The convention is used that τ_{ij} is positive if i and j have the same sign. Since pressure, p, is compressive, it represents a negative contribution to the normal stress. Making use of the Einstein convention,

$$\frac{\partial \tau_{ij}}{\partial x_j} = \frac{\partial \tau_{ix}}{\partial x} + \frac{\partial \tau_{iy}}{\partial y} + \frac{\partial \tau_{iz}}{\partial z} \tag{6.2}$$

For a Newtonian incompressible fluid (Schlichting, 1968)

$$\tau_{ij} = -\delta_{ij} p + \mu\left(\frac{\partial u_i}{\partial x_j} + \frac{\partial u_j}{\partial x_i}\right) \tag{6.3}$$

where δ_{ij} is a delta function which is unity when $i = j$ and zero otherwise. For $\mu = 0$, $\tau_{xx} = \tau_{yy} = \tau_{zz} = -p$ and all the shear stresses are zero, so (4.10) is recovered.

The two-dimensional system to be considered is flow in a rectangular channel which is infinitely wide. The flow is fully developed, so the mean velocity varies only in the y-direction. The equations describing an incompressible two-dimensional viscous fluid are the equation for conservation of mass

$$\frac{\partial u_x}{\partial x} + \frac{\partial u_y}{\partial y} = 0 \tag{6.4}$$

and the equations for conservation of momentum

$$\frac{Du_x}{Dt} = -\frac{1}{\rho}\frac{\partial p}{\partial x} + \frac{\mu}{\rho}\left(\frac{\partial^2 u_x}{\partial x^2} + \frac{\partial^2 u_x}{\partial y^2}\right) + g_x \tag{6.5}$$

$$\frac{Du_y}{Dt} = -\frac{1}{\rho}\frac{\partial p}{\partial y} + \frac{\mu}{\rho}\left(\frac{\partial^2 u_y}{\partial x^2} + \frac{\partial^2 u_y}{\partial y^2}\right) + g_y \tag{6.6}$$

where the x-axis is in the mean flow direction. Since the flow is fully developed, the average of (6.5) produces the following equation for the mean velocity in a non-turbulent flow:

$$0 = -\frac{1}{\rho}\frac{\partial \overline{p}}{\partial x} + \left(\frac{\mu}{\rho}\right)\frac{\partial^2 \overline{u}_x}{\partial y^2} + g_x \tag{6.7}$$

The velocity is given as the sum of the average and the fluctuating component imposed by a two-dimensional wave motion at the interface.

$$u_x = \overline{u}_x + u'_x \tag{6.8}$$

$$u_y = \overline{u}_y + u'_y \tag{6.9}$$

$$p = \overline{p} + p' \tag{6.10}$$

where $\overline{u}_y = 0$. Equations (6.8), (6.9) and (6.10) are substituted into (6.5), (6.6). Equation (6.7) is subtracted from (6.5). The imposed waves are assumed to have a small amplitude so second-order terms in the fluctuating quantities are neglected. The following equations for u'_x, u'_y and p' are obtained:

$$\frac{\partial u'_x}{\partial x} + \frac{\partial u'_y}{\partial y} = 0 \tag{6.11}$$

$$\frac{\partial u'_x}{\partial t} + \overline{u}_x\frac{\partial u'_x}{\partial x} + u'_y\frac{d\overline{u}_x}{dy} = -\frac{1}{\rho}\frac{\partial p'}{\partial x} + \left(\frac{\mu}{\rho}\right)\left(\frac{\partial^2 u'_x}{\partial x^2} + \frac{\partial^2 u'_x}{\partial y^2}\right) \tag{6.12}$$

$$\frac{\partial u'_y}{\partial t} + \overline{u}_x\frac{\partial u'_y}{\partial x} = -\frac{1}{\rho}\frac{\partial p'}{\partial y} + \left(\frac{\mu}{\rho}\right)\left(\frac{\partial^2 u'_y}{\partial x^2} + \frac{\partial^2 u'_y}{\partial y^2}\right) \tag{6.13}$$

The interfacial disturbance is defined by (4.25), (4.29) as a sinusoidal wave for which the displacement from the average location of the interface is given as $\eta = a \exp ik(x - ct)$. Since the system is linear, the solution is of the form

$$\frac{u'_x}{\hat{u}_x} = \frac{u'_y}{\hat{u}_y} = \frac{p'}{\hat{p}} = \exp ik(x - ct) \tag{6.14}$$

where $\hat{u}_x(y)$, $\hat{u}_y(y)$, $\hat{p}(y)$ are complex and proportional to the wave amplitude, a. The wave velocity c is complex; the wave number k is real. Thus (6.11) and (6.14) give

$$0 = ik\hat{u}_x \exp ik(x - ct) + \frac{d\hat{u}_y}{dy} \exp ik(x - ct) \tag{6.15}$$

The following relation between $\hat{u}_x$ and $\hat{u}_y$ is obtained from (6.15):

$$\hat{u}_x = \frac{i}{k} \frac{d\hat{u}_y}{dy} \tag{6.16}$$

since $i^2 = -1$. The substitution of (6.14) into (6.12) and (6.13) yields

$$-ikc\hat{u}_x + ik\overline{u}_x\hat{u}_x + \hat{u}_y \frac{d\overline{u}_x}{dy} = -i\frac{k}{\rho}\hat{p} + \frac{\mu}{\rho}\left(\frac{d^2\hat{u}_x}{dy^2} - k^2\hat{u}_x\right) \tag{6.17}$$

$$-ikc\hat{u}_y + ik\overline{u}_x\hat{u}_y = -\frac{1}{\rho}\frac{d\hat{p}}{dy} + \frac{\mu}{\rho}\left(\frac{d^2\hat{u}_y}{dy^2} - k^2\hat{u}_y\right) \tag{6.18}$$

The pressure amplitude is eliminated between (6.17) and (6.18) after taking the derivative of (6.17) with respect to y. The amplitude $\hat{u}_x$ is eliminated from the resulting equation by substituting (6.16). The following relation for $\hat{u}_y$ is obtained:

$$(\overline{u}_x - c)\left(\frac{d^2\hat{u}_y}{dy^2} - k^2\hat{u}_y\right) - \hat{u}_y \frac{d^2\overline{u}_x}{dy^2} = -\frac{i}{k}\frac{\mu}{\rho}\left(\frac{d^4\hat{u}_y}{dy^4} - 2k^2\frac{d^2\hat{u}_y}{dy^2} + k^4\hat{u}_y\right) \tag{6.19}$$

This is the Orr–Sommerfeld equation used to calculate the stability of viscous flow fields (Schlichting, 1968). The solution of (6.19) can be substituted into (6.16) and (6.17) to give $\hat{u}_x(y)$ and $\hat{p}(y)$.

6.3 Boundary conditions

The inclusion of the effects of viscosity requires a consideration of the terms on the right side of (6.19). Thus, the equation changes from second order to fourth order if the right side is not zero. The solution of the inviscid equation (Chapter 4) needs the specification of two boundary conditions: the assumption that $\hat{u}_y = 0$ at the stationary wall and the specification of $\hat{u}_y$ at the interface by using the kinematic condition. The solution of (6.19) requires two additional boundary conditions. One of these is the assumption of zero tangential velocity at the wall (the no slip condition)

$$u_x = 0 \quad \text{at} \quad y = -h_{\mathrm{L}} \tag{6.20}$$

where $y = 0$ is the average location of the interface. The other, to be satisfied at the interface, is the equality of the wave-induced variations of the shear stresses in the liquid to the wave-induced shear stresses in the gas and tangential forces due to the variation of tensile stresses, such as surface tension, in the interface (caused by temperature variations or by variation of the concentration of a surface active agent).

Gas flowing over a wave, defined by (4.25), exerts a normal stress, $-P$, and a tangential stress, T, on the liquid as indicated in Figure 6.1a. These are defined in a

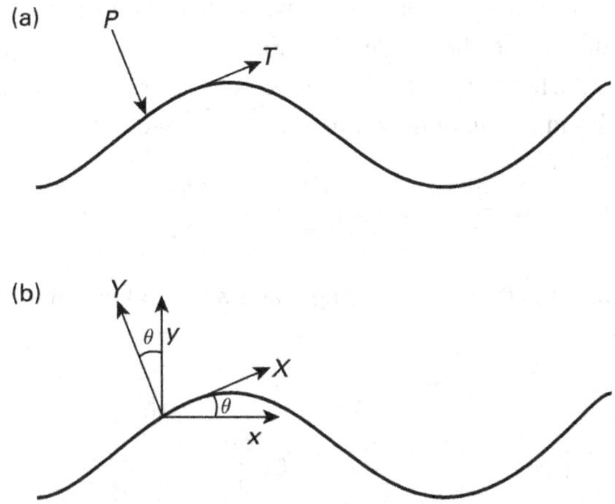

Figure 6.1 (a) Sketch of waves with vectors indicating pressure and shear stress at the surface. (b) Relation between boundary-layer coordinates and Cartesian coordinates.

coordinate system, X, Y, that is imbedded in the interface. Fluctuations of the imposed stresses and of the interfacial tension, σ, can be represented as given below, since these terms vary linearly with a.

$$T' = a\tilde{T} \exp ik(x - ct) \tag{6.21}$$

$$P' = a\tilde{P} \exp ik(x - ct) \tag{6.22}$$

$$\sigma' = a\tilde{\sigma} \exp ik(x - ct) \tag{6.23}$$

where the linear dependence of $\hat{T}$, $\hat{P}$ and $\hat{\sigma}$ on wave amplitude, a, is recognized and $\tilde{T}$, $\tilde{P}$ and $\tilde{\sigma}$ can be complex.

Using (4.28), the real parts of (6.21) and (6.22) are given as

$$T' = a\tilde{T}_R \cos k(x - c_R t) - a\tilde{T}_I \sin k(x - c_R t) \tag{6.24}$$

$$P' = a\tilde{P}_R \cos k(x - c_R t) - a\tilde{P}_I \sin k(x - c_R t) \tag{6.25}$$

where $\tilde{P}_R$, $\tilde{T}_R$ are components in phase (or 180° out of phase) with the wave height and $\tilde{P}_I$, $\tilde{T}_I$ are in phase with the wave slope.

The equality of the shear stresses and the tangential forces at the interface can be expressed as

$$T_i + \frac{\partial \sigma}{\partial X} = (\tau_{XY})_i \tag{6.26}$$

where $(\tau_{XY})_i$ is the shear stress at the interface of the liquid in the X,Y coordinate system. We want to express these liquid stresses in laboratory coordinates x,y as shown in Figure 6.1b. To do this, the transformation for a Cartesian tensor (Long, 1963) is used:

$$\tau_{XY} = \tau_{kl} a_{Xk} a_{Yl} \tag{6.27}$$

where $a_{Xk} = \cos(X, x_k)$ and $a_{Yl} = \cos(Y, x_l)$, with $x_1 = x$, $x_2 = y$ and $x_3 = z$. For a very small amplitude wave the angle θ in Figure 6.1b is small enough that $\cos(X,x) = \cos(Y,y) \approx 1$, where $\cos(X,y) \approx \partial\eta/\partial x$, $\cos(Y,x) \approx -\partial\eta/\partial x$. Thus, by summing with respect to k and l, equations (6.26) and (6.27) give

$$\tau_{XY} = -\tau_{xx}\frac{\partial\eta}{\partial x} + \tau_{xy} + \tau_{yy}\frac{\partial\eta}{\partial x} - \tau_{yx}\left(\frac{\partial\eta}{\partial x}\right)^2 \tag{6.28}$$

Represent the stresses in (6.26) as the sum of averages and wave-induced fluctuations and substitute (6.28)

$$\overline{T}_{\mathrm{i}} = \overline{\tau}_{xy} \tag{6.29}$$

$$\frac{\partial\sigma'}{\partial x} + T'_{\mathrm{i}} = -\overline{\tau}_{xx}\frac{\partial\eta}{\partial x} + \left(\tau'_{xy}\right)_{\mathrm{i}} + \overline{\tau}_{yy}\frac{\partial\eta}{\partial x} \tag{6.30}$$

where second-order terms in the fluctuations have been neglected. Again, for small-amplitude waves,

$$\overline{\tau}_{xx} = \overline{\tau}_{yy} = -\overline{p} \tag{6.31}$$

so

$$\frac{\partial\sigma'}{\partial x} + T' = \mu\left(\frac{\partial u'_x}{\partial y} + \frac{\partial u'_y}{\partial x}\right)_{\mathrm{i}} \tag{6.32}$$

where, from (6.3),

$$\tau'_{xy} = \mu\left(\frac{\partial u'_x}{\partial y} + \frac{\partial u'_y}{\partial x}\right) \tag{6.33}$$

For small-amplitude waves, (6.32) can be evaluated at $y = 0$ so

$$\frac{\partial\sigma'}{\partial x} + T' = \mu\left(\frac{\partial u'_x}{\partial y} + \frac{\partial u'_y}{\partial x}\right)_0 \tag{6.34}$$

By substituting (6.21), (6.23) and (6.14) and using (6.16) to eliminate $\hat{u}_x$ the following equation is obtained from (6.34):

$$0 = ik\tilde{\sigma} + \tilde{T} - \mu\left(\frac{i}{k}\frac{d^2\tilde{u}_y}{dy^2} + ik\tilde{u}_y\right)_{y=0} \tag{6.35}$$

6.4 Solution of the Orr–Sommerfeld equation

6.4.1 General solution

The solution of (6.19) for $\hat{u}_y$ is given by

$$\hat{u}_y = A\hat{u}_{y1} + B\hat{u}_{y2} + C\hat{u}_{y3} + D\hat{u}_{y4} \tag{6.36}$$

where $\hat{u}_{y1}, \hat{u}_{y2}, \hat{u}_{y3}, \hat{u}_{y4}$ are four independent solutions. The constants A, B, C, D can be obtained by using the four constraints on the system provided by the boundary conditions,

$$\hat{u}_y = 0 \qquad \frac{d\hat{u}_y}{dy} = 0 \quad \text{at} \quad y = -h_L \tag{6.37}$$

equation (6.35) and the kinematic condition, which relates the normal velocity in the liquid at the interface to the wave amplitude; see (4.42).

$$\hat{u}_{y0} = aik(\overline{u}_{x0} - c) \tag{6.38}$$

6.4.2 Dispersion relation

The principal interest is the development of a relation for the complex wave velocity, so the calculation of the velocity field is bypassed. As was shown in Section 4.3, this can be realized by relating the normal stresses on both sides of the interface

$$-P = (\tau_{YY})_{Y=\eta} - \sigma \frac{\partial^2 \eta}{\partial x^2} \tag{6.39}$$

or, for small-amplitude waves

$$-P \approx (\tau_{YY})_{Y=0} + \rho g \eta - \sigma \frac{\partial^2 \eta}{\partial x^2} \tag{6.40}$$

where P is the pressure at the interface. The term $\rho g \eta$ takes account of the pressure change from $y = \eta$ to $y = 0$ due to hydrostatic head. In Chapter 4, the pressures on both sides of the interface were related to the velocity fields by using the Bernoulli equation. This cannot be done if viscous effects are included.

The fluid stress τ_{YY} is related to stresses in the x–y laboratory coordinate system by using (6.27)

$$(\tau_{YY})_0 = -(\tau_{xy})_0 \frac{\partial \eta}{\partial x} + (\tau_{yy})_0 - (\tau_{yx})_0 \frac{\partial \eta}{\partial x} \tag{6.41}$$

From (6.3),

$$\tau_{yy} = -p + 2\mu \left(\frac{\partial u_y}{\partial y} \right) \tag{6.42}$$

Substituting (6.41) into (6.40) and using (6.42) gives the following equation for the wave-induced fluctuating quantities, where second-order terms have been neglected.

$$\rho_G g \eta - P' = -p'(0) + 2\mu \left(\frac{\partial u_y'}{\partial y} \right)_0 - 2\overline{T} \frac{\partial \eta}{\partial x} + \rho_L g \eta - \sigma \frac{\partial^2 \eta}{\partial x^2} \tag{6.43}$$

Substitute (6.22) for P' and evaluate $p'(0)$ with (6.17). Use (6.14) and, after some algebra, (6.43) gives

$$0 = -\left[a\tilde{P} + a(\rho - \rho_G)g + a\sigma k^2 - a2ik\tilde{T}\right] + \mu\left[\frac{1}{k^2}\frac{d^3\hat{u}_y}{dy^3} - 3\frac{d\hat{u}_y}{dy}\right]_0 \\ -\frac{i}{k}\rho(\overline{u}_{x_0} - c)\left(\frac{d\hat{u}_y}{dy}\right)_{y=0} + \frac{i}{k}\rho\hat{u}_y(0)\left(\frac{d\overline{u}_y}{dy}\right)_{y=0} \tag{6.44}$$

The conditions on the shear stress at the interface, (6.35), and the pressure, (6.44), at the interface can be written as

$$H_0(\hat{u}_y) = 0 \tag{6.45}$$

$$G_0(\hat{u}_y) = 0 \tag{6.46}$$

If (6.37), (6.45) and (6.46) are substituted into (6.36), the following equations for constants A, B, C, D are obtained:

$$0 = A(\hat{u}_{y1})_{-h_L} + B(\hat{u}_{y2})_{-h_L} + C(\hat{u}_{y3})_{-h_L} + D(\hat{u}_{y4})_{-h_L} \tag{6.47}$$

$$0 = A\left(\frac{d\hat{u}_{y1}}{dy}\right)_{-h_L} + B\left(\frac{d\hat{u}_{y2}}{dy}\right)_{-h_L} + C\left(\frac{d\hat{u}_{y3}}{dy}\right)_{-h_L} + D\left(\frac{d\hat{u}_{y4}}{dy}\right)_{-h_L} \tag{6.48}$$

$$0 = AH_0(\hat{u}_{y1}) + BH_0(\hat{u}_{y2}) + CH_0(\hat{u}_{y3}) + DH_0(\hat{u}_{y4}) \tag{6.49}$$

$$0 = AG_0(\hat{u}_{y1}) + BG_0(\hat{u}_{y2}) + CG_0(\hat{u}_{y3}) + DG_0(\hat{u}_{y4}) \tag{6.50}$$

The zero subscripts indicate that the terms are evaluated at the average location of the interface. Note that the above equations are homogeneous in that the left sides are zero. From the theory of determinants, a solution for A, B, C, D can exist only if the determinant of the coefficients in these equations equals zero; that is,

$$0 = \begin{vmatrix} (\hat{u}_{y1})_{-h_L} & (\hat{u}_{y2})_{-h_L} & (\hat{u}_{y3})_{-h_L} & (\hat{u}_{y4})_{-h_L} \\ \left(\frac{d\hat{u}_{y1}}{dy}\right)_{-h_L} & \left(\frac{d\hat{u}_{y2}}{dy}\right)_{-h_L} & \left(\frac{d\hat{u}_{y3}}{dy}\right)_{-h_L} & \left(\frac{d\hat{u}_{y4}}{dy}\right)_{-h_L} \\ H_0(\hat{u}_{y1}) & H_0(\hat{u}_{y2}) & H_0(\hat{u}_{y3}) & H_0(\hat{u}_{y4}) \\ G_0(\hat{u}_{y1}) & G_0(\hat{u}_{y2}) & G_0(\hat{u}_{y3}) & G_0(\hat{u}_{y4}) \end{vmatrix} \tag{6.51}$$

Equation (6.51) gives a relation between the complex wave velocity, $c = c_R + c_I$, and the wave number, k. The real and imaginary parts of the expansion produce two equations that can be solved for c_R and c_I. This calculation is simplified by using the assumption that $(c - \overline{u}_0)/(\mu/\rho)k$ is large.

Note that

$$H(\hat{u}_y) = \frac{i\mu}{k}\frac{d^2\hat{u}_y}{dy^2} + \left(i\mu k - \frac{\tilde{\sigma}}{(\overline{u}_{x0} - c)} + \frac{i\tilde{T}}{k(\overline{u}_{x0} - c)}\right)\hat{u}_y \tag{6.52}$$

$$G(\hat{u}_y) = \mu\frac{d^3\hat{u}_y}{dy^3} + \left[-3k^2\mu - ik\rho(\overline{u}_x - c)\right]\frac{d\hat{u}_y}{dy} \\ + \frac{ik\hat{u}_y}{(\overline{u}_{x0} - c)}\left[\tilde{P} + (\rho - \rho_G)g + \sigma k^2 - 2ik\overline{T} + \rho\left(\frac{d\overline{u}_x}{dy}\right)_0(\overline{u}_{x0} - c)\right] \tag{6.53}$$

where (6.38) is incorporated by using

$$\frac{\hat{u}_{y0}}{aik(\overline{u}_{x0} - c)} = 1 \tag{6.54}$$

6.5 Special solutions for constant liquid velocity

6.5.1 Solution of (6.19) for constant liquid velocity

Solutions of (6.19) for cases where $\overline{u}_x$ is not varying spatially can be obtained by substituting $\exp(ny)$. It is found that this provides four solutions since the expression for n is a quartic. These are $\exp(ky)$, $\exp(-ky)$, $\exp(my)$, $\exp(-my)$, where

$$m^2 = k^2 + \frac{i(\overline{u}_x - c)k}{\mu/\rho} \tag{6.55}$$

Thus $\hat{u}_y$ can be represented either as

$$\hat{u}_y = Ae^{ky} + Be^{-ky} + Ce^{my} + De^{-my} \tag{6.56}$$

or as

$$\hat{u}_y = A\coth(ky) + B\sinh(ky) + C\coth(my) + D\sinh(my) \tag{6.57}$$

6.5.2 Free-surface waves

The case of zero values of the mean gas and liquid flows will now be considered; that is, $\overline{u}_x = 0$, $T = 0$, $P = 0$. The interface is assumed to be clean so the surface tension is constant. The goal is to derive a relationship for wave damping due to viscosity.

The solution for a deep liquid is considered. Wave-induced disturbances decay to zero at large distances from the interface (large negative values of y). Thus, B and C in (6.56) may be taken as zero. The constants A and D are calculated using (6.52) and (6.53) with $\tilde{P}, \overline{T}, \tilde{T}, \tilde{\sigma}$ equal to zero. For the case considered, the shear condition, equation (6.35) or (6.52), gives

$$0 = AH_0(e^{ky}) + DH_0(e^{-my}) \tag{6.58}$$

where

$$H(\hat{u}_y) = \mu\left(\frac{i}{k}\right)\frac{d^2\hat{u}_y}{dy^2} + \mu ik\hat{u}_y \tag{6.59}$$

The condition on the normal stress gives

$$0 = AG_0(e^{ky}) + DG_0(e^{-my}) \tag{6.60}$$

The only way for a solution to exist is that the determinant of the coefficients *A* and *D* in (6.58) and (6.60) equals zero

$$0 = H_0(e^{-my})G_0(e^{ky}) - G_0(e^{-my})H_0(e^{ky}) \tag{6.61}$$

(This equation can also be obtained by solving (6.58) and (6.60) for *A/D* and equating the two results.)

The expansion of the secular determinant (6.61) gives

$$c^2\left[1 - \frac{2k(\mu/\rho)}{ic}\right] = c_0^2 - \left(\frac{\mu}{\rho}\right)\frac{c}{ik^2}\left[-2k^3\left(1 - \frac{2k(\mu/\rho)}{ic}\right) + 2k^2 m\left(\frac{2k(\mu/\rho)}{ic}\right)\right] \tag{6.62}$$

where

$$c_0^2 = \frac{g(\rho - \rho_G)}{\rho k} + \frac{\sigma}{\rho}k \tag{6.63}$$

Solving (6.62) for *c* gives

$$c = -i2k\frac{\mu}{\rho} \pm \left[c_0^2 + 4km\left(\frac{\mu}{\rho}\right)^2\right]^{1/2} \tag{6.64}$$

For small μ/ρ (or large $c/\upsilon k$, where υ is the kinematic viscosity), equation (6.64) simplifies further to

$$c = -i2k\frac{\mu}{\rho} \pm c_0 \tag{6.65}$$

where $c = c_R + ic_I$. Equation (6.65), therefore, gives

$$c_I = -\frac{2k\mu}{\rho} \tag{6.66}$$

$$c_R = \pm c_0 \tag{6.67}$$

Thus, from (6.66) and (6.67), the real part of (4.27) is

$$\eta = a\cos k(x - c_0 t)\exp(-2\upsilon k^2 t) \tag{6.68}$$

This represents a disturbance with wave number *k* propagating with velocity c_0 and decaying exponentially with time due to viscous damping.

The fluctuating normal velocity, u'_y, is given as

$$u'_y = \hat{u}_y \cos k(x - c_0 t)\exp(-2\upsilon k^2 t) \tag{6.69}$$

where, from (6.56),

$$\hat{u}_y = Ae^{ky} + De^{-my} \tag{6.70}$$

Terms e^{ky} and e^{-my} are, respectively, the inviscid and viscous parts of the solution. Both terms in (6.70) show an exponential decrease with decreasing y, where $y = 0$ is the average location of the interface. The characteristic length is $1/k$ for the inviscid solution. For small μ/ρ, the real part of m is

$$m_{\mathrm{R}} = \left[\frac{(c_{\mathrm{R}} - \overline{u})_x k}{\upsilon}\right]^{1/2} \frac{\sqrt{2}}{2} \tag{6.71}$$

The characteristic length of the viscous solution, for small υ, is

$$l = \left[\frac{\upsilon}{(c - \overline{u}_x)k}\right]^{1/2} \tag{6.72}$$

For large Reynolds numbers, the ratio of the viscous length to the inviscid length is small so that viscous effects are confined to a thin boundary layer close to the interface.

For a finite layer of height, h_{L}, it is convenient to use (6.57) as the solution. The dispersion relation is obtained by solving the complete determinant (6.51), for small viscosity. This yields

$$c_{\mathrm{R}}^2 = c_0^2 \tanh(kh_{\mathrm{L}}) \tag{6.73}$$

$$-c_{\mathrm{I}} = 2k\upsilon + \frac{(\upsilon k)^{1/2} c_0^{1/2} \tanh^{-3/4}(kh_{\mathrm{L}}) \operatorname{sech}^2(kh_{\mathrm{L}})}{(2)^{3/2}} \tag{6.74}$$

where c_0^2 is defined by (6.63). The two terms on the right side of (6.74), respectively, represent the damping in thin viscous boundary layers at the interface and at the wall.

For very deep liquid layers (large kh_{L}), $\operatorname{sech}(kh_{\mathrm{L}}) = 0$; equation (6.66) is recovered. As kh_{L} decreases $\operatorname{sech}^2(kh_{\mathrm{L}})$ and wave damping increase. Note that the second term on the right side of (6.74) varies as $\upsilon^{1/2}$ (rather than υ) so, for small υ, dissipation in the boundary layer at the wall can be dominant.

A derivation of (6.73) and (6.74) was given by Kim (1968). He verified these equations by generating two-dimensional waves with an oscillating dipper. Wave damping was studied by measuring the decrease in amplitude downstream of the dipper, that is, damping in space, rather than in time. Thus, the wave height is represented by

$$\eta = a \exp i(kx - ct) \tag{6.75}$$

where c is real and k is complex

$$k = k_{\mathrm{R}} + ik_{\mathrm{I}} \tag{6.76}$$

Thus,

$$\eta = a \exp i(k_{\mathrm{R}}x - ct) \exp(-k_{\mathrm{I}}x) \tag{6.77}$$

so that k_{I} is the spatial damping coefficient. As discussed in Chapter 4, Gaster (1962) has shown that k_{I} can be related to the temporal coefficient, $k_{\mathrm{R}}c_{\mathrm{I}}$, by using the equation

$$k_{\mathrm{I}} = \frac{k_{\mathrm{R}} c_{\mathrm{I}}}{c_{\mathrm{g}}} \tag{6.78}$$

where c_{I} is calculated with (6.74) and c_{g} is the group velocity, defined as

$$c_{\mathrm{g}} = k \frac{dc_{\mathrm{R}}}{dk} \tag{6.79}$$

Wave velocity c_{R} is given by (6.73). Thus

$$c_{\mathrm{g}} = \frac{(3\sigma k^2 + \rho g)\tanh(kh_{\mathrm{L}}) + (\rho g k + \sigma k^3)\,\mathrm{sech}^2(kh_{\mathrm{L}})}{2\rho c_{\mathrm{R}} k_{\mathrm{R}}} \tag{6.80}$$

Kim (1968) found good agreement between measurements of k_{I} and c_{R} with (6.73), (6.74), if (6.78) is used to relate spatial damping to temporal damping.

6.6 Wave generation by a gas flow

6.6.1 Dispersion relation for large Reynolds numbers

As discussed in Section 2.2.1, the first disturbances that appear at the interface when air flows over water are regular two-dimensional waves of the type shown in Figure 6.2. The liquid will not have a uniform velocity so that $d\overline{u}_x/dy \neq 0$.

The analysis presented in this subsection was given by Cohen (Cohen & Hanratty, 1965; Cohen, 1964). It uses the height of the liquid, h_{L}, and the velocity of the liquid at the interface, $\overline{u}_{x0}$, to make variables dimensionless. A low viscosity, that is, a large Reynolds number, is assumed. The dimensionless wave number is of the order of unity, so $(h_{\mathrm{L}}\rho c/\mu)(kh_{\mathrm{L}})$ is also large. The equation for the disturbed flow created by the interfacial waves has two inviscid and two viscous solutions, as shown in Section 6.5.

Figure 6.2 Photograph of two-dimensional waves generated by an air flow. Hanratty & Engen, 1957.

The inviscid solutions are obtained by considering only the left side of (6.19). For the case to be considered, $d\overline{u}_x/dy \neq 0$. However,

$$l = \frac{d^2\overline{u}_x/dy^2}{k^2 c} \tag{6.81}$$

is usually a small number so the two inviscid solutions are

$$u_{y1} = e^{ky} \quad \hat{u}_{y2} = e^{-ky} \tag{6.82}$$

As discussed earlier in this section, the viscous solutions can be simplified since viscosity is important only in thin boundary layers at the wall and near the interface, where the velocity profiles are, respectively,

$$\overline{u} = \left(\frac{d\overline{u}_x}{dy}\right)_{\mathrm{W}} (h_{\mathrm{L}} + y) \tag{6.83}$$

$$\overline{u} = \overline{u}_{x0} + \left(\frac{d\overline{u}_x}{dy}\right)_0 y \tag{6.84}$$

with $y = 0$, the average location of the interface. Asymptotic solutions of (6.19) for $\hat{u}_{y3}$ and $\hat{u}_{y4}$, in the limit of small viscous effects, have been presented by Cohen & Hanratty (1965) and by Craik (1966). These were obtained by scaling the distances from the wall and from the interface with viscous lengths characteristic of the sizes of the viscous boundary layers. The solutions of the scaled differential equations for small disturbances were determined by assuming that they can be represented by power series of the form

$$\hat{u}_y = \hat{u}_y^{(0)} + (k\mathrm{Re}_{\mathrm{L0}})^{-1/2}\hat{u}_y^{(1)} + (k\mathrm{Re}_{\mathrm{L0}})^{-1}\hat{u}_y^{(2)} + \cdots \tag{6.85}$$

As mentioned above, all terms in the solution are scaled with $\overline{u}_0$ and h_{L}. The expansion parameter is $k\mathrm{Re}_{\mathrm{L0}}$, where $\mathrm{Re}_{\mathrm{L0}}$ is a Reynolds number based on the mean velocity at the interface, $\overline{u}_{x0}$, the height of the liquid layer, and k is made dimensionless with h_{L}.

The main interest is to develop a dispersion relation that satisfies boundary conditions (6.45), (6.46) and (6.37). The viscous solutions at the wall and at the interface are chosen so that they decay very rapidly with distance away from the boundary. Therefore, the solutions for the viscous boundary layers at the wall and at the interface are, respectively, zero at the interface and at the wall. The determinant (6.51) simplifies to

$$0 = \begin{vmatrix} (\hat{u}_{y1})_{-h_{\mathrm{L}}} & (\hat{u}_{y2})_{-h_{\mathrm{L}}} & (\hat{u}_{y3})_{-h_{\mathrm{L}}} & 0 \\ \left(\dfrac{d\hat{u}_{y1}}{dy}\right)_{-h_{\mathrm{L}}} & \left(\dfrac{d\hat{u}_{y2}}{dy}\right)_{-h_{\mathrm{L}}} & \left(\dfrac{d\hat{u}_{y3}}{dy}\right)_{-h_{\mathrm{L}}} & 0 \\ H_0(\hat{u}_{y1}) & H_0(\hat{u}_{y2}) & 0 & H_0(\hat{u}_{y4}) \\ G_0(\hat{u}_{y1}) & G_0(\hat{u}_{y2}) & 0 & G_0(\hat{u}_{y4}) \end{vmatrix} \tag{6.86}$$

For the case being considered, all terms in the equations for the $H()$ operator and the $G()$ operator are considered except that $\tilde{\sigma}$ is set equal to zero.

The imaginary and real parts of the determinant (6.86) are separately set equal to zero. These two equations give the real and imaginary parts of the wave velocity, c_R and c_I, which, respectively, define the wave velocity and the critical condition for waves to appear. The imaginary part of (6.86) gives

$$c_I = 1 - \frac{(d\bar{u}_x/dy)_0}{2k}\tanh k \pm \left[\left(\frac{(d\bar{u}_x/dy)_0}{2k}\tanh k\right)^2 + \tanh k\left(k\mathrm{We} + \frac{1}{k\mathrm{Fr}^2} + \frac{\tilde{P}_R}{k}\right)\right]^{1/2} \tag{6.87}$$

Equation (6.87) is the same as was obtained for inviscid flows, except that a term involving $(D\bar{u}_x)_0$ appears, where $D = d/dy$. The Froude number and the Weber number in (6.87) are given as $\mathrm{Fr}^2 = \bar{u}_{x0}^2/gh_L$ and $\mathrm{We} = \sigma/\rho\bar{u}_{x0}^2 h_L$.

For neutral stability, $c_I = 0$, the real part of (6.86) then gives

$$\tilde{P}_I + \tilde{T}_R\left[\coth k + \frac{(D\bar{u}_x)_0}{k(c-1)}\right] = \frac{4(D\bar{u}_x)_0 k^2}{k\mathrm{Re}_{L0}} + \frac{4k^3(c-1)\coth k}{k\mathrm{Re}_{L0}} + \frac{k^2(c-1)^2(\coth^2 k - 1)}{(2kc\mathrm{Re}_{L0})^{12}} \tag{6.88}$$

Since $\tilde{P}_I$ and $\tilde{T}_R$ are functions of the gas velocity, (6.88) gives the gas flow needed to generate waves.

All of the terms in (6.87) and (6.88) have been made dimensionless with $\bar{u}_0$ and h_L, so the wave number k is the product of the dimensional wave number and h_L. The term multiplying $\tilde{T}_R$ is equal to the ratio of the amplitudes of the wave-induced streamwise and normal velocity fluctuations, $|\hat{u}_{x0}/\hat{u}_{y0}|$. If (6.88) is multiplied by $\hat{u}_{y0}$, the terms on the left side represent the energy transferred from the gas to the liquid, while those on the right side represent energy dissipated at the interface and at the wall. The shear stress amplitude, $\tilde{T}_R$, is expected to be an order of magnitude smaller than the pressure amplitude, $\tilde{P}_I$, because the gas viscosity is small. However, for very small heights of the liquid and/or large wavelengths, $\hat{u}_{x0} >> \hat{u}_{y0}$, so the contribution by the shear stress could then be dominant.

In order to examine the growth of surface waves, the wave velocity is considered to be complex. The time constant representing the growth rate, kc_I, is given by the following equation if $c_R >> c_I$ and c_I^2 is small

$$kc_I\left[2(c_R - 1)\coth k + \frac{D\bar{u}_{x0}}{k(c_R - 1)}\right] = \left[\tilde{P}_I + \tilde{T}_R\left(\coth k + \frac{D\bar{u}_{x0}}{k(c_R - 1)}\right) - \frac{4Du_{x0}}{\mathrm{Re}_{L0}} - \frac{2k^2(c_R - 1)\coth k}{\mathrm{Re}_{L0}} - \frac{k^{3/2}(c_R - 1)^2(\coth^2 k - 1)}{(2c_R\mathrm{Re}_{L0})^{1/2}}\right] \tag{6.89}$$

This would not be valid for gas velocities near that needed for a Kelvin–Helmholtz instability. In these cases, the growth rate of waves can be "explosively" large.

6.6.2 Influence of $\tilde{P}$ and $\tilde{T}$

The critical issue in using (6.87), (6.88) and (6.89) is the evaluation of the wave-induced variations of the pressure and the surface shear stress. Of particular interest is the prediction of the gas velocity at which waves first appear.

This problem was first investigated by Lord Kelvin, who considered fluids of infinite extent. Irrotational two-dimensional flows, with constant $\overline{u}_L$ and $\overline{u}_G$, were assumed (see Chapter 4). The predicted critical gas velocity is much larger than is observed for air–water systems. This led Jeffreys (1924, 1925) to develop his theory of sheltering for which the gas pressure in phase with the wave height is given by Kelvin–Helmholtz theory

$$\hat{P}_R = -a\rho_G k(\overline{u}_G - c_R)^2 \tag{6.90}$$

and for which the pressure amplitude in phase with the wave slope is given by

$$\hat{P}_I = as\rho_G k(\overline{u}_G - c_R)^2 \tag{6.91}$$

where s is the sheltering coefficient. A rationale of this equation is that the gas separates just behind the crest and reattaches near the trough. This creates a dead zone of low pressure behind the wave. From observations of the initiation of waves on a body of water, Jeffreys suggested that $s \approx 0.3$.

Several difficulties arise in using this interpretation. Since very small waves exist at transition, flow separation might not occur. The gas velocity varies with distance from the interface, so one needs to select an appropriate velocity. One choice is to use the gas velocity at $y = 1/k$. Another is to use a spatial mean. Articles by Benjamin (1959) and by Miles (1957) are milestones in that they show sheltering can occur without separation.

Cohen & Hanratty (1965) tested the Jeffreys equation by measuring the gas flow at which waves are initiated for flow of air and water (or a water–glycerine solution) in a rectangular channel with a height of 2.54 cm and a width of 30.5 cm. The wave velocity was greater than the liquid velocity, so the liquid did not receive energy from the average velocity profile in the liquid. That is, waves were not the result of a Tollmien–Schlichting instability in the liquid, as suggested by Feldman (1957). Since the wave velocity was much smaller than the average gas velocity, resonance with turbulence in the gas (Phillips, 1957; Charles & Lilleleht, 1965) could not be responsible for the observed waves. That is, components of the turbulent pressure fluctuations with lengths comparable to the waves at the interface would have velocities much higher than the wave velocities, since they would be comparable to the average gas velocity.

Thus, one could expect that the energy is supplied to the waves by air flow through pressure and shear forces at the interface, as described in (6.89). Cohen & Hanratty (1965) substituted (6.91) into (6.88), with $s = 0.3$, and defined the characteristic gas velocity as the spatial average. Since air has a small viscosity, the contribution of $\hat{T}_R$ was ignored. Approximate agreement was noted between the measured and the predicted critical air velocity for the initiation of waves.

6.6.3 Predictions of $\hat{P}$ and $\hat{T}$ using the quasi-laminar assumption

The approach proposed by Jeffreys cannot be rationalized by assuming separation at the initiation condition. Therefore, considerable attention has been given to the development of theoretical predictions of $\hat{P}_{\mathrm{I}}$ and $\hat{T}_{\mathrm{R}}$. The approach taken by Miles (1957,1959a,b,1962a,b) and Benjamin (1959) was to use an equation such as (6.14) to represent wave-induced variation of the gas velocity. Turbulence in the gas is taken into account only insofar as it defines the average velocity profile (a quasi-laminar approximation). The thin viscous layer in the gas, which exists close to the interface, gives rise to phase shifts in the wave-induced velocity fluctuations much the same as is found in Rayleigh's solution for the flow in a viscous fluid above an oscillating plate (Schlichting, 1968, p. 85). Consequently, the maximum in the velocity does not occur at the wave crest. This causes a phase shift from that predicted by Kelvin–Helmholtz theory and an interfacial shear stress profile which has a maximum upstream of the wave crest. Miles also considered viscous effects for a non-turbulent gas flow in the critical layer, where the gas velocity equals the wave velocity.

Cohen & Hanratty (1965) compared experimental results with a solution of the quasi-laminar disturbance equation in a Cartesian coordinate system (6.19). The pressure and the shear stress in the gas phase along the wave interface were obtained from this calculated velocity field. They used numerical methods developed by Miles (1962a,b) and measurements of the profile of the average gas velocity in a channel with smooth walls. The predicted critical gas-phase Reynolds number for the appearance of waves was somewhat larger than observed, but it was close enough to encourage further exploration of this linear theory.

6.6.4 Changes in the quasi-laminar theory to include wave-induced turbulence

The works of Miles and Benjamin have prompted a number of studies which have focused on the accuracy of the quasi-laminar assumption and on the influence of the choice of a coordinate system in which to formulate the disturbance equation.

For example, Benjamin (1959) used an orthogonal curvilinear coordinate system, ξ–η, for which $\eta = 0$ coincides with the wave profile when the waves have a small amplitude. Coordinates of constant η correspond to the streamlines determined in the inviscid analysis in Chapter 4. (However, it is noted that Benjamin argues that the influence of terms introduced by the use of curvilinear coordinates have small effects.) A conformal mapping technique (Caponi *et al.*, 1982; McClean, 1983) has been developed to map on a rectangular grid the region between a wavy boundary and a plane at η_{T}, located in the log-layer of a deep turbulent boundary layer. This mapping has also been adapted to the case where η is a rigid wavy wall (Frederick, 1986; Frederick & Hanratty, 1988). Thorsness *et al.* (1978) and Abrams & Hanratty (1985) have used a boundary-layer coordinate system for which η and ξ are, respectively, perpendicular and parallel to the wave surface. Solutions in boundary-layer coordinates are matched to an outer solution which is linearized around an average profile defined in a Cartesian coordinate system.

In what follows, a Cartesian coordinate system is used in order to demonstrate how the equation for small disturbances changes if the effects of turbulence are considered. In Cartesian coordinates, the equation of conservation of mass for an incompressible fluid is

$$\frac{\partial u_{\mathrm{G}j}}{\partial x_j} = 0 \tag{6.92}$$

where $u_{\mathrm{G}j}$ signifies the j-component of the gas velocity. If $u_{\mathrm{G}i}\left(\partial u_{\mathrm{G}j}/\partial x_j\right) = 0$ is added to the equation for conservation of momentum, equation (6.1) can be written as

$$\rho_{\mathrm{G}}\frac{\partial u_{\mathrm{G}i}}{\partial t} + \rho_{\mathrm{G}}\left(\frac{\partial u_{\mathrm{G}j}u_{\mathrm{G}i}}{\partial x_j}\right) = -\frac{\partial p}{\partial x_j} + \frac{\partial \tau_{ji}}{\partial x_j} \tag{6.93}$$

where τ_{ji} represents the effect of viscous stresses (see equation (6.3)).

The variables are represented as the sum of a mean and fluctuations due to turbulence. Thus

$$u_{\mathrm{G}i} = \overline{u}_{\mathrm{G}i} + u^{\mathrm{t}}_{\mathrm{G}i} \tag{6.94}$$

where, by definition, $\overline{u^{\mathrm{t}}_i} = 0$. Take an average of (6.93) to get

$$\rho_{\mathrm{G}}\left(\overline{u}_{\mathrm{G}j}\frac{\partial \overline{u}_{\mathrm{G}i}}{\partial x_j} + \frac{\partial \overline{u^{\mathrm{t}}_{\mathrm{G}i}u^{\mathrm{t}}_{\mathrm{G}j}}}{\partial x_j}\right) = -\frac{\partial \overline{p}}{\partial x_i} + \frac{\partial \overline{\tau}_{ji}}{\partial x_j} \tag{6.95}$$

The second term on the left side represents a net flow of momentum out of the control volume due to turbulence. It is usually brought to the right side, multiplied by the density and called a Reynolds stress. Define

$$R_{ij} = -\overline{u^{\mathrm{t}}_{\mathrm{G}i}u^{\mathrm{t}}_{\mathrm{G}j}} \tag{6.96}$$

For an incompressible flow over a flat surface,

$$\rho_{\mathrm{G}}\overline{u}_{\mathrm{G}j}\frac{\partial \overline{u}_{\mathrm{G}i}}{\partial x_j} = -\frac{\partial \overline{p}}{\partial x_i} + \frac{\partial \overline{\tau}_{ji}}{\partial x_j} + \rho_{\mathrm{G}}\frac{\partial (R_{ji})}{\partial x_j} \tag{6.97}$$

Two-dimensional interfacial waves induce two-dimensional variations of R_{ij}, as well as of $u_{\mathrm{G}i}$, so that

$$R_{ij} = \overline{R}_{ij} + r_{ij} \tag{6.98}$$

For small-amplitude waves, an equation analogous to (6.14) can be used to represent r_{ij} and $u'_{\mathrm{G}j}$.

$$r_{ij} = \hat{r}_{ij}\exp ik(x - ct) \tag{6.99}$$

The substitution of (6.98), (6.99) into (6.97) and the use of the assumption of small disturbances yields two equations analogous to (6.17) and (6.18) for disturbances introduced by the waves, with the exception that additional terms involving $\hat{r}_{xy}$, $\hat{r}_{xx}$, $\hat{r}_{yy}$, $\hat{r}_{yx}$ and their derivatives appear on the right side. The elimination of $\hat{p}$ between the x- and y-momentum balance equations yields

$$(\overline{u}_G - c)\left(\frac{d^2\hat{u}_{Gy}}{dy^2} - k^2\hat{u}_{Gy}\right) - \hat{u}_{Gy}\frac{d^2\overline{u}_{Gx}}{dy^2} = -\frac{i}{k}\upsilon_G\left(\frac{d^4\hat{u}_{Gy}}{dy^4} - 2k^2\frac{d^2\hat{u}_{Gy}}{dy^2} + k^4\hat{u}_{Gy}\right) - ik\mathfrak{R} \tag{6.100}$$

where

$$\mathfrak{R} = D^2\hat{r}_{xy} + ikD\hat{r}_{xx} + k^2\hat{r}_{yx} - ikD\hat{r}_{yy} \tag{6.101}$$

with $D = d/dy$. A comparison with (6.19) shows that the effect of wave-induced turbulence is represented by $\mathfrak{R}$ in Cartesian coordinates. The theoretical challenge is the modeling of $\mathfrak{R}$. A number of approaches have been explored.

6.7 Flow over solid wavy surfaces

6.7.1 Measurements of shear stress along a solid wavy surface

The testing of turbulence models by measuring stresses at a gas–liquid interface has not been realized. Therefore, studies of turbulent flow over solid waves have been pursued. Pressure profiles have been measured for air and water flows over solid waves by a number of investigators (Kendall, 1970; Hsu & Kennedy, 1971; Sigal, 1970; Beebe, 1972; Zilker, 1976; Buckles *et al.*, 1984; Kuzan, 1983). These are summarized by Abrams & Hanratty (1985). The drawback with these studies is that the realization of measureable pressure variations requires the use of waves with large amplitudes. Thus, a linear behavior is usually not observed.

This motivated studies of turbulent liquid flow over solid waves with very small amplitudes. Measurements of the profiles of wall shear stress were made by Cook (1970), Thorsness *et al.* (1978), Abrams & Hanratty (1985). In these studies, the wavy boundary formed one wall of a 2 in × 24 in rectangular channel through which an electrolyte was circulated. The waves were preceded by a 27.5 ft length of channel, with smooth walls, so that studies were made with a fully developed turbulent velocity field. Wall shear stresses were determined by using electrochemical techniques developed by Reiss & Hanratty (1963), Mitchell & Hanratty (1966), Son & Hanratty (1969). The basic idea is to measure the rate of mass transfer to small circular electrodes mounted flush with the wall. These electrodes form cathodes of electrochemical cells, which have an anode of much larger area. A large enough voltage is applied so that the current is controlled by the rate of mass transfer to the cathode and the concentration of the reactive species is zero at the surface of the cathode. A mass transfer coefficient, K, is defined and, from a mass balance equation, the wall shear stress is related to the mass transfer coefficient by the equation

$$\tau_W = \left(\frac{2K\Gamma(4/3)}{3}\right)\frac{4L_e\mu}{D^2} \tag{6.102}$$

where D is the diffusion coefficient, Γ is the gamma function and L_e is the equivalent length of the electrode, equal to 0.82 times the diameter of the electrode.

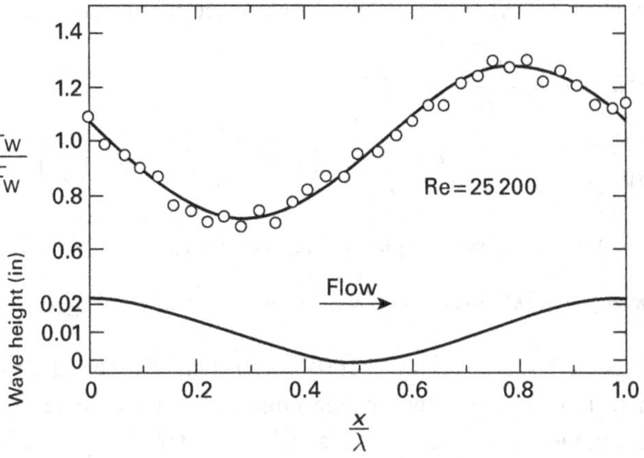

Figure 6.3 Profile of boundary shear stress. Thorsness *et al.*, 1978.

An example of measurements of the wall shear stress is given in Figure 6.3. This shows part of a train of ten sinusoidal waves with an amplitude of 0.0114 in and a wavelength of 2 in. The Reynolds number characterizing the flow was 25 200, where Re is based on the channel half-width and the bulk-averaged velocity.

6.7.2 Equations for flow over solid waves, in boundary-layer coordinates

The solid wave is not moving ($c = 0$), so that the possible influence of a critical layer, where the fluid velocity and the wave velocity are equal, need not be considered. The line drawn through the data in Figure 6.3 represents the best fit, in a least-squares sense, of a sinusoidal curve with a wavelength equal to that of the waves on the surface. The results suggest that the waves had small enough amplitudes for linear theory to be applicable; that is, the measurements contain only one harmonic. Further evidence for linearity is that the average shear stress over the wavy surface is the same as would be measured in a channel with flat walls. These measurements of τ_W can be represented by an amplitude and a phase angle (the number of degrees that the maximum is upstream of the crest).

They were interpreted by Thorsness *et al.* (1978) by using a boundary-layer coordinate system. The y-axis is perpendicular to the wave surface, with $y = 0$ corresponding to the surface; the x-axis is parallel to the wave surface. The velocity components in the x- and y-directions are given as

$$u_x = \langle \overline{u}_x \rangle + \hat{u}_x e^{ikx} \tag{6.103}$$

$$u_y = \hat{u}_y e^{ikx} \tag{6.104}$$

where $\overline{u}_x$ is the time-averaged velocity at a given y and $\langle \overline{u}_x \rangle$ is the average along one wavelength. From conservation of mass

$$-ik\hat{u}_x = \frac{d\hat{u}_y}{dy} \tag{6.105}$$

The following equation is derived for $\hat{u}_y$ by using conservation of momentum:

$$\begin{aligned}&\langle \overline{u}_x\rangle\left(\frac{d^2\hat{u}_y}{dy^2}-k^2\hat{u}_y\right)-\hat{u}_y\frac{d^2\langle \overline{u}_x\rangle}{dy^2}-ik^3\langle \overline{u}_x\rangle^2 a\\&=-\frac{i\upsilon}{k}\left(\frac{d^4\hat{u}_y}{dy^4}-2k^2\frac{d^2\hat{u}_y}{dy^2}+k^4\hat{u}_y\right)-\upsilon\left(a2k^2\frac{d^2\langle \overline{u}_x\rangle}{dy^2}-ak^4\langle \overline{u}_x\rangle\right)+\Re\end{aligned} \tag{6.106}$$

The wavelength average of $\overline{u}_x$ is set equal to the average for a flat plate and

$$\Re = aik^3\overline{R}_{xx}+a3k^2D\overline{R}_{xy}+ik\left(D\hat{r}_{xx}-D\hat{r}_{yy}\right)+k^2\hat{r}_{xy}+D^2\hat{r}_{xy} \tag{6.107}$$

A comparison of (6.106), (6.107) with (6.100), (6.101) shows that additional terms are introduced when formulating the disturbance equation in curvilinear coordinates. Of these changes, the inclusion of a centripetal acceleration term (the last term on the left side of equation (6.106)) could be important. All of the non-homogeneous terms are linear in the amplitude, a, as would be expected since $\hat{u}_y$ varies linearly with a because of the assumption of small-amplitude waves.

Equation (6.106) is solved numerically using the boundary conditions of zero velocity and zero velocity gradient at the wall

$$\hat{u}_y = 0 \qquad \frac{d\hat{u}_y}{dy}=0 \quad at\, y=0 \tag{6.108}$$

and the expectation that a velocity in the fluid far from the wave surface is the same as would be present for a flat surface:

$$\frac{d\hat{u}_y}{dy}=-\frac{d\overline{u}_x}{dy}ika \tag{6.109}$$

$$\hat{u}_y = -\overline{u}_x ika \tag{6.110}$$

The shear stress at the boundary is calculated from the solution for $\hat{u}_y$ as follows:

$$\frac{T}{\rho}=\frac{\langle T\rangle}{\rho}-\frac{\upsilon}{ik}\left(\frac{d^2\hat{u}_y}{dy^2}\right)_0 e^{ikx} \tag{6.111}$$

The pressure at the wavy boundary is obtained from the x-momentum equation evaluated at the wall.

$$\frac{P}{\rho}=\frac{\langle P\rangle}{\rho}+\frac{\upsilon}{k}\left(\frac{d^3\hat{u}_y}{dy^3}\right)_0 e^{ikx}-ai\upsilon k\left(\frac{d^2\langle \overline{u}_x\rangle}{dy^2}\right)_0 e^{ikx} \tag{6.112}$$

6.7.3 Evaluation of $\hat{r}_{ij}$ using a modified van Driest model

From (6.97) it is seen that the equations for a turbulent fluid require the specification of turbulent stresses

$$\tau_{ij}^{t} = -\rho\overline{u_i^t u_j^t} = -\rho R_{ij} \tag{6.113}$$

in addition to viscous stresses. The simplest approach is to represent R_{ij} by a Boussinesq approximation (Hinze, 1975, pp. 23–24), which is analogous to (6.3).

$$R_{ij} = -\frac{q^{t2}}{3}\delta_{ij} + \upsilon^{t}\left(\frac{\partial u_i}{\partial x_j} + \frac{\partial u_j}{\partial x_i}\right) \tag{6.114}$$

where υ^{t} is a turbulent kinematic viscosity, which can vary with spatial location, and

$$q^{t2} = \overline{u_x^{t2}} + \overline{u_y^{t2}} + \overline{u_z^{t2}} \tag{6.115}$$

The first term on the right side of (6.114) is the average pressure due to turbulence. Assume that the waves not only induce variations in the fluid velocity but also in the turbulent viscosity so that

$$\upsilon^{t} = \overline{\upsilon}^{t} + \hat{\upsilon}^{t}\exp ik(x - ct)t \tag{6.116}$$

Neglecting higher-order terms in the wave-induced fluctuations, equation (6.114) gives

$$R_{ij} = \overline{R}_{ij} + \overline{\upsilon}^{t}\left(\frac{\partial \hat{u}_i}{\partial x_j} + \frac{\partial \hat{u}_j}{\partial x_i}\right) + \hat{\upsilon}^{t}\left(\frac{\partial \overline{u}_i}{\partial x_j} + \frac{\partial \overline{u}_j}{\partial x_i}\right) \tag{6.117}$$

Thus

$$\hat{r}_{ij} = \hat{\upsilon}^{t}\left(\frac{\partial \overline{u}_i}{\partial x_j} + \frac{\partial \overline{u}_j}{\partial x_i}\right) + \overline{\upsilon}^{t}\left(\frac{\partial \hat{u}_i}{\partial x_j} + \frac{\partial \hat{u}_j}{\partial x_i}\right) \tag{6.118}$$

is substituted into (6.106), (6.107) to obtain an equation for $\hat{u}_y$ in boundary-layer coordinates. The theoretical issue in this Boussinesq formulation is the specification of υ^{t}.

The influence of turbulence can be seen by examining measurements for fully developed flow in a channel or a pipe with flat walls. Consider a horizontal volume of unit breadth attached to the wall. The force due to the existence of a pressure gradient is $-y_W\Delta p$. This is balanced by shear stresses at the top and bottom of the volume, $\tau\Delta x - \tau_W\Delta x$. Thus

$$(\tau - \tau_W) - y_W\frac{\Delta p}{\Delta x} = 0 \tag{6.119}$$

For a symmetric fully developed flow in a channel, $\tau = 0$ at the half-width, $y_W = B/2$, so

$$\tau_W = -\frac{B}{2}\frac{\Delta p}{\Delta x} \tag{6.120}$$

where $\Delta p/\Delta x$ is negative. The friction velocity, $v^* = (\tau_W/\rho)^{1/2}$, can be calculated if the pressure gradient is measured. The local fluid shear stress, which is obtained from (6.119), can be set equal to the sum of a turbulent and a viscous contribution

$$\tau_{yx} = \tau_{yx}^{\ell} + \tau_{yx}^{t} \tag{6.121}$$

where

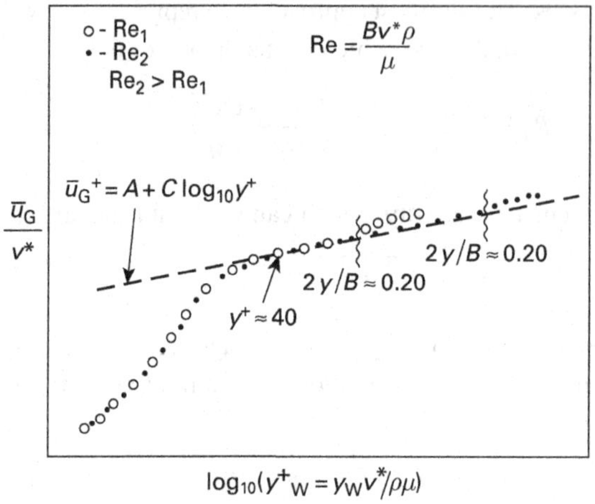

Figure 6.4 Sketch of a turbulent velocity profile where B is the half-height of the channel.

$$\tau_{yx}^{\ell} = \mu \frac{d\overline{u}_x}{dy} \tag{6.122}$$

$$\tau_{yx}^{t} = -\rho \overline{u_y^t u_x^t} \tag{6.123}$$

Measurements of turbulent velocity profiles are sketched in Figure 6.4, where the half-height of the channel is designated by $B/2$.

Velocities and distances from the wall have been made dimensionless with the friction velocity and the friction length υ/v^*. The region $y^+ < 5$ is called the laminar sub-layer since turbulent stresses are negligible compared to viscous stresses and the velocity is given by

$$u^+ = y^+ \tag{6.124}$$

For $y^+ > 40$ viscous stresses can be neglected. In the region $y^+ = 40$ to $2y/B = 0.20$, the velocity profile is given by

$$\overline{u}^+ = A + C \log_{10} y^+ \tag{6.125}$$

where $A = C = 5.6$. As sketched in Figure 6.4, the region beyond $2y/B = 0.20$ is a weak function of $\text{Re} = Bv^*/\upsilon$. The choice of $A = 5.5$ and $C = 5.75$ gives a good approximation over the whole pipe cross-section, with the exception of $y^+ < 40$ (see Schlichting, 1968). The region between $y^+ = 5$ and the inner edge of the log-layer, where both viscous and turbulent stresses are important, is known as the buffer layer.

A frequently used approach in specifying the turbulent stresses is the mixing-length theory of Prandtl. This is inspired by the kinetic theory of gases whereby the kinematic viscosity is shown to be proportional to the product of the molecular velocity and the molecular mean-free path. Thus, the kinematic turbulent viscosity is given as

$$\upsilon^{\mathrm{t}} \propto v^{\mathrm{t}} \ell \tag{6.126}$$

where v^{t} is a characteristic turbulent velocity and ℓ is the mixing length.

Prandtl assumed that $v^{\mathrm{t}} \propto \ell |d\bar{u}_x/dy|$. Thus, in a boundary layer

$$\frac{\tau^{\mathrm{t}}}{\rho} = \ell^2 \left|\frac{d\bar{u}_x}{dy}\right| \left(\frac{d\bar{u}_x}{dy}\right) \tag{6.127}$$

Consider a turbulent boundary layer on a flat surface where the pressure gradient is zero and the fluid stress, $\tau = \tau^{\mathrm{t}} + \tau^{\ell}$, is constant and equal to τ_{W}. Beyond $y^+ = 40$ equation (6.127) gives

$$v^{*2} = \ell^2 \left|\frac{d\bar{u}_x}{dy}\right| \left(\frac{d\bar{u}_x}{dy}\right) \tag{6.128}$$

where $v^* = (\tau_{\mathrm{W}}/\rho)^{1/2}$, since $\tau = \tau^{\mathrm{t}} = \tau_{\mathrm{W}}$. From (6.125) the velocity gradient in the log-layer is

$$\frac{d\bar{u}_x}{dy} = \frac{v^*}{\kappa y} \tag{6.129}$$

where $\kappa = 0.4$ is the von Karman constant. Substituting (6.129) into (6.128) gives

$$\ell = \kappa y \tag{6.130}$$

and from (6.130) and (6.126)

$$\upsilon^t = \kappa v^* y \tag{6.131}$$

Thus, both ℓ and υ^{t} vary linearly with y in the log-layer.

Van Driest (Hinze, 1975, p. 622) suggested the following relation for the mixing length

$$\ell = \kappa y \left[1 - \exp\left(-\frac{y v^*}{\upsilon A}\right)\right] \tag{6.132}$$

Note that, at large y, (6.132) gives (6.130). The term in brackets represents the role of viscosity in damping the mixing length in the buffer layer. The parameter A is a measure of the thickness of the viscous wall region. A value of 25 fits measurements of the velocity profile over a flat boundary, for which $dp/dx = 0$. Loyd *et al.* (1970) applied (6.132) to turbulent boundary layers with a range of imposed pressure gradients. The thickness of the viscous wall layer varies with pressure gradient so A is given by the function

$$A = \overline{A}\left(1 + \frac{dp}{dx}k_1 + \left(\frac{dp}{dx}\right)^2 k_2 + \cdots\right) \tag{6.133}$$

with $k_1 = -30$ and $k_2 = 1.54 \times 10^3$ for equilibrium flows for which the pressure gradient is not varying. For flow over a small-amplitude sinusoidal solid wave, the quadratic term in (6.133) can be neglected, so

$$A = \overline{A} + a\tilde{A}\,e^{ix} \tag{6.134}$$

The pressure variation is given as

$$p = \overline{p} + a\tilde{p}e^{ikx} \tag{6.135}$$

From equation (6.133) with $c = 0$ (since a solid wave is being considered)

$$a\tilde{A} = a\overline{A}k_1 ik\tilde{p}(0) \tag{6.136}$$

The substitution of (6.134), (6.136) into (6.132) provides an equation for the wave-induced variation of the mixing length.

Loyd *et al.* (1970) have suggested that for non-equilibrium conditions, such as exist for flow over waves, an effective pressure gradient should be used in (6.133), where

$$\frac{d\left(\frac{dp}{dx}\right)_{\text{eff}}}{dx} = \frac{\left(\frac{dp}{dx}\right) - \left(\frac{dp}{dx}\right)_{\text{eff}}}{k_{\text{L}}} \tag{6.137}$$

This introduces a lag between the imposition of a non-zero pressure gradient and a change of the scale, for which the following relation is obtained

$$A = \overline{A} + a\frac{k_1\overline{A}ik\tilde{p}(0)}{1 + ikk_{\text{L}}}e^{ikx} \tag{6.138}$$

Loyd *et al.* suggested that $k_{\text{L}} = 3000$.

6.7.4 Testing turbulence models against wall shear stress measurements

Thorsness *et al.* (1978) used measurements of the shear stress variation along small-amplitude solid waves, such as shown in Figure 6.3, to test different models for the Reynolds stress. The profiles of wall shear stress were represented by phase angles and amplitudes, plotted against the dimensionless wave number, as shown in Figures 6.5 and 6.6. The quasi-laminar approach (Section 6.6.3) involves the neglect of all terms in (6.106) and (6.107) which involve $\hat{r}_{ij}$ or its derivatives. The resulting equation differs from the formulation in a Cartesian system because of terms associated with the use of curvilinear coordinates.

Thorsness *et al.* found that the quasi-laminar assumption in Cartesian coordinates produces large errors. The use of the quasi-laminar assumption in boundary-layer coordinates overpredicts the magnitude of $\hat{T} = a\tilde{T}$ (Figure 6.6) and incorrectly predicts that the phase angle is relatively insensitive to changes in k^+ (Figure 6.5). This led Thorsness *et al.* to explore whether improved results could be obtained by considering wave-induced variations in the turbulence.

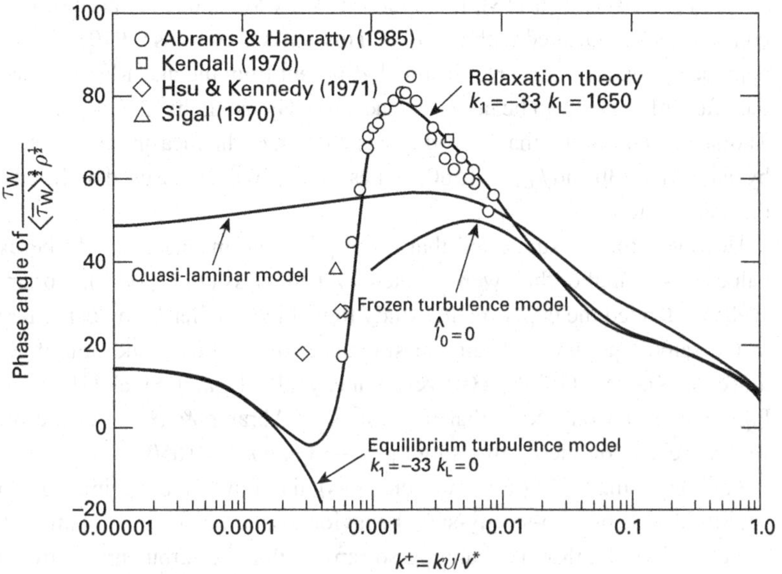

Figure 6.5 The phase angle characterizing the variation of the wall shear stress as a function of the dimensionless wave number. Abrams & Hanratty, 1985.

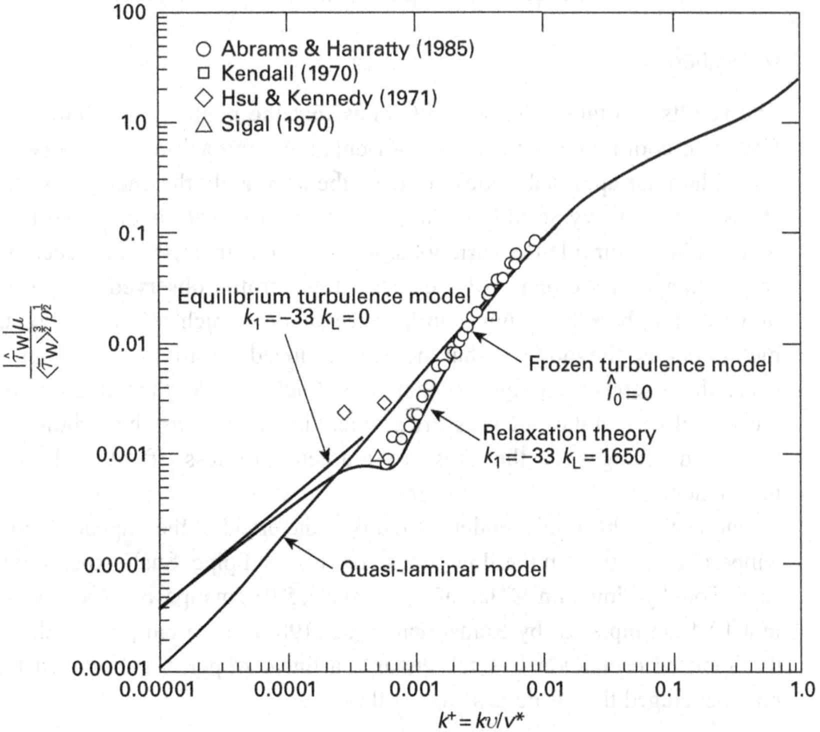

Figure 6.6 The amplitude characterizing the variation of the wall shear stress as a function of the dimensionless wave number. Abrams & Hanratty, 1985.

From (6.117) and (6.118), it is seen that the wave-induced variations of R_{ij} can have contributions associated with $\overline{\upsilon}^t$ and $\hat{\upsilon}^t$. Hussain & Reynolds (1970) assumed $\hat{\upsilon}^t = 0$. This approach did not produce results much different from the quasi-laminar approach. Thus, the effect of $\hat{\upsilon}^t$ needs to be taken into account. By using the model of Loyd *et al.* (1970), Thorsness *et al.* found that a good representation of the measurements could be obtained by using $k_1 = -30$ and $k_L = 1500$, values close to what were used by Loyd *et al.* for quite different systems.

However, this fit suggested that a sharp change in phase should be experienced at values of k^+ smaller than were studied by Thorsness *et al.* (1978). Abrams & Hanratty (1985) extended the experimental study to include smaller k^+ by using larger flow rates in the same equipment. Their measurements of the phase and amplitude, along with those of Kendall (1970), Hsu & Kennedy (1971) and Sigal (1970) are plotted in Figures 6.5 and 6.6. Note that the results of Abrams & Hanratty are predicted quite well by relaxation theory by using $k_1 = -33$ and $k_L = 1650$.

For very small k^+ (large wavelengths) the flow is changing so slowly that an equilibrium or pseudo-steady-state behavior exists. For large k^+ (small dimensionless wavelengths), the flow is changing so rapidly that the turbulence is frozen, $\hat{\mu}^t = 0$. At intermediate k^+, the flow is in a transition region between these two extremes.

6.8 Response of turbulence to imposed temporal oscillations

6.8.1 Motivation

The results in Figures 6.5 and 6.6 show that the neglect of turbulence effects in a Cartesian coordinate system (Miles–Benjamin approach) or in a curvilinear system (quasi-laminar approach) does not describe accurately the changes of the wall shear stress along a wavy solid boundary. Some improvement is obtained by considering wave-induced turbulence variations. Of particular interest is the accounting for the sharp change in the phase of the shear stress profile observed over a narrow range of wave numbers by using a turbulence model which allows for relaxation. The picture behind the model is that the wave-induced variations of the pressure gradient cause the viscous wall layer to change its thickness. As the wave number increases the spatial variation of the pressure gradient is too rapid for the turbulence to respond fully and changes in the pressure gradient are less effective in changing the turbulence.

One way to check this model is to look at the effect of the imposition of a sinusoidal temporal oscillation on the flow in a smooth walled pipe. Such experiments have been carried out by Finnicum & Hanratty (1988) in a 5.08 cm pipe, by Mao & Hanratty (1986) in a 19.4 cm pipe and by Ramaprian & Tu (1983) in a 5 cm pipe. If the amplitude of the imposed oscillation is small enough, a linear response is obtained for which the phase-averaged field is described as follows:

$$u_x = \overline{u}_x(y) + \hat{u}_x(y)\cos\left[\varpi t + \theta_u(y)\right] \tag{6.139}$$

$$\frac{dp}{dx} = \frac{d\bar{p}}{dx} + \frac{d\hat{p}}{dx}\cos \varpi t \tag{6.140}$$

with x being the distance in the flow direction, y the distance from the wall, ϖ the angular frequency, t the time and θ_u, the phase relative to the pressure gradient. The time-mean pressure gradient and the time-mean velocity of the undisturbed flow are designated by $d\bar{p}/dx$ and $\bar{u}_x(y)$. The oscillations of the wall shear stress are given by

$$\tau_W = \bar{\tau}_W + \hat{\tau}_W \cos(\varpi t + \theta_{\tau W}) \tag{6.141}$$

where $\theta_{\tau W}$ is the phase of τ_W relative to the pressure gradient. Thus, for the flows being considered, the velocity field is exposed to a time-varying pressure gradient, while flow over a solid wavy wall is exposed to a spatially varying pressure gradient. The geometric complications associated with a wavy wall are avoided.

6.8.2 Measurements

Figure 6.7 presents measurements of $-d\hat{p}/dx$ and of $\hat{\tau}_W/\bar{\tau}_W$ obtained in a 5.08 cm pipe for a Reynolds number of 17 100. The ratio of the oscillations of the centerline velocity to the mean centerline velocity, a, was 0.102. The frequency of the imposed oscillations was f = 0.50 Hz. The circular frequency, made dimensionless with wall parameters, was $\varpi^+ = 0.0093$. The curve through the data represents a best least-squares fit of a cosine function. Note that $\hat{\tau}_W$ lags $-d\hat{p}/dx$ by about 90°. Figure 6.7 shows a linear behavior in that the measurements are described by a single harmonic and the mean shear stress at the wall is unchanged.

Measurements of the amplitude, $\hat{\tau}_W$, and of the phase of τ_W relative to the pressure gradient, $\theta_{\tau W}$, in a 5.08 cm pipe and in a 19.4 cm pipe are plotted in Figures 6.8

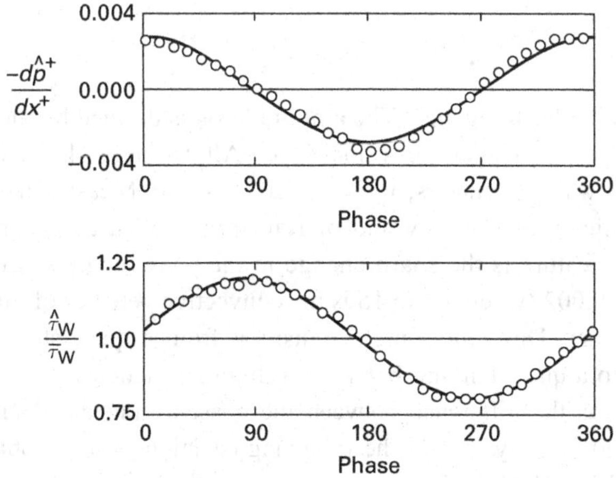

Figure 6.7 Phase-averaged values of the wall shear stress and the pressure gradient at Re = 17 100, f = 0.50 Hz and a = 0.102 centerline velocity. Finnicum & Hanratty, 1988.

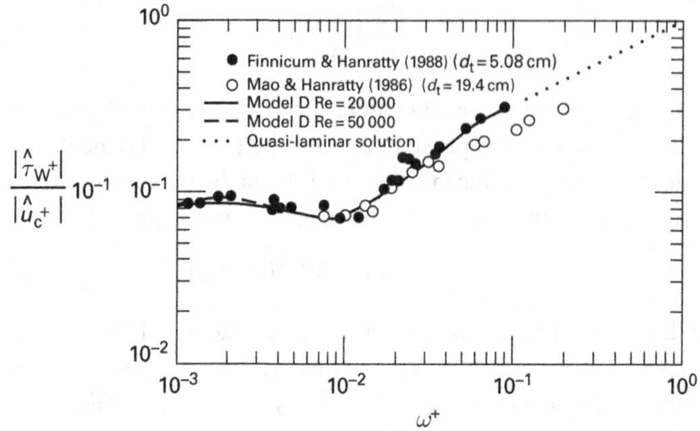

Figure 6.8 Amplitude of the wall shear stress variation. Finnicum & Hanratty, 1988.

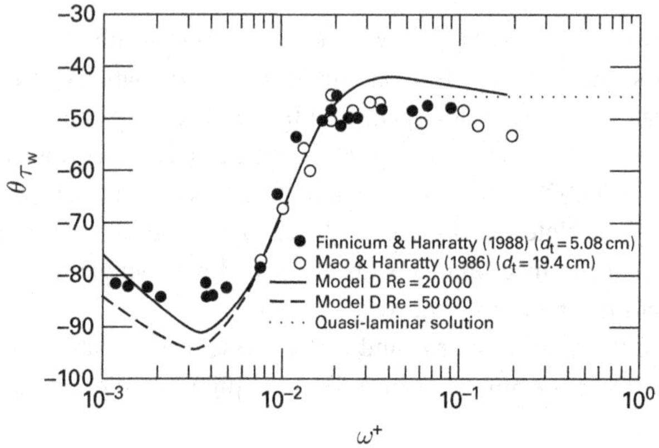

Figure 6.9 Phase lag (relative to the pressure gradient) of the wall shear stress variation. Finnicum & Hanratty, 1988.

and 6.9 against the circular frequency. The amplitude $\hat{\tau}_W$ is divided by the amplitude of the oscillations of the centerline velocity $\hat{u}_c$. All variables have been made dimensionless using wall parameters, $v^* = (\overline{\tau}_W/\rho)^{1/2}$ and υ. Measurements of Mao & Hanratty, of Finnicum & Hanratty and of Ramaprian & Tu agree, if plotted in this way. A striking feature is the sharp change in the phase angle when $\varpi^+/15$ is in the range 0.0005–0.002 (where $c^+ = 15$ is the convective velocity of turbulence in the viscous wall layer). This represents a transition from a quasi-steady behavior at low frequencies to a quasi-laminar behavior at high frequencies.

Designate $\tilde{\tilde{u}}_x(y,t)$ as the difference between the phase-averaged velocity, $u_x(y,t)$, and the time-averaged velocity, $\overline{u}_x(y)$. The following equation for $\tilde{\tilde{u}}_x$ is obtained from the x-momentum balance for flow in a pipe, where τ^t is the phase-averaged Reynolds stress and $\overline{\tau}^t$ is the time-averaged Reynolds stress:

$$\rho \frac{\partial \tilde{\tilde{u}}_x}{\partial t} = -\frac{\partial \tilde{\tilde{p}}}{\partial x} + \frac{1}{r}\frac{\partial}{\partial r}[r(\tau^{\mathrm{t}} - \overline{\tau}^{\mathrm{t}})] + \frac{\mu}{\rho}\frac{\partial}{\partial r}\left(r\frac{\partial \tilde{\tilde{u}}_x}{\partial r}\right) \tag{6.142}$$

In order to solve (6.142), the phase-averaged Reynolds stress, τ^{t}, needs to be specified. By using the eddy viscosity model and assuming small wave-induced fluctuations

$$\tilde{\tilde{\tau}}^{\mathrm{t}} = \tau^{\mathrm{t}} - \overline{\tau}^{\mathrm{t}} = \rho\left(\overline{\upsilon}^{\mathrm{t}}\frac{\partial \tilde{\tilde{u}}_x}{\partial r} + \tilde{\tilde{\upsilon}}^{\mathrm{t}}\frac{\partial \overline{u}_x}{\partial r}\right) \tag{6.143}$$

where the time-averaged stress, $\overline{\tau}^{\mathrm{t}}$, is given as

$$\overline{\tau}^{\mathrm{t}} = \rho\overline{\upsilon}^{\mathrm{t}}\frac{\partial \overline{u}_x}{\partial r} \tag{6.144}$$

6.8.3 Comparison with quasi-steady solution

In the limit of small ϖ the inertia term on the left side of (6.142) can be ignored. Then the relation between τ_{W} and the centerline velocity, u_{c}, is the same as obtained for steady flow

$$\tau_{\mathrm{W}} = \frac{1}{2} f \rho u_{\mathrm{c}}^2 \tag{6.145}$$

$$f = A\left(\frac{d_{\mathrm{t}} u_{\mathrm{c}}}{\upsilon}\right)^{-n} \tag{6.146}$$

The quasi-steady solution relating $\tilde{\tilde{\tau}}_{\mathrm{W}}$ to $\tilde{\tilde{u}}_{\mathrm{c}}$ is obtained by substituting $\tau_{\mathrm{W}} = \overline{\tau}_{\mathrm{W}} + \tilde{\tilde{\tau}}_{\mathrm{W}}$ and $u_{\mathrm{c}} = \overline{u}_{\mathrm{c}} + \tilde{\tilde{u}}_{\mathrm{c}}$ into (6.145) and (6.146). The following result was presented by Finnicum & Hanratty

$$\frac{\tilde{\tilde{\tau}}_{\mathrm{W}}^{+}}{\tilde{\tilde{u}}_{\mathrm{c}}} = 0.339\,\mathrm{Re}^{-0.125} \tag{6.147}$$

Equation (6.147) is a weak function of Reynolds number. Over the range of Reynolds numbers covered by the experiments (10000–50000), $\tilde{\tilde{\tau}}_{\mathrm{W}}^{+} = 0.095|\tilde{\tilde{u}}_{\mathrm{c}}^{+}|$ is a good approximation. In addition, quasi-steady behavior predicts that the phase of $\tilde{\tilde{\tau}}_{\mathrm{W}}$ would be the same as for $\tilde{\tilde{u}}_{\mathrm{c}}$ ($-90°$). Figures 6.8 and 6.9 indicate that for $\varpi^{+} \leq 0.004$ the quasi-steady prediction roughly approximates amplitudes and phases of the flow-induced variations of the magnitude and phase of the flow-induced turbulent velocity fluctuations. Note also that $\varpi^{+} = 0.004$ corresponds to the frequency at which the plug flow at the center of the pipe disappears and for which the relationship of the centerline velocity to the favorable pressure gradient differs from (6.149).

For almost all of the ϖ^{+} covered in Figures 6.8 and 6.9 viscous and turbulent stresses can be ignored in the center of the pipe, so the disturbed velocity profile is flat and described by

$$\rho\frac{\partial \tilde{\tilde{u}}_x}{\partial t} = -\frac{\partial \tilde{\tilde{p}}}{\partial x} \tag{6.148}$$

A solution of (6.148) gives

$$\varpi\rho|\hat{u}| = |\frac{\partial \hat{p}}{\partial x}| \tag{6.149}$$

and a phase angle relative to the favorable pressure gradient of

$$\theta_{\hat{u}} = \frac{1}{2}\pi \tag{6.150}$$

The region outside the core is influenced by turbulent and viscous stresses. As ϖ increases the region of plug flow increases. Eventually, it extends to the laminar sub-layer where a quasi-laminar model is valid

$$\rho\frac{\partial \tilde{\tilde{u}}_x}{\partial t} = -\frac{\partial \tilde{\tilde{p}}}{\partial x} + \frac{\mu}{\rho}\frac{\partial^2 \tilde{\tilde{u}}_x}{\partial y^2} \tag{6.151}$$

with y being the distance from the wall. The solution of (6.151) is presented by Uchida (1956). It gives a shear stress at the wall which leads the pressure gradient by 45° and has an amplitude given by

$$\frac{|\tilde{\tilde{\tau}}^+_{\mathrm{W}}|}{|\tilde{\tilde{u}}^+_{\mathrm{c}}|} = (\varpi^+)^{1/2} \tag{6.152}$$

The quasi-laminar solution is indicated by the dotted lines in Figures 6.8 and 6.9. Good agreement between predicted and measured amplitudes is observed for large ϖ^+ in Figure 6.8. Agreement is also noted for the phase angle at large ϖ^+ shown in Figure 6.9. The same type of relaxation observed for measurements of the spatial variation of τ_{W} induced by a wavy boundary over a range of dimensionless wave numbers of 0.0005–0.002 is shown in Figure 6.9. This suggests that a similar mechanism is operating in the two studies. These transitions can be compared numerically by replotting Figure 6.9 using ϖ^+/c^+ as the abscissa, where $c^+ = 15$ is the convection velocity characterizing turbulence in the viscous wall region.

Both Mao & Hanratty (1986) and Finnicum & Hanratty (1988) used the model described in Section 6.7.3 to capture the behavior at intermediate frequencies. The main difference is that the relaxation equation (6.137) is rewritten as

$$\frac{d\left(\frac{dp^+}{dx}\right)_{\mathrm{eff}}}{c^+ dt^+} = \frac{\left(\frac{dp^+}{dx^+}\right) - \left(\frac{dp^+}{dx^+}\right)_{\mathrm{eff}}}{k_{\mathrm{L}}} \tag{6.153}$$

Equation (6.142) has been solved numerically by using the analytical solution of (6.151) by Uchida (1956) as the initial condition and the boundary condition that the velocity is zero at the wall. Calculations by Finnicum & Hanratty (1988) used $k_1 = -25$ and $k_{\mathrm{L}} = 3500$ (Model D). A good fit to the measured amplitudes is shown in Figure 6.8. A satisfactory fit to the phase angle data is shown in Figure 6.9. The relaxation model does a reasonable job in representing the response of turbulent shear stresses both to a wavy boundary and to temporally induced oscillations. However, the physics behind the model has not been established. Clearly, more work on this problem is needed.

Note that the amplitudes measured by Mao & Hanratty at high frequencies are below the quasi-laminar solution. Their experiments were done in a larger pipe than used by Finnicum & Hanratty. Mao pointed out that, in a 19.4cm pipe, the frequency of the imposed oscillations is close to that of the turbulent velocity fluctuations. This suggests that there could be a direct interaction between the imposed flow oscillations and the turbulence.

6.9 Prediction of wave generation

Section 2.2 describes studies of the initiation of waves by an air flow (Cohen & Hanratty, 1965) in an enclosed horizontal channel with a height of 2.54cm and a width of 30.5 cm. The height of the liquid layer was large enough for the waves to receive their energy from induced gas-phase pressure variations in phase with the wave slope. The critical gas-phase Reynolds number, defined with the height of the gas space and the spatially averaged gas velocity, was measured as $Re_G = 3600$ for a liquid layer with a viscosity of 3.9cP and a height of 0.0147ft flowing along the bottom of the channel. Observed waves extended over the whole width of the channel. At the same liquid flow, these two-dimensional waves were observed up to $Re_G = 9600$. The waves became three-dimensional above this gas flow.

Figure 6.10 shows loci of $c_I = 0$ calculated by Cohen & Hanratty (1965) using the quasi-laminar assumption in a Cartesian coordinate system that was explored by Miles and by Benjamin (see Section 6.6.3). Stable conditions ($c_I < 0$) are predicted to exist outside the curve labeled "Miles–Benjamin." Note that the gas Reynolds number below which unstable conditions do not exist is predicted to be about $Re_G = 4100$. The points represent wave numbers of observed two-dimensional waves. The dashed curve is the fastest-growing wave number predicted by linear theory.

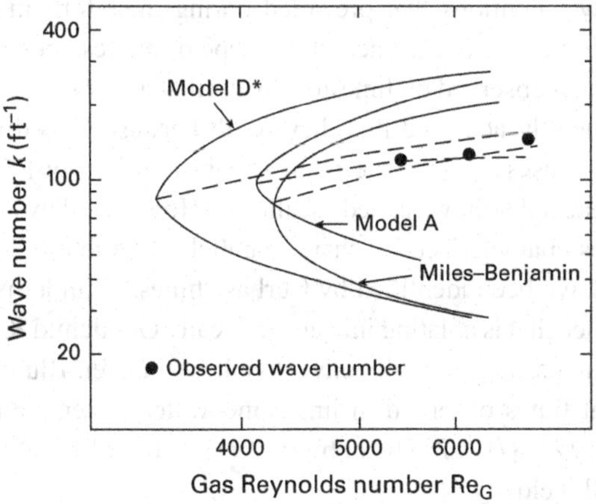

Figure 6.10 Calculation of neutral stability curve and loci of maximum growth for water–glycerine ($H = 0.0147$ ft, $\mu = 3.9$ cP).

Frederick (1982) also calculated the behavior in a boundary-layer coordinate system for two cases: (1) where wave-induced variations of the turbulence are ignored (Model A in Figure 6.10); (2) where wave-induced variations of the turbulence are calculated (Model D*) with the relaxation theory outlined in Section 6.7.2, using $k_l = -55$ and $k_L = 3000$. The wave velocities observed in the experiments were small, $c^+ = 1 - 3$. Therefore, the influence of c was ignored; that is, the equations developed for a wavy solid boundary were used. (Frederick showed that $\tilde{P}_I$ is not sensitive to changes in c_R when the wave velocity is small.)

As seen in Figure 6.10, the calculations are not highly dependent on the model that is used to calculate $\tilde{P}_I$. The chief limitation is that there are small differences in the prediction of the critical gas Reynolds number. The use of a relaxation model for the wave-induced turbulence produces a value of the critical gas Reynolds number close to observation, Re = 3600.

6.10 Dissolution patterns

A consideration of the variation of mass transfer along a small-amplitude wavy surface is an interesting extension of the theory outlined in Section 6.7. If a turbulent fluid flows over a soluble planar surface, material need not be removed uniformly along the surface. As outlined by Hanratty (1981) and by Thorsness & Hanratty (1979b), a growing two-dimensional wave can emerge if the function describing the variation of the convective mass transfer rate along a small-amplitude sinusoidal waveform has a maximum somewhere in the trough region. Depending on whether the maximum is on the downstream side of the trough, the upstream side of the trough or right at the bottom of the trough, the wave will propagate in the direction of flow, will propagate in a direction opposite to the flow, or will remain stationary.

The chief motivation for carrying out such a stability analysis is to explain the origin of wavelike patterns that are caused by diffusional processes and to relate the wavelength and wave velocity to flow conditions that prevailed during their formation. Hanratty (1981) provides a discussion of the occurrence of this type of process in nature.

Wavelike patterns are often observed on limestone walls that are or have been exposed to flowing water. It is generally accepted that they result because CO_2 diffuses to the surface and bicarbonate diffuses back to the main stream (Curl, 1966). The patterns usually appear as closely packed hollows called "scallops." However, they also appear as two-dimensional structures characterized by sharp, parallel crests that lie transverse to the flow direction. These have been identified by Curl as "flutes." Similar patterns have been observed on snow or ice that is ablating into an air stream. Goodchild & Ford (1971) argue that these patterns are associated with convective heat transfer. Blumberg & Curl (1966) have indicated that flutes observed in limestone–water systems and in ice–air systems are characterized by $\lambda U_B/\upsilon \approx 23\,000$ or by $\lambda v^*/\upsilon = 2200$, where v^* is the friction velocity and U_B is the bulk velocity.

Waveforms similar to flutes have been observed on the underside of ice covers on rivers or canals (Ashton & Kennedy, 1972; Hsu *et al.*, 1979). These are described as regular, long-crested, downstream-migrating waves which develop under conditions in which ice is melting. Controlled experiments by Hsu & Kennedy (1971) give $\lambda v^*/\upsilon = 3180$.

To consider the stability of a dissolving surface, define a disturbance at the interface of the following form:

$$h(t) = \overline{h}(t) + a(t)\cos(kx - kct) \tag{6.154}$$

Term $\overline{h}(t)$ is the average height of the surface at any time, and term $a(t)$ is the amplitude of the disturbance wave. A mass balance for a surface which is dissolving solely because of convective transfer is

$$-\rho_S \frac{dh}{dt} = N \tag{6.155}$$

where ρ_S is the density of the solid and N is the local rate of convective transfer per unit area from the solid to the flowing fluid.

The surface flux can be decomposed into mean and fluctuating components defined in the following way:

$$N = \overline{N} \pm a|\tilde{n}|\cos[(kx + \theta) - kct] \tag{6.156}$$

where the minus sign is to be used when convective transfer is occurring from the fluid to the solid. The phase angle, θ, is the number of degrees by which the maximum in the mass-transfer rate precedes the maximum of the wave profile. By using the trigonometric relation for the sum of two angles, (6.156) can be written in the following form:

$$N = \overline{N} \pm a|\tilde{n}|\cos\theta\cos(kx - ct) \mp a|\tilde{n}|\sin\theta\sin(kx - ct) \tag{6.157}$$

If (6.156) and (6.154) are substituted into (6.155) the following relations are obtained for a dissolving surface:

$$-\rho_S \frac{d\overline{h}}{dt} = \overline{N} \tag{6.158}$$

$$-\rho_S \frac{da}{dt} = a|\tilde{n}|\cos\theta \tag{6.159}$$

$$c = \frac{|\tilde{n}|\sin\theta}{k\rho_S} \tag{6.160}$$

Equation (6.159) can be solved to get

$$a = a_0 e^{\varpi t} \tag{6.161}$$

with

$$\varpi = -|\tilde{n}|\frac{\cos\theta}{\rho_S} \tag{6.162}$$

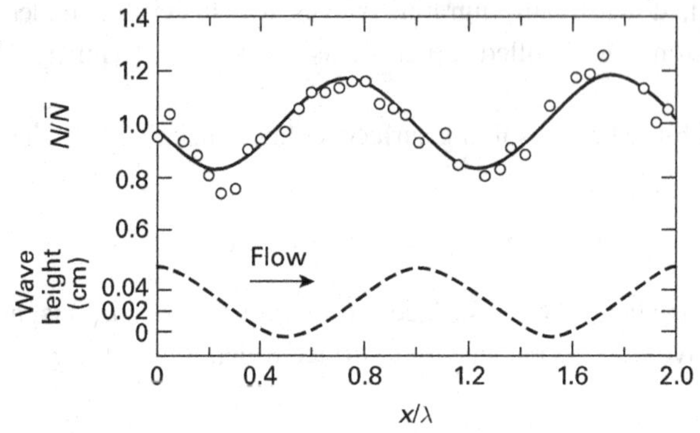

Figure 6.11 Variation of the mass transfer rate along a wavy surface for Re = 20 900. Thorsness & Hanratty, 1979.

If ϖ is positive the surface waves will grow and if ϖ is negative they will decay. For a dissolution process, a positive ϖ requires $\cos\theta$ to be negative so that $\pi/2<\theta<3\pi/2$. Thus, the understanding of the stability of a dissolving surface depends on predicting the phase of the mass transfer profile on a wavy surface.

Thorsness & Hanratty (1979a) used electrochemical techniques to measure the spatial variation of N along a small-amplitude solid wavy wall which is exchanging mass with a turbulently flowing liquid. The Schmidt number characterizing the system was equal to 729. A rectangular channel with a flat top and a wavy bottom was used. A typical profile of N is shown in Figure 6.11 for Re $=$ 20 900 where Re is based on the height of the channel and the bulk velocity. The wave profile is the dashed curve in the figure. The solid curve is the least-squares fit to the data. Note that the maximum in N is upstream of the crest and in the trough region.

This, to a large extent, reflects the phase shift of the wall shear stress due to viscous effects (see Figure 6.3). In order to predict instability it is necessary to predict the additional shift due to the mass transfer process.

The shift of the profile due to mass transfer is associated with wave-induced variations of the velocity field and of the turbulence. It is found that the first of these effects is not sufficient to produce phase angles of $\pi/2<\theta<3\pi/2$. One needs to consider wave-induced variations of the turbulence. The papers by Thorsness & Hanratty (1979a, b) assume that the turbulent diffusivity equals the turbulent kinematic viscosity, described in Section 6.7.2. Instability of a dissolving surface (Schmidt number 729) is predicted for $3\times10^{-4}<\alpha^+<3\times10^{-3}$, in rough agreement with the measurements of Thorsness & Hanratty (1979a).

Ashton & Kennedy (1972) and Hsu *et al.* (1979) measured the wavelength and velocity of small-amplitude waves at a water–ice interface. These covered a range of water velocities of 9.1 cm/s to 83.2 cm/s. Wavelengths were found to decrease with increasing velocity from 106.4 cm to 13.1 cm. These are represented quite well by $\lambda v^*/\upsilon = 3180$. This corresponds to a value of $\alpha^+ = 0.002$, in good agreement with a calculation for maximum growth (Hanratty, 1981). Furthermore, the predicted wave velocity also agrees with measurements. These ice waves are controlled by heat transfer from the water to the ice. The process is characterized by a Prandtl number of 13.7.

References

Abrams, J. & Hanratty, T.J. 1985 Relaxation effects observed for turbulent flow over a wavy surface. *J. Fluid Mech.* 151, 443–455.

Ashton, G.D. & Kennedy, J.F. 1972 Ripples on the underside of river ice covers. *J. Hydraulics Div., Proc. ASCE* 96, 1603–1624.

Beebe, P.S. 1972 *Turbulent flow over a wavy boundary.* Ph.D. thesis, Department of Civil Engineering, Colorado State University.

Benjamin, T.B. 1959 Shearing flow over a wavy boundary. *J. Fluid Mech.* 6, 161–205.

Blumberg, P.M. & Curl, R.L. 1974 Experimental and theoretical studies of dissolution roughness. *J. Fluid. Mech.* 65, 735–751.

Buckles, J.J., Adrian, R.J. & Hanratty, T.J. 1984 Turbulent flow over large-amplitude wavy surfaces. *J. Fluid Mech.* 140, 47.

Caponi, E.A., Fornberg, B., Knight, D.D., McLean, J.W., Saffman, P.G. & Yuen, Y. 1982 Calculation of laminar viscous flow over a moving wavy surface. *J. Fluid Mech.* 124, 347–363.

Charles, M.E. & Lilleleht, L.U. 1965 An experimental investigation of stability and interfacial waves in co-current flow of two liquids. *J. Fluid Mech.* 22, 217–224.

Cohen, L.S. 1964 *Interaction between turbulent air and a flowing liquid film.* Ph.D. thesis, Department of Chemical Engineering, University of Illinois.

Cohen, L.S. & Hanratty, T.J. 1965 Generation of waves in the concurrent flow of air and a liquid. *AIChE JL* 11, 138–144.

Cook, G.W. 1970 Turbulent flow over solid wavy surfaces. Ph.D. thesis, Department of Chemical Engineering, University of Illinois.

Craik, A.D.D. 1966 Wind-generated waves in thin liquid films. *J. Fluid Mech.* 26, 369–392.

Curl, R.L. 1966 Scallops and flutes. *Trans. Cave Res. Group, Gt. Britain* 7, 121–160.

Feldman, S. 1957 On the hydrodynamic stability of two viscous, incompressible flluids in parallel uniform shearing motion. *J. Fluid Mech.* 2, 343–370.

Finnicum, D.S. & Hanratty, T.J. 1988 Influence of imposed flow oscillations on turbulence. *Phys-Chem. Hydrodynamics* 10, 585–598.

Frederick, K.A. 1982 Wave generation at a gas–liquid interface. M.Sc. thesis, Department of Chemical Engineering, University of Illinois.

Frederick, K.A. 1986 Velocity measurements for a turbulent non-separated flow over solid waves. Ph.D. thesis, University of Illinois.

Frederick, K.A. & Hanratty, T.J. 1988 Velocity measurements for a turbulent non-separated flow over solid waves. *Exper. Fluids* 6, 477–486.

Gaster, M. 1962 A note on the relation between temporally-increasing and spatially-increasing disturbances in hydrodynamic stability. *J. Fluid Mech.*, 14, 222–224.

Goodchild, M.F. & Ford, D.C. 1971 Analysis of scallop patterns by simulation under controlled conditions. *J. Geol.* 79, 52–62.

Hanratty, T.J. 1981 Stability of surfaces that are dissolving or being formed by convective diffusion. *Ann. Rev. Fluid Mech.* 13, 231–252.

Hanratty, T. J. & Engen, J. M. 1957 Interaction between a turbulent air stream and a moving water surface. *AIChE Jl* 3, 299–304.

Hinze, J.O. 1975 *Turbulence*. New York: McGraw-Hill.

Hsu, S. & Kennedy, J.F. 1971 Turbulent flow over wavy pipes. *J. Fluid Mech.* 47, 481–502.

Hsu, K.S., Locher, F.A. & Kennedy, J.F. 1979 Forced convection heat transfer from irregular melting boundaries, Report for the Iowa Institute of Hydraulic Research, University of Iowa, Iowa City, Iowa.

Hussain, A. K. M. F. & Reynolds, W.C. 1970 The mechanics of a perturbation wave in turbulent shear flow. Thermosciences Division Report RA-6, Stanford University, California.

Jeffreys, H. 1924 On the formation of water waves by wind. *Proc. R. Soc. Lond.* A107, 189–206.

Jeffreys, H. 1925 On the formation of water waves by wind. *Proc. R. Soc. Lond.* A110, 341–347.

Kendall, J.M. 1970 The turbulent boundary-layer over a wall with progressive surface waves. *J. Fluid Mech.* 41, 249–281.

Kim, Y.Y. 1968 Wave motion on viscoelastic fluids. M.Sc. thesis, Department of Chemical Engineering, University of Illinois.

Kuzan, J.D. 1983 Separated flow over a large amplitude wavy surface. M.Sc. thesis, Department of Chemical Engineering, University of Illinois, Urbana.

Long, R.R. 1963 *Engineering Science Mechanics*. Englewood Cliffs, NJ: Prentice Hall.

Loyd, R.J., Moffat, R.J. & Kays, W.M. 1970 The turbulent boundary-layer on a porous plate: An experimental study of the fluid dynamics with strong favorable pressure gradients and blowing. Report HMT13, Thermosciences Division, Department of Mechanical Engineering, Stanford University, California.

Mao, Z. & Hanratty, T.J. 1986 Studies of the wall shear stress in a turbulent pulsating pipe flow. *J. Fluid Mech.* 170, 545–564.

McLean, J.W. 1983 Computation of turbulent flow over a moving wavy surface. *Phys. Fluids* 26, 2065–2073.

Miles, J.W. 1957 On the generation of surface waves by shear flow. *J. Fluid Mech.* 3, 185–204.

Miles, J.W. 1959a On the generation of surface waves by shear flow. Part 2. *J. Fluid Mech.* 6, 568–582.

Miles, J.W. 1959b On the generation of surface waves by shear flows. Part 3 Kelvin–Helmholtz instability. *J. Fluid Mech.* 6, 583–598.

Miles, J.W. 1962a A note on the inviscid Orr–Sommerfeld equation, *J. Fluid Mech.* 13, 427–432.

Miles, J.W. 1962b On the generation of surface waves by shear flows. Part 4. *J. Fluid Mech.* 13, 433–448.

Mitchell, J.E. & Hanratty, T.J. 1966 A study of turbulence at a wall using an electrochemical wall shear stress meter. *J. Fluid Mech.* 26, 199–221.

Phillips, G.M. 1957 On the generation of waves by turbulent wind. *J. Fluid Mech.* 2, 417–445.

Ramaprian, B.R. & Tu, S.W. 1983 Fully-developed periodic turbulent pipe flow. Part 2. The detailed structure of the flow. *J. Fluid Mech.* 137, 59–81.

Reiss, L.P. & Hanratty, T.J. 1963 An experimental study of the unsteady nature of the viscous sublayer. *AIChE Jl* 9, 154–160.

Schlichting, H. 1968 *Boundary-layer Theory*, 6th edn. New York: McGraw-Hill.

Sigal, A. 1970 *An experimental investigation of the turbulent boundary- layer over a wavy wall.* Ph.D. thesis, Department of Aeronautical Engineering, California Institute of Technology.

Son, J.S. & Hanratty, T.J. 1969 Velocity gradients at the wall for flow around a cylinder at Reynolds numbers 5×10^3 to 10^5. *J. Fluid Mech.* 35, 353–365.

Thorsness, C.B. & Hanratty, T.J. 1979a Mass transfer between a flowing fluid and a solid wavy surface. *AIChE Jl* 25, 686–698.

Thorsness, C.B. & Hanratty, T.J. 1979b Stability of dissolving and depositing surfaces. *AIChE Jl* 25, 697–701.

Thorsness, C.B., Morrisroe, P.E. & Hanratty, T.J. 1978 A comparison of linear theory with measurements of the variation of shear stress along a solid wave. *Chem. Eng. Sci.* 33, 579–592.

Uchida, S. 1956 The pulsating viscous flow superposed on the steady laminar motion of incompressible fluid in a circular pipe. *Z. angew. Math. Phys.* 7, 403–431.

Zilker, D.P. 1976 Flow over wavy surfaces. Ph.D. thesis, Department of Chemical Engineering, University of Illinois.

7 Large-wavelength waves; integral equations

7.1 Prologue

As shown in Section 4.4, a shallow-wave assumption can be made when the height of the liquid is small compared with the wavelength, that is, $k\bar{h}_L$ is a small number. Under this condition, velocity fluctuations associated with wave motion extend throughout the liquid layer; shear stresses at the interface or at the wall are directly felt over the entire liquid layer. This chapter considers such flows. Integral equation (7.7), which describes the variation of the average velocity, is developed.

The principal application of this equation is the development of (7.11) and (7.12), which describe the velocity field associated with the propagation of large-wavelength waves at the interface. These show that the wave behavior is closely related to the wave-induced variation of the shear stress at the wall. A pseudo-steady state is assumed to capture this relation, equation (7.25).

An examination of laminar free flow on inclined planes provides an opportunity to test the accuracy of (7.11) and (7.12) against analytical solutions. Periodic roll waves are considered for free turbulent flows down inclined planes. Equations (7.11) and (7.12) are used to predict the velocity of these waves and the channel angle required for their appearance. These periodic roll waves can be modeled by considering their fronts to be hydraulic jumps.

Intermittent waves with steep fronts are also observed at the interface of gas–liquid layers in horizontal channels that are moved along by interfacial stresses. These are modeled by applying equations (7.11) and (7.12) to the gas phase. That is, the gas space is small enough compared with the wavelength for a large-wavelength analysis to be applied to the disturbed gas flow. The critical gas velocity is correctly predicted at large liquid flows.

These flow surges in the liquid layer have been called roll waves because their appearance is similar to what has been observed in flows down inclined planes. However, they differ because their fronts are not hydraulic jumps.

Ripples with steep fronts are always observed on the wall layers associated with annular flows. These are also analyzed by using equations (7.11) and (7.12). This is an interesting case for which the film is so thin that the use of a large-wavelength analysis requires a consideration of the effect of surface tension.

7.2 Integral equation used to analyze the behavior of liquid layers

7.2.1 Approach

Detailed calculations of the velocity field in the liquid are not given. Instead, the spatially averaged velocity, u_a, is considered. Equations for conservation of mass and conservation of momentum, which define the spatial and temporal variation of u_a, are developed in (7.3), (7.5). These are used to derive (7.11) and (7.12), which define the variation of u_a associated with the appearance of sinusoidal waves at the interface. These two equations define the real and imaginary parts of the complex wave velocity, $c = c_R + ic_I$. The conditions for which $c_I = 0$ define the initiation of waves.

7.2.2 Dispersion relation

The two-dimensional gas–liquid flow, sketched in Figure 7.1, is considered. An average velocity in the liquid layer is defined as

$$u_a = \frac{1}{h}\int_0^h u\,dy \tag{7.1}$$

where, for simplification, h is used instead of h_L. Conservation of mass can be expressed as

$$\frac{\partial}{\partial x}(hu_a) + \frac{\partial h}{\partial t} = 0 \tag{7.2}$$

or as

$$h\frac{\partial u_a}{\partial x} + u_a\frac{\partial h}{\partial x} + \frac{\partial h}{\partial t} = 0 \tag{7.3}$$

If the wavelengths of the interfacial disturbances are large compared with the height of the liquid, the shallow-liquid assumption can be made whereby pressure changes in the

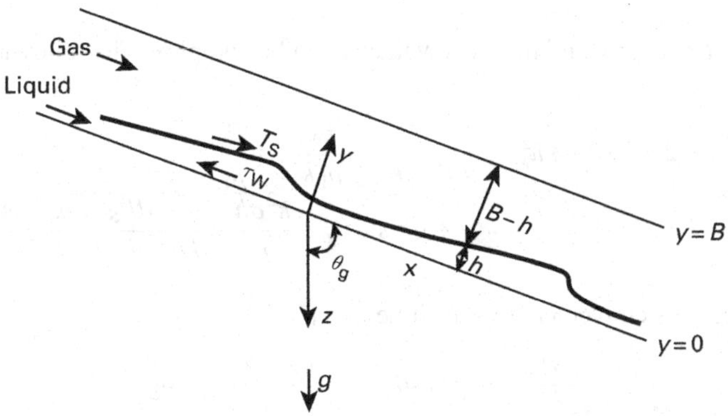

Figure 7.1 Sketch of flow system.

y-direction vary only because of hydrostatic head. Therefore, the pressure in the layer is given as

$$p = P_S + (h - y)\rho_L\, g \sin\theta_g - \sigma\frac{\partial^2 h}{\partial x^2} \tag{7.4}$$

where θ_g is the angle between the flow direction and the direction of gravity, P_S is the pressure at the surface and the last term represents the change of pressure across the interface because of surface tension effects. A formal derivation of this result can be found in a paper by Alekseenko *et al.* (1985), which also presents a justification for using integral methods to calculate waves on thin films.

The momentum balance equation can be written as

$$\frac{\partial}{\partial t}\int_0^h u\,dy + \frac{\partial}{\partial x}\int_0^h u^2\,dy = \frac{T_S}{\rho_L} - \frac{\tau_W}{\rho_L} + gh\cos\theta_g - \frac{h}{\rho_L}\frac{\partial P_S}{\partial_x} + gh\sin\theta_g\frac{\partial h}{\partial x} + \frac{\sigma}{\rho_L}\frac{\partial^2 h}{\partial x^2} \tag{7.5}$$

Conservation of mass, equation (7.3), is added to (7.5) and a velocity profile shape parameter is defined as

$$\Gamma_S = \frac{1}{h\overline{u}_a^2}\int_0^h u^2\,dy \tag{7.6}$$

The following equation is obtained:

$$\begin{aligned}&\frac{\partial u_a}{\partial t} + (2\Gamma_S - 1)u_a\frac{\partial u_a}{\partial x} + (\Gamma_S - 1)\frac{u_a^2}{h}\frac{\partial h}{\partial x} + u_a^2\frac{\partial \Gamma_S}{\partial x}\\ &= \frac{T_S}{h\rho_L} - \frac{\tau_W}{h\rho_L} + g\cos\theta_g - \frac{1}{\rho_L}\frac{\partial P_S}{\partial x} - g\sin\theta_g\frac{\partial h}{\partial x} + \frac{\sigma}{\rho_L}\frac{\partial^2 h}{\partial x^2}\end{aligned} \tag{7.7}$$

Represent all the terms in (7.7) by the sum of an average and a fluctuating component introduced by waves at the interface. The fluctuations are small enough for quadratic terms to be ignored. They are defined by a wave equation

$$h' = \frac{u'}{\tilde{u}_a} = \frac{\Gamma'}{\tilde{\Gamma}} = \frac{P_S'}{\tilde{P}_S} = \frac{T_S'}{\tilde{T}_S} = \frac{\tau_W'}{\tilde{\tau}_W} = a\exp ik(x - ct) \tag{7.8}$$

where $\tilde{u}_a, \tilde{\Gamma}_S, \tilde{P}_S, \tilde{T}_S, \tilde{\tau}_W$ and the wave velocity can be complex. The following equation is obtained:

$$\begin{aligned}c^2 + \overline{u}_a^2\overline{\Gamma}_S - 2\overline{u}_a\overline{\Gamma}_S c - h\overline{u}_a^2\tilde{\Gamma}_S &= i\frac{\tilde{T}_S}{\rho_L k} - i\frac{\tilde{\tau}_W}{\rho_L k} + \overline{h}\frac{\tilde{P}_S}{\rho_L}\\ &\quad + g\overline{h}\sin\theta_g + \frac{k^2\sigma\overline{h}}{\rho_L} - \frac{i}{k\rho_L}\frac{d\overline{P}_S}{dx} + \frac{ig\cos\theta_g}{k}\end{aligned} \tag{7.9}$$

In some cases, it is convenient to substitute into (7.7)

$$-\frac{\partial\overline{P}_S}{\partial x} + \rho_L g\cos\theta_g = -\frac{\overline{T}_S - \overline{\tau}_W}{\overline{h}} = -\hat{g} \tag{7.10}$$

where $\hat{g}$ is defined in equation (3.20) (see Asali & Hanratty, 1993; Craik, 1968). The quantities on the left side of (7.9) are the inertia terms.

For plug flow in the film, $\overline{\Gamma}_{\mathrm{S}}=1$ and $\tilde{\Gamma}_{\mathrm{S}}=0$, so the inertia terms are equal to $(\overline{u}_{\mathrm{a}}-c)^2$ and, consequently, are always destabilizing. For $\overline{u}_{\mathrm{a}}=c$, the effects of inertia vanish for a plug flow. If the velocity varies linearly with distance from the wall, $\overline{\Gamma}_{\mathrm{S}}=4/3$. Then inertia effects vanish if the wave velocity equals the liquid velocity at the interface, $2\overline{u}_{\mathrm{a}}$.

The real and imaginary parts of (7.9) are, respectively,

$$-c_{\mathrm{I}}^2+c_{\mathrm{R}}^2-2\overline{\Gamma}_{\mathrm{S}}c_{\mathrm{R}}\overline{u}_{\mathrm{a}}+\overline{\Gamma}_{\mathrm{S}}\overline{u}_{\mathrm{a}}^2-\overline{h}\overline{u}_{\mathrm{a}}^2\tilde{\Gamma}_{\mathrm{SR}}=\frac{\tilde{T}_{\mathrm{SI}}}{\rho_{\mathrm{L}}k}+\frac{\tilde{\tau}_{\mathrm{WI}}}{\rho_{\mathrm{L}}k}+\frac{k\overline{h}\tilde{P}_{\mathrm{SR}}}{\rho_{\mathrm{L}}k}+g\overline{h}\sin\theta_{\mathrm{g}}+\frac{k^2\sigma\overline{h}}{\rho_{\mathrm{L}}} \tag{7.11}$$

$$-\overline{h}\overline{u}_{\mathrm{a}}^2\tilde{\Gamma}_{\mathrm{SI}}+2c_{\mathrm{I}}\left(c_{\mathrm{R}}-\overline{\Gamma}_{\mathrm{S}}\overline{u}_{\mathrm{a}}\right)=\frac{\tilde{T}_{\mathrm{SR}}}{\rho_{\mathrm{L}}k}-\frac{\tilde{\tau}_{\mathrm{WR}}}{\rho_{\mathrm{L}}k}-\frac{1}{\rho_{\mathrm{L}}k}\frac{d\overline{P}_{\mathrm{S}}}{dx}+\frac{g\cos\theta_{\mathrm{g}}}{k}+\frac{k\overline{h}\tilde{P}_{\mathrm{SI}}}{\rho_{\mathrm{L}}k} \tag{7.12}$$

The term kc_{I} represents the growth rate of unstable waves and $c_{\mathrm{I}}=0$ defines neutral stability. Since $\tilde{\tau}_{\mathrm{WR}}$ is strongly related to c_{R}, equation (7.12) may be viewed as defining c_{R} under neutral stability conditions. This velocity is the kinematic wave velocity defined by Lighthill & Whitham (1955). In this context, (7.11) with $c_{\mathrm{I}}=0$ defines the dynamic conditions needed for neutral stability.

Note that in (7.11) and (7.12), as $k\overline{h}$ becomes smaller, for a fixed gas velocity, $\tilde{T}_{\mathrm{SR}}$ and $\tilde{T}_{\mathrm{SI}}$ become more important relative to $\tilde{P}_{\mathrm{SR}}$ and $\tilde{P}_{\mathrm{SI}}$. The reason for this, as pointed out by Cohen & Hanratty (1965), is that pressure and shear stress fluctuations in the gas, feed energy into the film through normal velocity fluctuations and tangential velocity fluctuations in the liquid, respectively. As $k\overline{h}\to 0$, the ratio of the normal to the tangential velocity fluctuations in the liquid approaches zero. Waves appearing under these circumstances are kinematic (see Section 7.4) and called slow waves by Craik (1966).

Non-zero values of c_{I} indicate growing or decaying waves. In order to calculate the fastest-growing wave it is necessary to develop relations for $\tilde{T}_{\mathrm{S}}$, $\tilde{P}_{\mathrm{S}}$, $\tilde{\tau}_{\mathrm{W}}$, and $\tilde{\Gamma}_{\mathrm{S}}$. Terms $\tilde{\tau}_{\mathrm{W}}$ and $\tilde{\Gamma}_{\mathrm{S}}$ are evaluated by using a pseudo-steady state approximation.

7.2.3 Relation between the wall shear stress and u_{a}

Chapter 3 on film flows provides relations between the volumetric flow and the wall shear stress for both laminar and turbulent flows. For laminar flow, a solution of the Navier–Stokes equations gives

$$\tau_{\mathrm{W}}=\frac{2\mu_{\mathrm{L}}u_{\mathrm{a}}}{h}-\frac{h\hat{g}}{3} \tag{7.13}$$

$$\Gamma_{\mathrm{S}}=\frac{4}{3}+\frac{1}{270}\frac{\hat{g}^2h^4}{u_{\mathrm{a}}^2\mu_{\mathrm{L}}^2}+\frac{1}{18}\frac{h^2\hat{g}}{\mu_{\mathrm{L}}u_{\mathrm{a}}} \tag{7.14}$$

$$\hat{g}=\frac{dP_{\mathrm{S}}}{dx}-\rho_{\mathrm{L}}g\cos\theta_{\mathrm{g}}=\frac{T_{\mathrm{S}}-\tau_{\mathrm{W}}}{h} \tag{7.15}$$

From Chapter 3, the following relation can be used for a sheared turbulent flow:

$$\frac{u}{v_c^*} = f\left(\frac{yv_c^*}{\upsilon_L}\right) \tag{7.16}$$

with

$$v_c^* = \left(\frac{\tau_c}{\rho_L}\right)^{1/2} \tag{7.17}$$

$$\tau_c = \frac{2}{3}\tau_W + \frac{1}{3}\tau_S + \frac{1}{3}\hat{g} \tag{7.18}$$

The integration of (7.16) from 0 to h^+ yields

$$\frac{hv_c^*}{\upsilon_L} = \gamma(\mathrm{Re_L}) \tag{7.19}$$

$$\gamma = \left[\left(1.414\mathrm{Re_L^{0.5}}\right)^{2.5} + \left(0.132\mathrm{Re_L^{0.9}}\right)^{2.5}\right]^{0.4} \tag{7.20}$$

where $\mathrm{Re_L} = 4\Gamma\rho_L/\mu_L$ and Γ is the volumetric flow per unit spanwise length. Hanratty & Hershman (1961) show that equations (7.16–7.20) give

$$\tau_W = \frac{2\mu_L u_a}{h} B(\mathrm{Re_L}) - \frac{1}{3}h\hat{g} \tag{7.21}$$

$$B(\mathrm{Re_L}) = \frac{\gamma^2(\mathrm{Re_L})}{2(\mathrm{Re_L})} \tag{7.22}$$

For a laminar flow, $B(\mathrm{Re_L}) = 1$. From (7.6), the following equation for $\Gamma_S(y^+)$ is obtained:

$$\Gamma_S = \frac{h^+}{\mathrm{Re_L^2}}\int_0^{h^+} f^2(y^+)dy^+ \tag{7.23}$$

The function $f(y^+)$ has been evaluated by Henstock & Hanratty (1976) by using the van Driest mixing-length model with $A = 26$, $\kappa = 0.4$. The calculated dependence of h^+, B and Γ_S on $\mathrm{Re_L}$ is given by equations (7.19), (7.20) and Figures 7.2, 7.3, 7.4.

Equations for $\tilde{\tau}_W$ and $\tilde{\Gamma}$ are obtained by assuming that the relations of τ_W and Γ_S to local values of u_a and h are the same as would exist for an undisturbed flow (the pseudo-steady-state approximation). The sum of a time-average and a fluctuating component is substituted for the variables in (7.21), (7.22) and (7.23). The fluctuations are represented by (7.8) and $\tilde{u}_a$ is eliminated with

$$\overline{h}\tilde{u}_a = c - \overline{u}_a \tag{7.24}$$

obtained from conservation of mass, equation (7.3). Thus,

$$\tilde{\tau}_{WR} = \frac{3\mu_L\overline{u}_a B(\mathrm{Re_L})}{\overline{h}^2}\left[\frac{c}{\overline{u}_a} - 2 + \frac{c}{\overline{u}_a}\frac{\mathrm{Re_L}}{\overline{B}}\frac{d\overline{B}}{d\,\mathrm{Re_L}}\right] - \frac{1}{2}\tilde{\tau}_{SR} \tag{7.25}$$

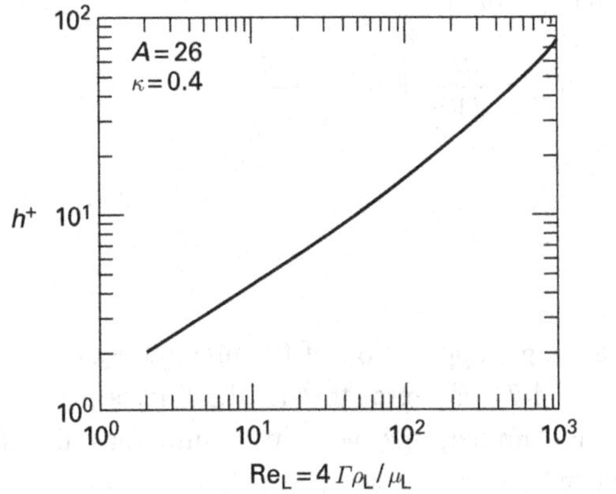

Figure 7.2 Calculation of film height. Henstock & Hanratty, 1976.

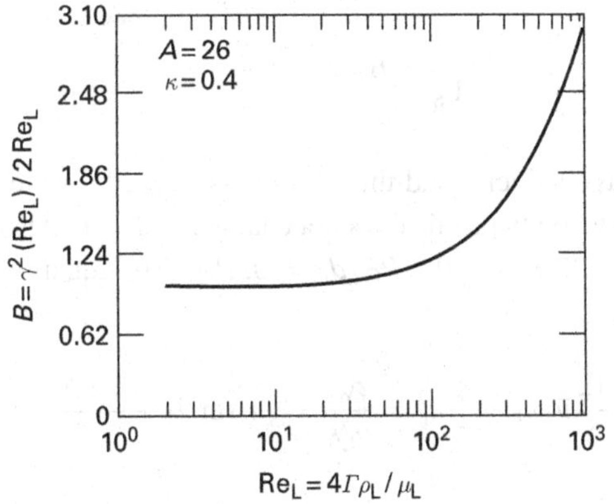

Figure 7.3 Calculation of the wall shear parameter as a function of the liquid film Reynolds number. Henstock & Hanratty, 1976.

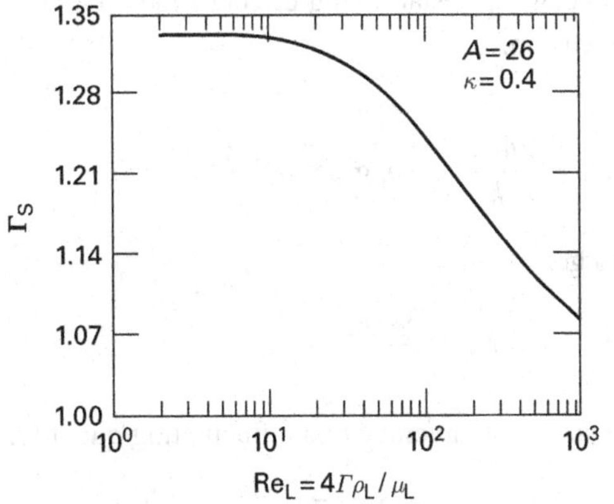

Figure 7.4 Calculation of the velocity profile shape parameter as a function of the liquid film Reynolds number. Henstock & Hanratty, 1976.

where $\tilde{\tau}_{\mathrm{WI}} = 0$, $\tilde{T}_{\mathrm{SI}} = 0$. Similarly, from (7.23),

$$\tilde{\Gamma}_{\mathrm{S}} = \frac{d\Gamma_{\mathrm{S}}}{d\,\mathrm{Re}_{\mathrm{L}}} \mathrm{Re}_{\mathrm{L}} \frac{c}{\overline{u}_{\mathrm{a}} - \overline{h}} \tag{7.26}$$

7.3 Laminar free flow

Laminar free flow provides a simple application of the integral equations (7.11) and (7.12). The pressure gradient and T_{S} are zero; that is, the film is pulled along only by gravity. Since the flow is laminar, the velocity profile has the shape of a parabola. From a force balance,

$$\tau_{\mathrm{W}} = \rho_{\mathrm{L}} h g \cos\theta_{\mathrm{g}} \tag{7.27}$$

and from (7.6)

$$\Gamma_{\mathrm{S}} = \frac{6}{5} \tag{7.28}$$

The goal is to predict the wave velocity and the conditions needed for the initiation of waves. Since the shape of the velocity profile does not change, $\tilde{\Gamma}_{\mathrm{SR}} = 0$. Also, since there is no gas flow, $\tilde{T}_{\mathrm{SI}} = 0$, $\tilde{T}_{\mathrm{SR}} = 0$, $\tilde{P}_{\mathrm{SR}} = 0$, $d\overline{P}_{\mathrm{G}}/dx = 0$, $\tilde{P}_{\mathrm{SI}} = 0$. Equations (7.11) and (7.12) can then be written as

$$-c_{\mathrm{I}}^2 + c_{\mathrm{R}}^2 - \frac{12}{5}\overline{u}_{\mathrm{a}}\, c_{\mathrm{R}} + \frac{6}{5}\overline{u}_{\mathrm{a}}^2 = \frac{\tilde{\tau}_{\mathrm{WI}}}{\rho_{\mathrm{L}} k} + g\overline{h}\,\sin\theta_{\mathrm{g}} + \frac{k^2\sigma\overline{h}}{\rho_{\mathrm{L}}} \tag{7.29}$$

$$2c_{\mathrm{I}}\left(c_{\mathrm{R}} - \frac{6}{5}\overline{u}_{\mathrm{a}}\right) = -\frac{\tilde{\tau}_{\mathrm{WR}}}{k\rho_{\mathrm{L}}} + \frac{g\,\cos\theta_{\mathrm{g}}}{k} \tag{7.30}$$

For neutral stability, $c_{\mathrm{I}} = 0$. Equation (7.30) defines the wave velocity, c_{R}. Equation (7.29) relates c_{I} to the imbalance between the destabilizing effect of inertia and the stabilizing effects of gravity and surface tension.

From (7.13) and (7.15),

$$\tau_{\mathrm{W}} = \frac{2\mu_{\mathrm{L}} u_{\mathrm{a}}}{h} + \left(\rho_{\mathrm{L}} g\,\cos\theta_{\mathrm{g}}\right)\frac{h}{3} \tag{7.31}$$

Substitute (7.27) into (7.31) to get

$$\tau_{\mathrm{W}} = \frac{3\mu_{\mathrm{L}} u_{\mathrm{a}}}{h} \tag{7.32}$$

Represent the variables as the sum of an average and a fluctuating quantity:

$$\tau_{\mathrm{W}} = \overline{\tau}_{\mathrm{W}} + \tau'_{\mathrm{W}} = \frac{3\mu_{\mathrm{L}}(\overline{u}_{\mathrm{a}} + u'_{\mathrm{a}})}{\overline{h} + h'} \tag{7.33}$$

Multiply the numerator and denominator by $(\overline{h} - h')$ and neglect second-order terms in the fluctuations:

$$\tau'_{W} = \frac{3\mu_L u'_a}{\overline{h}} - \frac{3\mu_L h' \overline{u}_a}{\overline{h}^2} \tag{7.34}$$

Represent the disturbance by (7.8), so that

$$a\tilde{\tau}_W = \frac{3a\mu_L \tilde{u}_a}{\overline{h}} - \frac{3\mu_L a \overline{u}_a}{\overline{h}^2} \tag{7.35}$$

From conservation of mass, equation (7.3),

$$\tilde{u}_a = \frac{c - \overline{u}_a}{\overline{h}} \tag{7.36}$$

This is substituted into (7.35), to give

$$\tilde{\tau}_{WR} = \frac{3(c_R - \overline{u}_a)\mu_L}{\overline{h}^2} - \frac{3\mu_L \overline{u}_a}{\overline{h}^2} \tag{7.37a}$$

Also, from (7.35) and (7.36),

$$\tilde{\tau}_{WI} = 0 \tag{7.37b}$$

since, at neutral stability, $c_I = 0$. Thus, (7.30) and (7.37a) give

$$0 = -\frac{3(c_R - \overline{u}_a)\mu_L}{\overline{h}^2 k \rho_L} + \frac{3\mu_L \overline{u}_a}{\overline{h}^2 k \rho_L} + \frac{g \cos\theta_g}{k} \tag{7.38}$$

at neutral stability. Eliminate $g \cos\theta_g$ with (7.27). The following equation is obtained by using (7.32) to eliminate τ_W:

$$\frac{c_R}{\overline{u}_a} = 3 \tag{7.39}$$

By substituting (7.39) and (7.37b) into (7.29), one obtains the following criterion for the initiation of an instability ($c_I = 0$):

$$\frac{g\overline{h} \sin\theta_g}{\overline{u}_a^2} = 3 - \frac{k^2 \sigma \overline{h}}{\rho_L \overline{u}_a^2} \tag{7.40}$$

Equation (7.40) defines the angle of the plane above which an instability will occur for a given liquid velocity.

Benjamin (1957) presented a more rigorous analysis of free laminar flow down an inclined plane by solving the Orr-Sommerfeld equation (6.19) with a series solution in y (instead of using an integral approach):

$$\psi = A_0 + A_1 y + A_2 y^2 + A_3 y^3 + \cdots \tag{7.41}$$

where ψ is the stream function defined for a two-dimensional flow by using conservation of mass: $u_x = -\partial\psi/\partial y$, $u_y = \partial\psi/\partial x$. This solution is valid for laminar flow for which the velocity profile is given by a power series. For large $\left(k\overline{h}\right)\left(\overline{h}\overline{u}_a\rho_L/\mu_L\right)$, it is not useful since it does not converge rapidly enough. Benjamin considered only four terms. The constants A_i are determined using the kinematic relation at the interface, the shear stress condition at the interface, and the conditions $\psi = 0$, $\partial\psi/\partial y = 0$ at the lower wall (see equation (6.20)). The solution is then substituted into the normal stress condition to obtain $c = f(k)$. The constants A_0 to A_3 are of order $(k\,\mathrm{Re}_L)$. The constants A_4 to A_7 are of order $(k\,\mathrm{Re}_L)^2$. Therefore, if the series is truncated at y^3, the solution ignores terms of order $(k\,\mathrm{Re}_L)^3$. For small $k\overline{h}$, the Benjamin analysis gives

$$\frac{c_R}{\overline{u}_a} = 3 \tag{7.42}$$

$$\frac{g\overline{h}\sin\theta_g}{\overline{u}_a^2} = \frac{18}{5} - \frac{k^2\sigma\overline{h}}{\rho_L\overline{u}_a^2} \tag{7.43}$$

These compare favorably with the results from the integral analysis, equations (7.39) and (7.40).

For a vertical wall, $\sin\theta_g = 0$ so (7.40) gives the following stability conditions:

$$\frac{k^2\sigma\overline{h}}{\rho_L\overline{u}_a^2} = 3 \tag{7.44}$$

or

$$\left(k\overline{h}\right)^2 = 3\frac{\rho_L\overline{u}_a^2\overline{h}}{\sigma} \tag{7.45}$$

An increase in $\overline{u}_a$ is predicted to cause a decrease in the wavelength at neutral stability. Thus, as pointed out by Benjamin (1957), a critical $\overline{u}_a$ is not predicted for a very long flow on a vertical wall. Binnie (1957) studied the initiation of waves on a water film flowing down the outside of a vertical tube. He found that transition occurs at Re = 4.4 for waves with $c_R/\overline{u}_a = 3.5$ and $\lambda/\overline{h} = 102$. For Re = 4.4 equation (7.45) predicts $\lambda/\overline{h} = 74$. Thus, approximate agreement is realized between the integral analysis and laboratory observations for vertical films. (See Hanratty & Hershman (1961) for a discussion of these results.)

7.4 Kinematic waves

Lighthill & Whitham (1955) and Whitham (1974) have demonstrated that the concept of a kinematic wave is useful in analyzing many phenomena. As indicated by the name, kinematic waves are different from classical wave motions (discussed in Chapter 4) encountered in dynamical systems, in that they are defined from the equation of conservation of mass. From (7.2)

$$\frac{\partial q}{\partial x} + \frac{\partial h}{\partial t} = 0 \tag{7.46}$$

where $q = h\overline{u}_a$ is the volumetric flow per unit breadth. Suppose a pseudo-steady-state approximation can be made where

$$q = f(h) \tag{7.47}$$

which is of the same form as exists under steady conditions

$$\overline{q} = f(\overline{h}) \tag{7.48}$$

Then, from (7.46),

$$\frac{\partial q}{\partial h}\frac{\partial h}{\partial x} + \frac{\partial h}{\partial t} = 0 \tag{7.49}$$

For a disturbance propagating at a velocity c

$$c\frac{\partial h}{\partial x} + \frac{\partial h}{\partial t} = 0 \tag{7.50}$$

Therefore the equating of (7.49) and (7.50) gives

$$c_{\mathrm{KW}} = \frac{\partial q}{\partial h} \tag{7.51}$$

For laminar flow down an inclined plane, the averaged volumetric flow and the average velocity are

$$\overline{q} = \frac{g\overline{h}^3 \cos\theta_{\mathrm{g}}}{3(\mu_{\mathrm{L}}/\rho_{\mathrm{L}})} \tag{7.52}$$

$$\overline{u}_{\mathrm{a}} = \frac{\rho_{\mathrm{L}} g\overline{h}^2 \cos\theta_{\mathrm{g}}}{3\mu_{\mathrm{L}}} \tag{7.53}$$

where θ_{g} is defined in Figure 7.1. Make a pseudo-steady-state assumption. From (7.52),

$$q = \frac{gh^3 \cos\theta_{\mathrm{g}}}{3(\mu_{\mathrm{L}}/\rho_{\mathrm{L}})} \tag{7.54}$$

$$c_{\mathrm{KW}} = \frac{dq}{dh} = \frac{3gh^2 \cos\theta_{\mathrm{g}}}{3(\mu_{\mathrm{L}}/\rho_{\mathrm{L}})} = 3\overline{u}_{\mathrm{a}} \tag{7.55}$$

where $\overline{u}_{\mathrm{a}} = q/\overline{h}$ is given by (7.53). The result of a stability analysis (7.39) can also be obtained by assuming that c_{R} is the kinematic wave velocity. Thus, Lighthill & Whitham point out that neutral stability can be defined as the condition for which

$$c_{\mathrm{KW}} = c \tag{7.56}$$

where c is the dynamic wave velocity. For free laminar flow this is given by (7.29) with $c_{\mathrm{I}} = 0$ and $\tilde{\tau}_{\mathrm{WI}} = 0$.

7.5 Roll waves

7.5.1 Outline

Roll waves have been discussed in Chapter 2. Equations (7.11) and (7.12) are now used to describe these waves for free turbulent flows down inclined planes and for gas–liquid flows.

7.5.2 Free turbulent flows on an inclined plane

Roll waves are observed for turbulent flows down inclined planes, such as spillways of dams and run-off channels. They are periodic and give the appearance of rollers that are observed at a beach. (However, as pointed out by Dressler (1949, 1952), their underlying mechanism is quite different, since they depend critically on the wall resistance.) A famous photograph of roll waves presented by Cornish (1934) is reproduced in Figure 7.5 and in Figure 11.33 of *Water Waves* by Stoker (1957). These waves are believed to evolve from large-wavelength regular waves. Thus, a prediction of their appearance can be obtained by considering the stability of a free

Figure 7.5 Roll waves looking downstream. Cornish, 1934.

turbulent flow on an inclined plane. The following equations are obtained from (7.11) and (7.12), if $\Gamma_S = 1$:

$$-c_I^2 + (c_R - \overline{u}_a)^2 = \frac{\tilde{\tau}_{WI}}{\rho_L k} + g\overline{h} \sin\theta_g \tag{7.57}$$

$$2c_I(c_R - \overline{u}_a) = -\frac{\tilde{\tau}_{WR}}{\rho_L k} + \frac{g \cos\theta_g}{k} \tag{7.58}$$

Here, the influence of surface tension is ignored because very large wavelengths are considered. Also the influence of the gas flow on the liquid flow can be ignored, so $\tilde{T}_{SI} = \tilde{T}_{SR} = \tilde{P}_{SR} = \tilde{P}_{SI} = \partial P/\partial x = 0$.

The wall shear stress for a smooth plane can be represented by

$$\tau_W = \frac{1}{2}\rho_L f_W u_a^2 \tag{7.59}$$

$$f_W = 0.046\ \mathrm{Re}_L^{0.2} \tag{7.60}$$

However, for the applications cited above, the bottom boundary may be considered to be completely rough. Thus, the friction factor varies as the ratio of the length scale, k_S, characterizing the roughness, to the height of the liquid

$$f_W \propto \frac{k_S}{h} \tag{7.61}$$

Thus,

$$\tau_W = \frac{1}{2}\frac{\rho_L u_a^2 k_S}{h} \tag{7.62}$$

Substitute for τ_W and u_a

$$\tau_W = \overline{\tau}_W + \tau'_W \quad u_a = \overline{u}_a + u'_a, \quad \text{where } \tau'_W = a\tilde{\tau}_W \tag{7.63}$$

From conservation of mass, equation (7.3),

$$u'_a = \tilde{u}_a a = \frac{a(c - \overline{u}_a)}{\overline{h}} \tag{7.64}$$

Neglect second-order terms in the fluctuating quantities and assume neutral stability, $c_I = 0$. From (7.62), (7.63) and (7.64), the following equation is obtained

$$\tilde{\tau}_W = \frac{1}{2}\frac{2\rho_L k_S}{\overline{h}}\frac{\overline{u}_a c_R}{\overline{h}} - \frac{\rho_L k_S}{\overline{h}^2}\frac{\overline{u}_a^2}{2} - \frac{1}{2}\frac{k_S}{\overline{h}}\rho_L\frac{2\overline{u}_a^2}{\overline{h}} \tag{7.65}$$

From (7.58), since $c_I = 0$ for neutral stability

$$0 = -\frac{\tilde{\tau}_{WR}}{k\rho_L} + \frac{g \cos\theta_g}{k} \tag{7.66}$$

Substitute (7.65) into (7.66). From a force balance, $g\cos\theta_g/k = \overline{\tau}_W/k\rho_L\overline{h}$. From (7.62), $\overline{\tau}_W = (1/2)\rho_L f_W \overline{u}_a^2$. Equation (7.66) provides the result

$$\frac{c_R}{\overline{u}_a} = 2 \tag{7.67}$$

This agrees with measurements made by Cornish on the waves shown in Figure 7.5. From (7.57), with $\tilde{\tau}_{WI} = 0$ and c_R given by (7.67), the following condition is obtained for neutral stability:

$$\frac{g\overline{h}}{\overline{u}_a^2}\sin\theta_g = 1 \tag{7.68}$$

Equation (7.68) predicts the liquid flow-rate at which roll waves will appear on a declined spillway. This defines the minimum angle at which roll waves will be observed.

7.5.3 Generation of roll waves by a gas flow

Figure 7.6 shows waves observed by Hanratty & Engen (1957) for air and water flowing in a rectangular enclosed channel with a large aspect ratio. Because of the similarity with the free waves described in Section 7.5.2, Engen & Hanratty called these large-wavelength disturbances roll waves. They differ from what is observed for free flows on inclined walls in that the liquid is moved by interfacial stresses associated with the air flow rather than by gravity. The conditions for neutral stability have been calculated by Hanratty & Hershman (1961), by Lin & Hanratty (1986), by Andreussi *et al.* (1985) and by Hanratty (1983).

Neutral stability is defined by (7.11) and (7.12) with $c_I = 0$. Since a fully developed horizontal flow is considered, $\sin\theta_g = 1$, $\cos\theta_g = 0$ and $d\overline{h}/dx = 0$. Thus, if $\overline{\Gamma} \approx 1$ and $\tilde{\Gamma} \approx 0$,

Figure 7.6 Roll observed by Hanratty and Engen for air–water flow in an enclosed rectangular channel. Hanratty & Engen, 1957.

$$(c - \bar{u}_a)^2 = g\bar{h} - \frac{\tilde{T}_{SI}}{\rho_L k} + \frac{\tilde{\tau}_{WI}}{\rho_L k} + \frac{k\bar{h}\tilde{P}_{SR}}{\rho_L k} \tag{7.69}$$

$$0 = \tilde{T}_{SR} - \tilde{\tau}_{WR} - \frac{d\bar{P}_S}{dx} + k\bar{h}\tilde{P}_{SI} \tag{7.70}$$

where surface tension effects are not considered and (7.25) describes $\tilde{\tau}_W$. Equation (7.70) defines the wave velocity and equation (7.69) defines the initiation of waves as occurring when inertia and gas-phase pressure variation overcome the stabilizing effect of gravity.

In the analyses by Hanratty & Hershman (1961) and by Lin & Hanratty (1986), an equation for $\tilde{P}_S$ is obtained by considering a momentum balance for the gas. Because the wavelength is considered to be large compared with the height of the gas space, $B - h$, integral equations of the type used for the liquid layer can be developed. Since $\Gamma_{SG} \approx 1$, conservation of mass and momentum for the gas flow are written as

$$-\frac{\partial h}{\partial t} + \frac{\partial}{\partial x}[(B - h)u_{Ga}] = 0 \tag{7.71}$$

$$\frac{\partial}{\partial t}[(B - h)u_{Ga}] + \frac{\partial}{\partial x}\left[(B - h)u_{Ga}^2\right] = -\frac{(B - h)}{\rho_G}\left(\frac{\partial P_S}{\partial x} + g\sin\theta_g\frac{\partial h}{\partial x}\right) - \frac{1}{\rho_G}(T_S + \tau_B) + g(B - h)\cos\theta_g \tag{7.72}$$

where u_{Ga} is the average velocity of the gas.

The interaction of the gas with the liquid is represented by $\tilde{T}_{SR}$, $\tilde{T}_{SI}$, $\tilde{P}_{SR}$, $\tilde{P}_{SI}$. The pressures $\tilde{P}_{SR}$ and $\tilde{P}_{SI}$ are obtained from (7.71) and (7.72) by using procedures outlined in this chapter to analyze the flow in a liquid layer perturbed according to (7.8). The gas phase is fully turbulent, so it is approximated by a plug flow with velocity u_{Ga}. Gravitational effects can be ignored.

$$\tilde{P}_{SR} = \frac{\rho_G}{(B - \bar{h})}\left[-(\bar{u}_{Ga} - c_R)^2 - \frac{1}{k\rho_G}\left(\tilde{T}_{SI} + \tilde{\tau}_{BI}\right)\right] \tag{7.73}$$

$$\tilde{P}_{SI} = \frac{1}{k(B - \bar{h})}\left[\frac{\overline{T}_S + \overline{\tau}_B}{(B - \bar{h})} + \left(\tilde{T}_{SR} + \tilde{\tau}_{BR}\right)\right] \tag{7.74}$$

Because of the assumption of a large-wavelength disturbance, the disturbed shear stresses at the interface and the top wall can be calculated by using the pseudo-steady-state approximation, whereby the instantaneous stress is related to flow variables by the equation derived for undisturbed flow in the gas:

$$T_S = \frac{1}{2}\rho_G f_S(u_{Ga} - c_R)^2 \tag{7.75}$$

$$\tau_B = \frac{1}{2}\rho_G f_B u_{Ga}^2 \tag{7.76}$$

$$f_B = f_S = 0.042\,\mathrm{Re}_G^{0.2} \tag{7.77}$$

Substitute $T_S = \overline{T}_S + T'_S$, $u_{Ga} = \overline{u}_{Ga} + u'_{Ga}$, $f_S = \overline{f}_S + f'_S$ and neglect terms which are second order in the fluctuations to obtain

$$\tilde{T}_S = \overline{T}_S \left[\frac{\tilde{f}_S}{\overline{f}_S} + 2 \frac{\tilde{u}_{Ga}}{(\overline{u}_{Ga} - c)} \right] \tag{7.78}$$

From conservation of mass

$$\tilde{u}_{Ga} = \frac{(u_{Ga} - c)}{(B - h)} \tag{7.79}$$

So (7.78) can be written as

$$\tilde{T}_{SR} = \overline{T}_S \left[\frac{\tilde{f}_S}{\overline{f}_S} + \frac{2}{(B - \overline{h})} \right] \tag{7.80}$$

$$\tilde{T}_{SI} = 0 \tag{7.81}$$

Also, from (7.76) and (7.79),

$$\tilde{\tau}_{BR} = \overline{\tau}_B \left[2 \frac{(\overline{u}_{Ga} - c)}{(B - \overline{h})\overline{u}_{Ga}} \right] \approx \overline{\tau}_B \left[\frac{2}{(B - \overline{h})} \right] \tag{7.82}$$

$$\tilde{\tau}_{BI} = 0 \tag{7.83}$$

The fluctuations in the friction factor, f'_S, are approximated by Andreussi *et al.* (1985) as

$$\frac{f'_S}{f_S} = \frac{a\tilde{f}_S}{f_S} = \frac{1}{\overline{f}_S} \left(\frac{\partial \overline{f}_S}{\partial \mathrm{Re}_L} \mathrm{Re}'_L + \frac{\partial f_S}{\partial \mathrm{Re}_G} \mathrm{Re}'_G \right) \tag{7.84}$$

From conservation of mass, $\mathrm{Re}'_G = 0$. Also,

$$\frac{\mathrm{Re}'_L}{\overline{\mathrm{Re}_L}} = \frac{c}{\overline{u}_a} \frac{h'}{\overline{h}} \tag{7.85}$$

Thus

$$\frac{\tilde{f}_S}{\overline{f}_S} = \frac{1}{\overline{f}_S} \frac{\partial \overline{f}_S}{\partial \mathrm{Re}_L} \frac{\overline{\mathrm{Re}_L} c}{\overline{u}_a \overline{h}} \tag{7.86}$$

So (7.80) gives, for $c_I = 0$,

$$\tilde{T}_{SR} = \overline{T}_S \left[\frac{2}{(B - h)} + \frac{\overline{\mathrm{Re}_L}}{f_S} \frac{\partial \overline{f}_S}{\partial \mathrm{Re}_L} \frac{c_R}{\overline{u}_a} \frac{1}{\overline{h}} \right] \tag{7.87}$$

The second term on the right side of (7.87) recognizes that the liquid layer is on the bottom of the channel and that the stress at the interface could depend on the flow of the liquid.

Equation (7.25) can be used to represent $\tilde{\tau}_{WR}$. By assuming $\tilde{T}_{SI} = \tilde{\tau}_{WI} = \tilde{\tau}_{BI} = 0$, equation (7.69), for neutral stability, becomes

$$(c_R - \bar{u}_a)^2 = g\bar{h} + \frac{\bar{h}\tilde{P}_{SR}}{\rho_L} \tag{7.88}$$

and (7.73) gives

$$\tilde{P}_{SR} = -\left(\frac{\rho_G}{\bar{B} - \bar{h}}\right)(\bar{u}_{Ga} - c_R)^2 \tag{7.89}$$

Equation (7.89) represents a Bernoulli effect whereby the squeezing of gas-phase streamlines at a wave crest causes an increase in velocity and, consequently, a decrease in gas-phase pressure at the crest. This is a destabilizing effect.

The wave velocity is not given by classical theory developed in Chapter 4 for an inviscid flow. As already mentioned, it is given by the equation for a kinematic wave, whose behavior is, to first order, dependent on the viscous drag at the wall. It is calculated by using (7.72), with $\tilde{T}_{SR}$, $\tilde{\tau}_{WR}$, $\tilde{P}_{SI}$ defined by (7.87), (7.25), (7.74). Equation (7.74) requires the specification of $\bar{\tau}_B$, $\bar{T}_S$, $\tilde{\tau}_{BR}$. These are estimated by using (7.75), (7.76), (7.82).

By substituting (7.89) into (7.88), the following stability condition is obtained:

$$(c_R - \bar{u}_a)^2 = g\bar{h} - \left(\frac{\rho_G}{B - \bar{h}}\right)(u_{Ga} - c_R)^2\frac{\bar{h}}{\rho_L} \tag{7.90}$$

The term on the left side represents the stabilizing of liquid inertia to an interfacial disturbance. The first term on the right side represents the stabilizing effect of gravity. The second term on the right side is the destabilizing effect of the wave induced variation of the gas phase pressure. The numerical solution of (7.90) to obtain the critical gas velocity was pursued by Lin & Hanratty (1986), Hanratty (1983) and by Asali *et al.* (1985). The influence of liquid inertia is not large, so a simplification of the calculation of $(c - u_a)$ is warranted. Thus, Hanratty (1983) used $c_R/u_a = 2$ and compared (7.90) with measurements by Hanratty & Hershman (1961), Woodmansee & Hanratty (1969), Miya *et al.* (1971). This comparison is reproduced in Figure 7.7. Good agreement is noted at large liquid Reynolds numbers. At very small liquid Reynolds numbers, transition is approximately independent of gas velocity and a critical liquid Reynolds number can be defined. Andreussi *et al.* (1985) suggest that it is associated with relaxation phenomena whereby τ_i and τ_W do not respond to changes in film height as is predicted by pseudo-steady-state relations. This explanation has not, however, been substantiated. Another possibility is to assume that a necessary condition is that the wall film needs to be able to sustain turbulence for a disturbance wave to appear. Thus, the film Reynolds number must be larger than a critical value at large gas velocities.

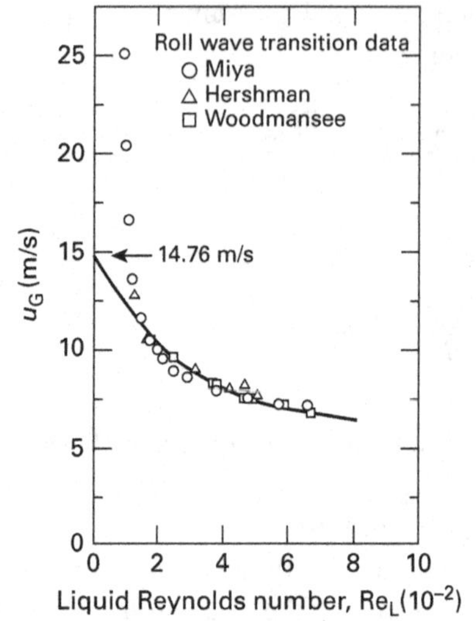

Figure 7.7 Comparison of an observed roll wave transition with calculations using $\hat{\tau}_{SI} = 0$. Meyer, 1983.

7.6 Model for a fully developed roll wave

Dressler (1949) presented a model for fully developed roll waves, in a declined channel, which uses non-linear shallow liquid theory. This work motivated a study for gas–liquid flow by Miya *et al.* (1971).

The equations of conservation of mass (7.2) and momentum (7.5) are written by assuming steady flow in a coordinate system moving with the velocity of the waves, c; that is,

$$X = x - ct \tag{7.91}$$

so that $\partial/\partial t = -c(d/dx)$. The integral form of conservation of mass is

$$h(c - u_a) = q \tag{7.92}$$

where the integration constant is called the progressive discharge by Dressler (1949). From conservation of momentum (7.7), one gets

$$\begin{aligned} &\frac{dh}{dX}\left[c^2 - \left(2\Gamma_S + \mathrm{Re_L}\frac{d\Gamma_S}{d\mathrm{Re_L}}\right)u_a c + \Gamma_S u_a^2 - gh\sin\theta_g\right] \\ &= \frac{1}{\rho_L}(\tau_W - T_S) + \frac{h}{\rho_L}\frac{dP_G}{dX} - gh\cos\theta_g \end{aligned} \tag{7.93}$$

where $\partial u_a/\partial x$ is eliminated by using conservation of mass. For the case analyzed by Dressler, free flow down an inclined wall, $T_S = 0$, $dP_G/dX = 0$, $\Gamma_S = 1$.

Equation (7.93) then simplifies to

$$\frac{dh}{dX}\left[\left(c - u_a^2\right) - gh\,\sin\theta_g\right] = \frac{\tau_W}{\rho_L} - gh\,\cos\theta_g \tag{7.94}$$

where τ_W is given by (7.62). Thus (7.94) is of the form

$$M_1 \frac{dh}{dX} = M_2 \tag{7.95}$$

where M_1 and M_2 are functions of h. This is integrated, numerically, to give the variation of h along the wave, from an initial value, h_0, to a peak value of h_P, where a hydraulic jump returns it to h_0. Thus, the solution is discontinued when

$$\frac{h}{h_0} = \frac{h_P}{h_0} = \frac{1}{2}\left[\left\langle 1 + \frac{8(c-u_0)^2}{gh_0}\right\rangle^{1/2} - 1\right] \tag{7.96}$$

Term M_1 reaches a value of zero at

$$(c-u_a)^2 = gh\sin\theta_g \tag{7.97}$$

Dressler arranged for M_2 to equal 0 at this critical point by selecting parameters for the solution. In this way, the predicted wave profile is continuous.

Gas–liquid flow in a channel is fundamentally different from flow down an inclined plane in that the front of the roll wave is not a hydraulic jump. This is demonstrated in Miya *et al.* (1971) by showing that the measured h_P/h_0 is smaller than predicted by (7.96).

For this flow, $\sin\theta_g = 1$ and $\cos\theta_g = 0$; the pressure gradient in the gas is obtained from (7.72) as

$$\frac{dP_G}{dX} = \frac{-\rho_G(u_{G0}-c)^2(B-h_0)^2}{(B-h)^3}\frac{dh}{dX} - \frac{(T_S+\tau_B)}{(B-h)} = 0 \tag{7.98}$$

where the effect of gravity is ignored because of the low gas density. Again, taking $\Gamma_S = 1$, equations (7.93) and (7.98) give

$$\frac{dh}{dX}\left[(c-u_a)^2 - gh + Gh\right] = \frac{1}{\rho_L}(\tau_W - T_S) - \frac{h}{\rho_G(B-h)}(T_S+\tau_B) \tag{7.99}$$

$$G = \frac{\rho_G(u_{G0}-c)^2(B-h_0)^2}{\rho(B-h)^3} \tag{7.100}$$

The term $G\,(dh/dX)$ represents the pressure gradient needed to balance inertial changes in the gas (a Bernoulli effect). In the situation being considered, it has the same effect as a negative gravitational force. For relatively high gas velocities $(-g+G)$ is positive or has a very small negative value. Therefore, the range of variables of interest is such that a solution of (7.99) is sought for which M_1 is always positive. The sign of the slope then depends on the sign of M_2. The term $[h/\rho_G\,(B-h)](T_S+\tau_B)$ appearing in M_2 is a force on a unit length of the liquid layer due to the gas-phase pressure gradient and is usually small compared with T_S. Therefore, the sign of the slope depends largely on the sign of $(\tau_W - T_S)$. This term is positive in the back of the wave and it becomes negative at the front due to relaxation of the wall shear stress. At $h = h_0$ and at $h = h_P$, $dh/dX = 0$; that is,

$$0 = \frac{1}{\rho_L}(\tau_W - T_S) - \frac{h}{\rho_G(B-h)}(T_S + \tau_B) \tag{7.101}$$

At these locations there is a balance between forces at the interface and at the solid boundary.

A comparison of the above analysis with experiment is given by Miya *et al.* (1971). A critical issue is the modeling of the relaxation of the wall shear stress in the front of the roll wave. Miya *et al.* speculate that the flow laminarizes in this region.

7.7 Ripples generated on a liquid film at high gas velocities

The wave structure at the interface for a high gas velocity (>20 m/s) has received considerable attention because of its importance in understanding interfacial stresses in annular flows (Hewitt & Hall-Taylor, 1970; Asali & Hanratty, 1993; Asali *et al.* 1985; Hanratty, 1991). Waves appear at arbitrarily small liquid flows, both in vertical and horizontal conduits.

The film is covered by long-crested slow-moving ripples, shown in Figure 7.8. These ripples have a steep front and a low ratio of the amplitude (10–20 μm) to wavelength (2–3 mm). Between ripples the surface is smooth and the flow appears laminar. At sufficiently high liquid rates, *disturbance* or *roll* waves appear on the film (see Figure 3.3). These have a much larger velocity than the ripples and very large spacing between successive waves. (They might be described as a collection of ripple waves.)

In this section, an analysis of ripple waves is presented. The goals are to identify the physical processes responsible for their appearance and to predict the characteristic length between them. This length scale is of importance in characterizing interfacial stresses in vertical gas–liquid annular flows. For example, Andritsos & Hanratty (1987) and Bontozoglou & Hanratty (1989) have argued that in separated flows the ratio of the

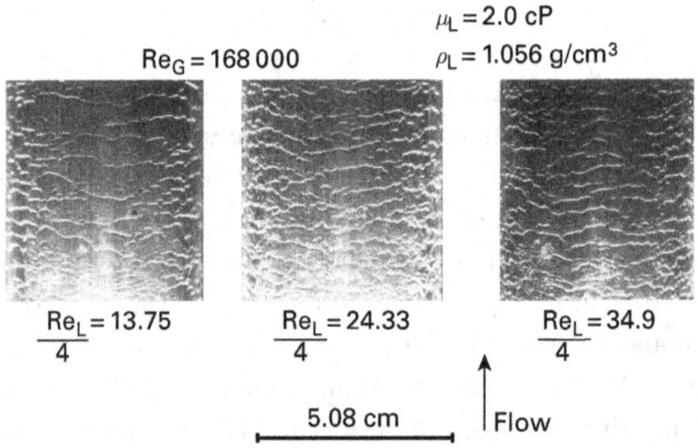

Figure 7.8 Ripple waves on a film with a viscosity of 2 cP. Asali & Hanratty, 1993.

interfacial drag to that for a smooth surface should scale with $\Delta h/\lambda$, where Δh is the wave height and λ is the wavelength.

The approach is to solve the linear momentum equations to determine the growth of small-amplitude two-dimensional wavelike disturbances at the interface. It is argued that the fastest growing wave is the precursor of the ripples (Taylor, 1963).

As mentioned above, the thicknesses of the films on which these ripple waves occur are quite small. For example, at a liquid Reynolds number of 36 and a gas Reynolds number of 78 000, a water film on the wall of a 4.2 cm pipe would have a height of 147 μm. This dictates that the analysis considers wavelengths which are large enough for a shallow-liquid assumption to be made, yet small enough for surface tension effects to be important. Thus, equations (7.9) to (7.12) are applicable. The initial growth of unstable waves is governed by an imbalance between the destabilizing effects of inertia, the component of surface stress in phase with the wave slope, the component of pressure 180° out of phase with the wave height and the stabilizing effect of surface tension.

Calculations for the growth rate of ripples are presented by Asali & Hanratty (1993). The analysis differs from that for roll waves in that surface tension is important, the flow is laminar and the wavelength is small compared with the height of the gas space. The large-wavelength analyses of the gas flow that have been used to calculate the wave-induced variation of the pressure and interfacial stress (7.73, 7.74, 7.75) are not valid. Thus, the analysis described in Sections 6.6 and 6.7 was used by Asali & Hanratty.

For gas flows over very thin films at high gas velocities, a number of simplifications in the analysis can be made. Since the flow is laminar, the average velocity at the interface, $\bar{u}_S$, is equal to $2\bar{u}_a$, so $\bar{\tau}_S \approx \bar{\tau}_W = 2\bar{u}\mu_L/h$. This means that $\bar{\tau}_S/\rho\bar{u}_a^2 = 2/\text{Re}_L$. From (7.10), $-\frac{\partial P_S}{\partial x} + \rho_L \cos\theta_g = 0$, since $\overline{T}_S \approx \bar{\tau}_W$. Asali & Hanratty (1993) show that equation (7.9) simplifies to

$$\left(\frac{c}{\bar{u}_S}-1\right)^2 + \frac{3}{4}\left(\frac{c}{\bar{u}_S}-1\right) + 3i\left(k\bar{h}\frac{\bar{h}\bar{u}_S}{\nu_L}\right)^{-1}\left(\frac{c}{\bar{u}_S}-1\right) = \frac{k^2\sigma\bar{h}}{\rho_L\bar{u}_S^2} + \frac{\bar{h}\tilde{P}_S}{\rho_L\bar{u}_S^2} + \frac{g\bar{h}}{\bar{u}_S^2}\sin\theta_g + \frac{3}{2}i\frac{\tilde{\tau}_S}{\rho_L k\bar{u}_S^2} \tag{7.102}$$

where $\bar{u}_S$ is the liquid velocity at the interface. Craik (1966) solved the problem by using the series approach developed by Benjamin (1957). For small $k\bar{h}$, he obtained

$$\frac{6}{5}\left(\frac{c}{\bar{u}_S}-1\right)^2 + \frac{7}{8}\left(\frac{c}{\bar{u}_S}-1\right) + 3i\left(k\bar{h}\frac{\bar{h}\bar{u}_S}{\nu_L}\right)^{-1}\left(\frac{c}{\bar{u}_S}-1\right) = \frac{k^2\sigma\bar{h}}{\rho_L\bar{u}_S^2} + \frac{\bar{h}\tilde{P}_S}{\rho_L\bar{u}_S^2} + \frac{g\bar{h}}{\bar{u}_S^2}\sin\theta_g + \frac{3}{2}i\frac{\tilde{T}_S}{\rho_L k\bar{u}_S^2} \tag{7.103}$$

where terms of order $(k\bar{h})^2$ are ignored. By comparing (7.102) and (7.103), it can be seen that the integral approach and the series approach produce similar results.

The complex wave velocity, $c = c_R + ic_1$, is substituted into (7.102). The real and imaginary parts of the resulting expression provide two equations that can be solved for c_R and c_I. Asali & Hanratty (1993) provide results obtained from such a calculation.

Studies of the properties of ripple waves have been carried out by Würz (1977), Taylor (1963), Shearer & Nedderman (1965). Asali & Hanratty compared their analysis with the study by Würz in a 2.54 cm horizontal channel, as well as with their own measurements in a 4.2 cm vertical pipe. Figure 7.9 shows calculations of the wave growth, kc_1, for different liquid Reynolds numbers where $\mathrm{Re}_{\mathrm{LF}}/4 = \overline{h}\overline{u}_{\mathrm{a}}\rho_{\mathrm{L}}/\mu_{\mathrm{L}}$. The maxima in the curves are the maximum rates of growth. Figure 7.10 shows that the ratio of the wavelength to

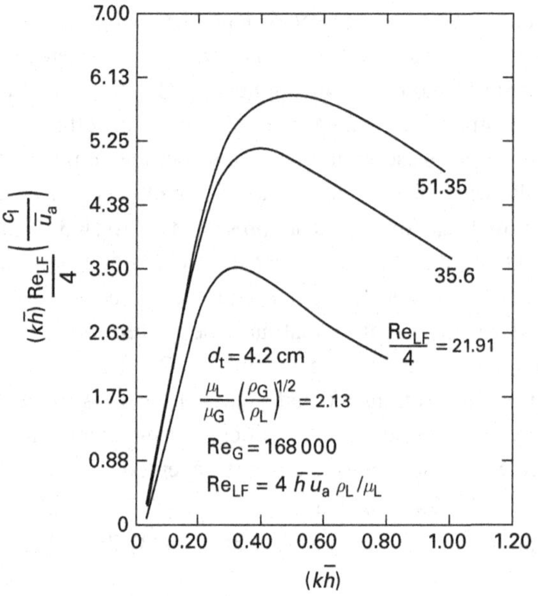

Figure 7.9 Examples of calculated growth rates of disturbances. Asali & Hanratty, 1993.

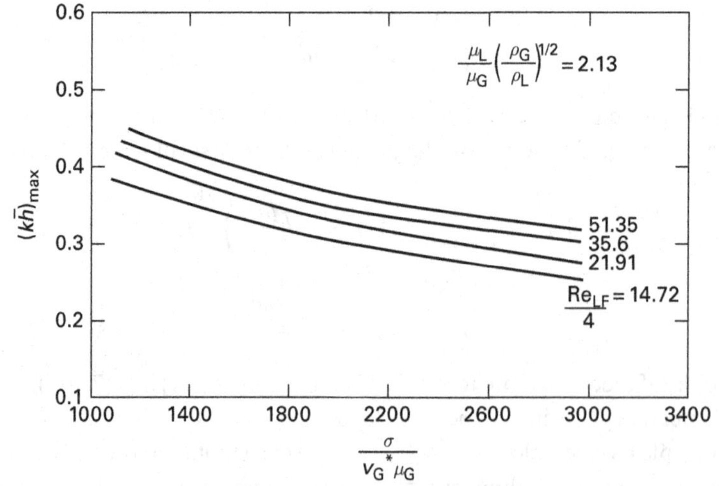

Figure 7.10 Calculated wave numbers for maximum growth. Asali & Hanratty, 1993.

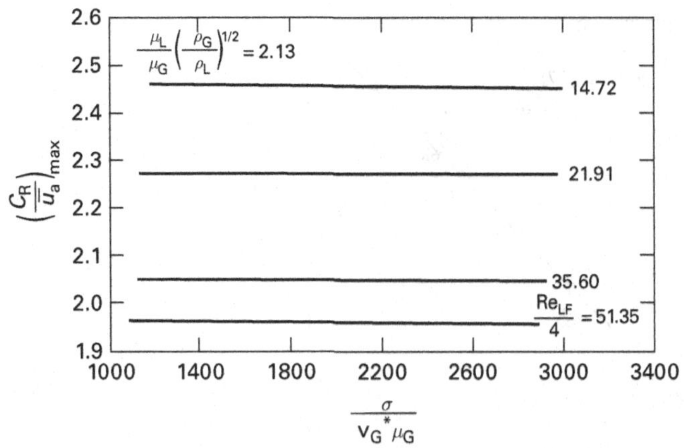

Figure 7.11 Calculated velocities of the fastest growing waves on a water film. Asali & Hanratty, 1993.

the film height for maximum growth is weakly dependent on the Reynolds number of the liquid film. Figure 7.11 shows the calculated velocity of the fastest growing waves. The dimensionless wave velocity, $c_R/\bar{u}_a$ is calculated to be approximately equal to 2. It is noted in the discussion before equations (7.11) and (7.12) that one should expect liquid inertia to have a small effect on the stability of the film. Instability occurs when the destabilizing effects of wave induced variations of the gas-phase pressure and shear stress at the interface overcome stabilizing effects of gravity and surface tension. Thus, the modeling of $\tilde{T}_{SI}$ and $\tilde{P}_{SR}$ is critical.

The calculations presented in the paper by Asali & Hanratty (1993) employed models for turbulent flow over wavy surfaces discussed in Sections 6.6 and 6.7 to calculate $\tilde{T}_{Si}$ and $\tilde{P}_{SR}$. They used a slight modification of the Model D explored by Thorsness *et al.* (1978). This approach is the relaxation analysis of Abrams & Hanratty (1984) or Model D* in Abrams (1984). It uses a boundary-layer coordinate system embedded in the wave surface and a modification of the van Driest model outlined by Loyd *et al.* (1970).

The main result is that the observed wavelength is twice the value of the predicted wavelength of the ripple waves at maximum growth. This is illustrated in Figure 7.12. Theory predicts the observed trends in the measurements. However, there is not an exact agreement between predicted and measured wavelengths. This should not be surprising because the observed waves, although broad-crested, are not two-dimensional. Asali & Hanratty (1993) provide the possible explanation that ripples evolve from a two-dimensional wave with a wavelength equal to the fastest-growing wave predicted by linear theory. As these waves grow in amplitude, they become unstable to a spanwise disturbance, as shown in Figure 7.13. According to this picture the two-dimensional waves are altered in such a way that the crests break into a group of isolated long-crested three-dimensional waves. Such an instability has been explored by McLean *et al.* (1981) for gravity waves on deep fluids.

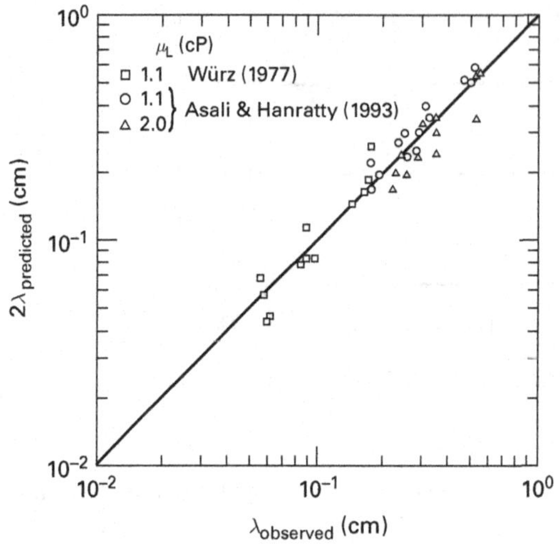

Figure 7.12 Comparison of calculated spacing of the most rapidly growing waves with measurements. Asali & Hanratty, 1993.

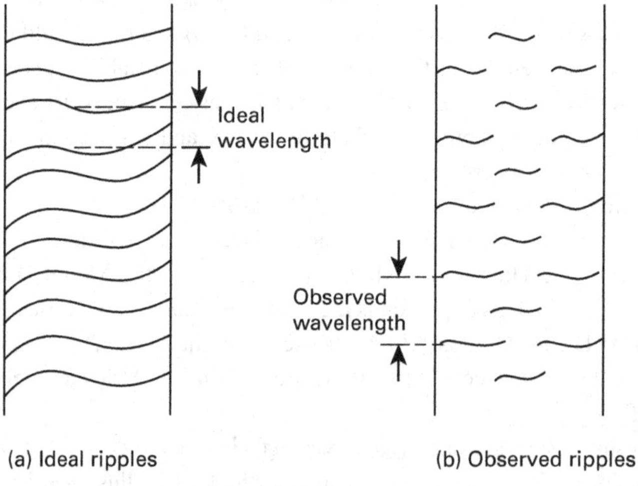

Figure 7.13 Postulated growth of infinitesimal two-dimensional ripples into a finite-amplitude three-dimensional wave pattern.

7.8 Roll waves in pipelines

The analysis presented in Section 7.5.2 for a rectangular channel was applied to a circular pipeline by Lin & Hanratty (1986). The simplified configuration for a stratified flow, presented in Chapter 5, was used. This was followed with analyses by Hurlburt & Hanratty (2002) and by Soleimani and Hanratty (2003). These show that viscous large-wavelength theory predicts the appearance of roll waves. At low gas velocities, these roll waves grow into slugs because the height of the stratified layer is such that slugs would be stable, if generated (see Chapter 2).

Soleimani & Hanratty (2003) present a detailed description of the disturbances in the pseudo-slug region, which are identified as roll waves. A striking feature is the relative insensitivity of liquid holdup (time-mean height) to changes in the liquid flow rate in the pseudo-slug regime. For example, at a superficial gas velocity of 8m/s, h/d_t varies from 0.17 to 0.18 as the superficial liquid velocity increases from 0.15 m/s to 0.25 m/s. This is explained by an increase in the velocity and frequency of roll waves with increasing superficial liquid velocity. The formation of slugs is initiated by the coalescence of roll waves to form stable slugs. The transition is predicted to occur when the holdup is such that propagating slugs would be stable.

References

Abrams, J. 1984 Turbulent flow over small amplitude solid waves. Ph.D. thesis, Univerity of Illinois.

Abrams, J. & Hanratty, T.J. 1984 Relaxation effects observed for turbulent flow over a wavy surface. *J. Fluid Mech.* 151, 443–455.

Alekseenko, S.V., Nakoryakov, V.E. & Pokusavev, B.G. 1985 Wave formation on vertical falling liquid films. *Int. J. Multiphase Flow* 11, 607–627.

Andreussi, P., Asali, J.C. & Hanratty, T.J. 1985 Initiation of roll waves in gas–liquid flows. *AIChE Jl* 31, 119–126.

Andritsos, N. & Hanratty, T.J. 1987 Influence of interfacial waves in stratified gas–liquid flows. *AIChE Jl* 33, 444–454.

Asali, J.C. & Hanratty, T.J. 1993 Ripples generated on a liquid film at high gas velocities. *Int. J. Multiphase Flow* 19, 229–243.

Asali, J.C., Andreussi, P. & Hanratty, T.J. 1985 Interfacial drag and film height for vertical annular flow. *AIChE Jl* 31, 895–902.

Benjamin, T.B. 1957 Wave formation in laminar flow down an inclined plane. *J. Fluid Mech.* 2, 554–574.

Binnie, A.M. 1957 Experiments on the onset of wave formation on a fluid flowing down a vertical plane. *J. Fluid Mech.* 2, 551–553.

Bontozoglou, V.L. & Hanratty, T.J. 1989 Wave height estimation in stratified gas–liquid flows. *AIChE Jl* 35, 1346–1350.

Cohen, L.S. & Hanratty, T.J. 1965 Generation of waves in concurrent flow of air and a liquid. *AIChE Jl* 11, 138–143.

Cornish, V. 1934 *Ocean Waves and Kindred Phenomena*. Cambridge: Cambridge University Press.

Craik, A.D.D. 1966 Wind generated waves in thin films. *J. Fluid Mech.* 26, 369–392.

Craik, A.D.D. 1968 Wind generated waves in contaminated liquid films. *J. Fluid Mech.* 31, 141–162.

Dressler, R.F. 1949 Mathematical solution of the problem of roll waves in inclined open channels. *Comm. Pure Appl. Math.* 2, 149–194.

Dressler, R.F. 1952 Stability of uniform flow and roll wave formation. *US Natl Bur. Std. Circular* 521, 237–241.

Hanratty, T.J. 1983 Interfacial instabilities caused by air flow over a thin layer. In *Waves on Fluid Interfaces,* ed. R.E. Meyer. New York: Academic Press, pp. 221–259.

Hanratty, T.J. 1991 Separated flow modeling and interfacial transport phenomena. *Appl. Scient. Res.* 48, 353–390.

Hanratty, T.J. & Engen, J.M. 1957 Interaction between a turbulent air stream and a moving water surface. *AIChE Jl* 3, 299–304.

Hanratty T.J. & Hershman, A. 1961 Initiation of roll waves. *AIChE Jl* 7, 489–497.

Henstock, W.H. & Hanratty, T.J. 1976 The interfacial drag and height of the wall layer in annular flows. *AIChE Jl* 7, 489–497.

Hewitt, G.F. & Hall-Taylor, N.S. 1970 *Annular Two-Phase Flow.* London: Academic Press.

Hurlburt, E.T. & Hanratty, T.J. 2002 Prediction of transition from stratified to slug and plug flows for long pipes. *Int J. Multiphase Flow* 28, 707–729.

Lighthill, M.J. & Whitham, G.B. 1955 Flood movement in long rivers. *Proc. R. Soc. Lond.* A229, 291.

Lin, P.Y. & Hanratty, T.J. 1986 Prediction of the initiation of slugs with linear stability theory. *Int. J. Multiphase Flow* 12, 79–98.

Loyd, R.J., Moffat, R.J. & Kays, W.M. 1970 The turbulent boundary on a plate: an experimental study of the fluid dynamics with strong favorable pressure gradients and blowing. Report No. HMT-13, Stanford University California.

McLean, J.W., Ma, Y.C., Martin, D.V. & Saffman, P.C. 1981 Three-dimensional instability of finite amplitude water waves. *Phys. Rev. Lett.* 46, 817–820.

Meyer, R.E. (ed.) 1983 *Waves on Fluid Interfaces.* New York: Academic Press, pp. 221–259.

Miya, M., Woodmansee, D.E. & Hanratty, T.J. 1971 A model for roll waves in gas–liquid flow. *Chem. Eng. Sci.*, 26, 1915–1931.

Shearer, C.J. & Nedderman, R.M. 1965 Pressure gradient and liquid film thickness in concurrent upwards flow of gas/liquid mixtures: application to film cooler design. *Chem. Eng. Sci.* 20, 679–683.

Soleimani, A. & Hanratty, T.J. 2003 Critical liquid flows for the transition from the pseudo-slug and stratified patterns to slug flow. *Int. J. Multiphase Flow* 29, 707–729.

Stoker, J.J. 1957 *Water Waves.* New York: Interscience.

Taylor, G.I. 1963 Generation of ripples by wind blowing over a viscous liquid. In *The Scientific Papers of Sir Geoffrey Ingram Taylor.* ed. G.K. Batchelor. Cambridge: Cambridge University Press, pp. 244–254.

Thorsness, C.B., Morrisoe, P.H. & Hanratty, T.J. 1978 A comparison of linear theory with measurements of the variation of shear stress along a solid wave. *Chem. Eng. Sci.* 33, 579–592.

Whitham, G.B. 1974 *Linear and Non-linear Waves.* New York: John Wiley.

Woodmansee, D.E. & Hanratty, T.J. 1969 Base film over which roll waves propagate. *AIChE Jl* 15, 712–715.

Würz, D. E. 1977 Flüssigkeits – Filmströmung unter einwirkung einer Uber-lufströmung. Thesis, Univ. Karlsruhe, Inst. für thermische strömung maschinen.

8 Bubble dynamics

8.1 Prologue

A central issue to be addressed in analyzing the behavior of bubbles in a gas–liquid flow is understanding the free-fall velocity of a spherical solid particle and the rise velocity of a spherical bubble in an infinite stationary fluid. Analytical solutions for these systems are available for very low particle Reynolds numbers (Stokes law and the Hadamard equation). A derivation of Stokes law is presented in the first part of this chapter.

Experiments show that Stokes law is valid for particle Reynolds numbers less than unity. For larger Re_P, empirical correlations of the drag coefficient are used. The description of the rise velocity of bubbles is complicated by possible contamination of the interface. Measurements of the rise velocity of single bubbles are usually presented as plots of U_S versus the bubble size for a given system. The structure of these plots reflects changes in the shape and behavior of the bubbles. Very large bubbles take the shape of a cap. A prediction of the rise velocity of these cap bubbles, developed by Batchelor, is presented in Section 8.7.

Other very large bubbles encountered in gas–liquid flows are the Taylor bubble, which is a part of slug flows in vertical pipes, and the Benjamin bubble used to model the shedding of liquid from slugs in horizontal pipes. These are described in Sections 8.8 and 8.9.

Bubbly flows are situations for which a swarm of bubbles is considered. A feature of these flows is that the rise velocities of bubbles are less than would be observed for single bubbles. In this sense, the bubble motion is hindered. There is a kinship of bubbly flows to sedimenting and fluidizing liquid–solid systems, where the critical issue is to relate the sedimenting and the fluidizing velocities to the free-fall velocity and the volume fraction of solids. These systems are described in Section 8.14.

8.2 Free-fall velocity of a solid sphere

A sphere released into a fluid would experience forces due to gravity, equal to $+\rho_P V_P g$, where $V_P = \pi d_P^3/6$ is the volume of the sphere and the plus sign indicates a force in the direction of gravity. The resisting force, F_D, is related to the magnitude of the relative velocity between the sphere and the fluid. The usual method for describing F_D is to define a drag coefficient, C_D.

$$F_D = -C_D \frac{1}{2}\rho_F |U_S|^2 S \tag{8.1}$$

where $|U_S|^2$ is the square of the magnitude of the relative velocity between the particle and the fluid and S is the projected area of the particle, $S = \pi d_P^2/4$.

The pressure in a stagnant fluid changes because of the variation of hydrostatic head. This gives rise to a buoyancy force F_B, which is related to the volume of fluid displaced by the particle, V_P,

$$F_B = -\rho_F V_P g \tag{8.2}$$

The acceleration of the sphere is given as $m_P dU_P/dt$ where $m_P = \rho_P V_P$ and U_P is the particle velocity. A particle moving through a fluid drags a volume of fluid with it. Thus, the force needed to accelerate the particle is given as $(\rho_P V_P + \rho_F V_P f)(dU_P/dt)$, where $V_P \rho_P$ is the mass of the particle, $\rho_F V_P f$ is the effective mass of the fluid dragged by the sphere and f is usually taken as a constant. Thus, from Newton's law of motion,

$$(\rho_P V_P + \rho_F f V_P)\frac{dU_P}{dt} = +\rho_P g V_P - \rho_F g V_P - C_D \frac{1}{2}\rho_F |U_S|^2 S \tag{8.3}$$

Under free-fall conditions, the mean acceleration is zero, so that, from (8.3), the mean free-fall velocity is

$$U_S^2 = \frac{V_P g(\rho_P - \rho_F)}{S\rho_F \frac{1}{2} C_D} \tag{8.4}$$

8.3 Stokes law for a solid sphere

The force F_D of a fluid on a falling sphere can be calculated exactly for the limit of very small particle Reynolds numbers, $\mathrm{Re}_P = d_P U_S \rho_F/\mu < 1$.

This is done by using the laws of conservation of mass and momentum. These are given for a Cartesian coordinate system by equations (4.7), (6.1) and (6.3). Their representation in a spherical polar coordinate system is given in Tables 3.4–1, 3.4–4, 3.4–7 of the book by Bird *et al.* (1960). The relations of fluid stresses to components of the velocity gradient tensor are given in Bird *et al.* (1960), Table 3.4–7, for a Newtonian fluid.

The following analysis leading to Stokes law is also taken from Bird *et al.* A sketch of a sphere settling with a velocity U_S and the definition of θ are given in Figure 8.1a. For a symmetric flow, $u_\phi = 0$, so there is no variation of the velocity in the ϕ-direction. For an incompressible fluid, the density is constant. Thus, the following equations define u_r, u_θ and p:

$$\frac{1}{r^2}\frac{\partial(r^2 u_r)}{\partial r} + \frac{1}{r\sin\theta}\frac{\partial(u_\theta \sin\theta)}{\partial\theta} = 0 \tag{8.5}$$

$$\rho\left(\frac{\partial u_r}{\partial t}+u_r\frac{\partial u_r}{\partial r}+\frac{u_\theta}{r}\frac{\partial u_r}{\partial \theta}-\frac{u_\theta^2}{r}\right)$$
$$=-\frac{\partial p}{\partial r}+\mu\left[\frac{1}{r^2}\frac{\partial^2}{\partial r^2}\left(r^2u_r\right)+\frac{1}{r^2\sin\theta}\frac{\partial}{\partial\theta}\left(\sin\theta\frac{\partial u_r}{\partial\theta}\right)\right]+\rho g_r \tag{8.6}$$

$$\rho\left(\frac{\partial u_\theta}{\partial t}+u_r\frac{\partial u_\theta}{\partial r}+\frac{u_\theta}{r}\frac{\partial u_\theta}{\partial \phi}+\frac{u_r u_\theta}{r}\right)$$
$$=-\frac{1}{r}\frac{\partial p}{\partial\theta}+\mu\left[\frac{1}{r^2}\frac{\partial}{\partial r}\left(r^2\frac{\partial u_\theta}{\partial r}\right)+\frac{1}{r^2}\frac{\partial}{\partial\theta}\left(\frac{1}{\sin\theta}\frac{\partial}{\partial\theta}(u_\theta\sin\theta)\right)+\frac{2}{r^2}\frac{\partial u_r}{\partial\theta}\right]+\rho g_\theta \tag{8.7}$$

As seen in Figure 8.1, boundary conditions on the surface of the sphere are $u_r=-U_S\cos\theta$, $u_\theta=-U_S\sin\theta$. The velocities are zero at large distances from the sphere since the fluid surrounding the sphere is stagnant.

The solution is more easily obtained by using a coordinate system which is moving, with the sphere, at a velocity U_S (see Figure 8.1b). An observer in the moving coordinate system sees a steady flow around the sphere, so the transient terms $\partial u_r/\partial t$ and $\partial u_\theta/\partial t$ are set equal to zero. The boundary conditions are that

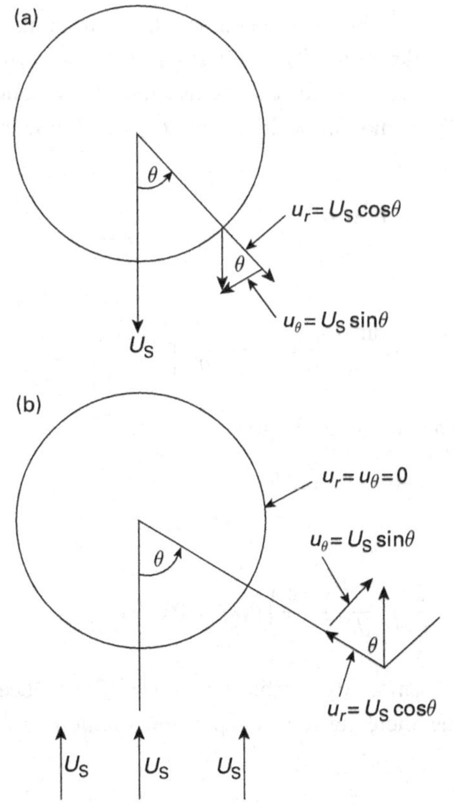

Figure 8.1 Solid sphere moving at a velocity through an infinite stagnant fluid: (a) as seen by a stationary observer, (b) as seen by an observer moving with the sphere.

$$u_r = u_\theta = 0 \tag{8.8}$$

at the surface of the sphere and that

$$u_r = U_S \cos\theta \tag{8.9}$$

$$u_\theta = U_S \sin\theta \tag{8.10}$$

at $r = \infty$.

The three equations to be solved have three unknowns u_r, u_θ and p. A stream function, $\psi(r, \theta)$, can be defined from the equation of conservation of mass, (8.5).

$$u_r = -\frac{1}{r^2 \sin\theta} \frac{\partial \psi}{\partial \theta} \tag{8.11}$$

$$u_\theta = \frac{1}{r \sin\theta} \frac{\partial \psi}{\partial r} \tag{8.12}$$

$$u_\phi = 0 \tag{8.13}$$

Its validity can be tested by substituting (8.11) and (8.12) into (8.5). The function ψ describes the current lines, that is, the amount of fluid that flows between two curves of constant ψ.

The substitution of (8.11) and (8.12) into the momentum equations eliminates u_r and u_θ. The pressure is eliminated by calculating $\partial^2 p/\partial\theta\partial r$ from (8.6) and $\partial^2 p/\partial r\partial\theta$ from (8.7). These are equated to obtain a relation for ψ. The assumption is made that fluid inertia can be neglected in a coordinate moving with velocity U_S. The following equation is obtained for the stream function, ψ:

$$E^4 \psi = 0 \tag{8.14}$$

where the operator is defined as

$$E^4 = \left[\frac{\partial^2}{\partial r^2} + \frac{\sin\theta}{r^2} \frac{\partial}{\partial \theta}\left(\frac{1}{\sin\theta} \frac{\partial}{\partial \theta}\right)\right]^2 \tag{8.15}$$

The boundary conditions suggest a solution of the form

$$\psi = f(r) \sin^2\theta \tag{8.16}$$

This satisfies (8.14) if

$$\left(\frac{d^2}{dr^2} - \frac{2}{r^2}\right)\left(\frac{d^2}{dr^2} - \frac{2}{r^2}\right) f(r) = 0 \tag{8.17}$$

The term r^n is the solution of this equation. Its substitution into (8.17) produces a fourth-order algebraic equation in n. Thus, there are four independent solutions: r^{-1}, r^1, r^2, r^4. The solution of (8.17) is given as

$$f(r) = \frac{A}{r} + Br + Cr^2 + Dr^4 \tag{8.18}$$

The constants A, B, C, D are specified so as to satisfy the boundary conditions:

$$A = -(1/4)U_S r_P^2 \quad B = (3/4)U_S r_P \quad C = -(1/2)U_S \quad D = 0$$

where r_P is the radius of the sphere. The velocity components are found by substituting (8.16) into (8.11) and (8.12) with $f(r)$ given by (8.18):

$$u_r = U_S \left[1 - \frac{3}{2}\left(\frac{r_P}{r}\right) + \frac{1}{2}\left(\frac{r_P}{r}\right)^3 \right] \cos\theta \tag{8.19}$$

$$u_\theta = U_S \left[1 - \frac{3}{4}\left(\frac{r_P}{r}\right) - \frac{1}{4}\left(\frac{r_P}{r}\right)^3 \right] \sin\theta \tag{8.20}$$

The fluid exerts a normal force due to the variation of the pressure around the sphere (the form drag) and a frictional force due to shear stresses exerted on the sphere (the skin friction). The pressure can be calculated with (8.7), (8.19) and (8.20). The local frictional stress at the surface of the sphere, $\tau_{r\theta}$, is calculated from the velocity field since

$$\tau_{r\theta} = \mu \left[r \frac{\partial}{\partial r}\left(\frac{u_\theta}{r}\right) \right] \quad \text{at} \quad r = r_P$$

The total force due to skin friction and form drag can be calculated by integrating p and $\tau_{r\theta}$ over the surface of the sphere. Bird *et al.* (1960) show that the total force is

$$F_D = 2\pi\mu r_P U_S + 4\pi\mu r_P U_S \tag{8.21}$$

where the first term is the skin friction and the second term is the form drag.

Stokes law is given as

$$F_D = 6\pi\mu r_P U_S \tag{8.22}$$

or as

$$C_D = \frac{24}{\mathrm{Re}_P} \tag{8.23}$$

Where C_D is the drag coefficient defined by (8.1). If (8.23) is substituted into (8.4) a free-fall velocity of

$$U_S = \frac{g(\rho_P - \rho_F)d_P^2}{18\mu} \tag{8.24}$$

is predicted for Reynolds numbers at which Stokes law is applicable. Thus U_S varies as d_P^2 at very low Re_P. Typical inertial and viscous terms appearing in the Navier–Stokes equations are $\rho u_r(\partial u_r/\partial r)$ and $\mu/r^2\left[\partial^2/\partial r^2(r^2 u_r)\right]$. If u_r scales with U_S and r scales with d_P, the ratio of the inertia forces to the viscous forces scales as

$$\frac{\rho U_S^2}{d_P} \bigg/ \frac{\mu_S U_S}{d_P^2} = \frac{d_P U_S \rho}{\mu} = \mathrm{Re}_P$$

Thus, one can expect that the neglect of inertia terms is valid for very small Re_P.

8.4 Rise velocity of bubbles at low $\mathrm{Re_P}$

The calculation of the rise velocity of a bubble in a stationary fluid at low $\mathrm{Re_P}$ is similar to that for the free-fall velocity of a solid particle. If ρ_P is taken as the density of the gas in the bubble and ρ_F as the density of the surrounding fluid, the sum of the gravitational force and the buoyancy force is given as $-(\rho_P - \rho_F)\, gV_P$, where a positive sign indicates a force or velocity opposite the direction of gravity.

Note that, in using (8.3), the mean inertia of the system is dominated by the mass of the liquid dragged along by the bubble; that is, $\rho_F V_P f >> \rho_P V_P$ in (8.3). As for a solid sphere, an equation for C_D can be developed at small Reynolds numbers if inertia terms in the Navier–Stokes equations are ignored. A spherical bubble differs from a spherical solid sphere in that the inside of the sphere can be in motion and the fluid velocity at the boundary need not be the same as for a solid sphere. Hadamard (1911) developed an equation for the velocity fields both inside and outside the bubble. The following relation was obtained for the terminal velocity in a stationary fluid:

$$U_S = \frac{d_{PS}^2 g(\rho_P - \rho_F)}{18\mu_F} \frac{(3\mu_P + 3\mu_F)}{(3\mu_P + 2\mu_F)} \tag{8.25}$$

where μ_P is the viscosity of the gas in the bubble. For $\mu_P = \infty$, (8.25) gives (8.24), the equation for a solid sphere at low $\mathrm{Re_P}$. For $\mu_P = 0$, (8.25) gives

$$U_S = \frac{d_{PS}^2 g(\rho_F - \rho_P)}{12\mu_F} \tag{8.26}$$

Thus, because of motion inside the bubble the terminal velocity is predicted to be 3/2 of the value for a rigid sphere (if the interface is clean).

8.5 Measurements of C_D

8.5.1 Solid spheres

Values of C_D for flow around solid spheres have been obtained in a number of experiments. These are presented in plots of C_D versus $\mathrm{Re_P}$ (Bird *et al.*, 1960; Schlichting, 1955; Soo, 1967; Crowe *et al.*, 1998). The agreement among different experiments is not perfect. This is because they were done for different global flow conditions (Soo, 1967; Crowe *et al.*, 1998).

The plots by Bird *et al.* and by Schlichting agree with Stokes law at very low Reynolds numbers. They show that (8.23) provides a good representation of the data for $\mathrm{Re_P} < 1$. Bird *et al.* suggest that $C_D \approx 18.5/\mathrm{Re}_P^{0.6}$ for $2 < \mathrm{Re_P} < 5 \times 10^2$. A rough representation for the drag coefficient at large $\mathrm{Re_P}$ is Newton's law, C_D = constant. Bird *et al.* give $C_D \approx 0.44$ for $5 \times 10^2 < \mathrm{Re_P} < 2 \times 10^5$. A sharp drop in C_D is found at very large $\mathrm{Re_P}$. This can be associated with a sudden transition to turbulence in the boundary layer, which causes a decrease in the low-pressure wake region. The substitution of $C_D = 0.5$ into (8.4) gives the following relation for the free-fall velocity of a solid sphere:

$$U_S = \left[\frac{2}{3}\frac{d_P(\rho_P - \rho_F)g}{\rho_F}\right]^{1/2} \tag{8.27}$$

Equations (8.24) and (8.27) indicate that $U_S \propto d_P^2$ at very low Reynolds numbers and that $U_S \propto d_P^{1/2}$ at large Reynolds numbers.

8.5.2 Gas bubbles

Studies of the rise velocity of single spherical bubbles in water at small Reynolds numbers usually give values of U_S close to what is predicted for a solid sphere, rather than (8.25). Levich (1962) explains this finding as due to contamination of the interface. He points out that gradients of the concentration of contaminants along the interface cause changes in surface tension which result in forces that can oppose the motion of the interface.

The size of a bubble can be represented by its volume, V_P, or by an equivalent radius, r_P, defined from the volume, that is, $V_P = (4/3)\pi r_P^3$. For $\mathrm{Re}_P = 1$, the equivalent radius is approximately 0.2 mm for air bubbles in water. A summary of measurements of the rise velocity of air bubbles in filtered or distilled water is given in Figure 8.2 (see also Hetsroni, 1982, Figure 1.3.24; Soo, 1967, Figure 3.20). For bubble radii less than 0.2 mm, the rise velocity is found to increase with the square of the bubble radius, as

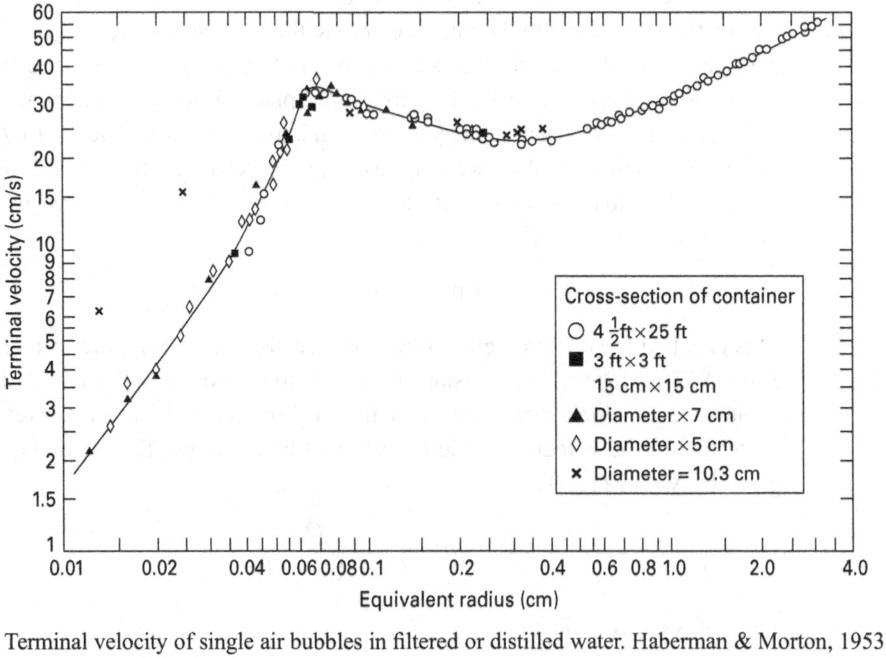

Figure 8.2 Terminal velocity of single air bubbles in filtered or distilled water. Haberman & Morton, 1953.

predicted by (8.24). The rise velocity shows a sharp local maximum of about 33 cm/s for $r_P \approx 0.7$ mm and a broad local minimum of about 24 cm/s for $r_P \approx 3$ mm.

At low Re_P, bubbles injected into the field keep a spherical shape up to $r_P \approx 0.7$ mm because surface tension effects are much greater than inertia effects. The bubbles move in straight lines. For equivalent radii of 0.7 to 3 mm, the bubbles behave differently. Their shape is in the form of a flattened ellipsoid whose minor axis is in the direction of mean motion; the bubbles vibrate and move in a spiral motion (Grace & Weber, 1982). For radii greater than 3 mm, the bubbles assume a turbulent motion. For radii of 1 to 10 mm, the rise velocity in water is roughly constant (approximately equal to 30 cm/s). For very large radii, say greater than 10 mm, the bubble size increases with $r_P^{0.5}$. This is the cap bubble regime described in Section 8.7. Studies of the effect of changes of the liquid properties do not produce definitive results. One of the reasons for this is that contamination of the interface can have an important effect (Haberman & Morton, 1953).

A dimensionless group characterizing the effect of surface tension is $U_S/(g\sigma/\rho_F)^{0.25}$. The following equation, proposed by Harmathy (1960), roughly represents measurements for large bubbles over a range of conditions:

$$U_S = 1.53\left(\frac{g\Delta\rho\sigma}{\rho_F^2}\right)^{0.25} \tag{8.28}$$

8.6 Gaseous volume fraction for bubbly flow

The gaseous volume fraction in a one-dimensional system, α, is defined as the ratio of the volumetric flow of gas to the total volumetric flow, given as $(u_{GS} + u_{LS})A$. This parameter is considered below for the case where variations of the gas density can be ignored.

One way to characterize the slip between the phases is to define a slip ratio, S, as equal to the ratio of the local gas velocity to the local liquid velocity (equation (1.67)). Equation (1.70) then defines α. Another way, used in the separated flow model (described in Section 1.4), is to consider the difference between the velocity of the gas and the velocity of the mixture, $(u_{GS} + u_{LS})$:

$$V_{\text{Drift}} = u_G - C_0(u_{GS} + u_{LS}) \tag{8.29}$$

This is called the drift velocity of the gas (see Nicklin *et al.*, 1962; Zuber & Findlay, 1965; Wallis, 1969). The constant C_0 was introduced by Zuber & Findlay. It has a value of 1.2, which recognizes that the nose of the bubble moves relative to the centerline of the velocity profile in turbulent liquid slugs. The following equation is given by Wallis (1969):

$$\alpha = \frac{Q_G}{Q_L + Q_G + AV_{\text{Drift}}} \tag{8.30}$$

where A is the column area.

8.7 Cap bubbles

Very large cap-shaped entities are characteristic of bubbles that contain a volume of gas larger than 5 cm^3. Batchelor (1967) presents photographs and describes these bubbles as "looking like an umbrella." He depicts them as a slice off a sphere of radius R and provides an analysis in a framework which is moving with the velocity of the bubble, U_S. The front of the bubble is, then, a stagnation point. It has a radius of curvature R. The bubble is large enough for effects of surface tension and viscosity to be neglected. A spherical polar coordinate system is used. The idealized bubble is symmetric in the ϕ-direction. The front face is steady, smooth and close to spherical in shape. The Bernoulli equation

$$\frac{p}{\rho} + \frac{q^2}{2} + gz = \text{const.} \tag{8.31}$$

is used to describe the flow along a streamline, where z is the distance upward in the vertical direction. The goal is to develop an equation for the rise velocity, U_S.

Consider a streamline embedded in the surface of the bubble. The velocity of the fluid at the nose of the bubble is zero in a moving coordinate system, so the constant in (8.31) is $p_S/\rho + gz_S$. Thus

$$\frac{p}{\rho} + \frac{q^2}{2} + gz = \frac{p_S}{\rho} + gz_S \tag{8.32}$$

where z_S is the location of the stagnation point. Since the region near a stagnation point is considered, the pressure is constant, that is, $p = p_S$, so

$$q^2 = 2g(z_S - z) \tag{8.33}$$

Equation (8.33) describes the velocity variation at the surface of the bubble in a coordinate system moving with the velocity of the bubble.

The problem is to relate q to U_S. This is done by solving for the velocity field. The fluid approaching the bubble has a uniform velocity and may be considered inviscid. Therefore, an irrotational flow is assumed. As shown in Chapter 4, the velocity vectors are then described as the gradient of a potential function. Conservation of mass for an incompressible fluid dictates that the potential function is defined by the Laplace equation (4.9).

The front of the bubble has a spherical shape, so flow around a sphere is considered. The Laplace equation in spherical coordinates is given in Table A.7–3 of Bird *et al.* (1960):

$$\frac{\partial}{\partial r}\left(r^2 \frac{\partial \phi}{\partial r}\right) + \frac{1}{\sin\theta}\frac{\partial}{\partial \theta}\left(\sin\theta \frac{\partial \phi}{\partial \theta}\right) = 0 \tag{8.34}$$

for a symmetric flow, where

$$u_r = -\frac{\partial \phi}{\partial r} \qquad u_\theta = -\frac{1}{r}\frac{\partial \phi}{\partial \theta} \tag{8.35}$$

Equation (8.34) is solved in a frame of reference moving with velocity U_S using the boundary conditions that u_r is zero at the boundary of the sphere and that U_S is the velocity at large distances from the sphere.

A solution of (8.34) is obtained by a separation of variables technique

$$\phi = f(r)\cos\theta \tag{8.36}$$

This satisfies (8.34) if

$$r^2\frac{d^2f}{dr^2} + 2r\frac{df}{dr} - 2f = 0 \tag{8.37}$$

Substitution of r^n into (8.37) gives two solutions with $n = -2$, $n = 1$ so

$$f = \frac{A}{r^2} + Br \tag{8.38}$$

To satisfy the condition that the velocity at a large distance from the surface is U_S

$$B = U_S \tag{8.39}$$

For a viscous flow, both u_θ and u_r are zero at the surface. However, for an inviscid flow only one of these conditions, $u_r = 0$, can be satisfied, so

$$A = \frac{R^3 U_S}{2} \tag{8.40}$$

Thus, $u_\theta \neq 0$ at the bubble surface. The solution

$$\phi = \frac{U_S R^3}{2r^2}\cos\theta + U_S r\cos\theta \tag{8.41}$$

is given in Streeter (1948, p. 67). The velocity at any point on the surface of the sphere $r = R$ is

$$-\frac{1}{r}\frac{\partial\phi}{\partial\theta} = q = \frac{3}{2}U_S\sin\theta \tag{8.42}$$

The following equation for the rise velocity is obtained by substituting (8.42) into (8.33):

$$2Rg(1 - \cos\theta) = \frac{9}{4}U_S^2\sin^2\theta \tag{8.43}$$

$$\frac{U_S^2}{gR} = \frac{8}{9}\frac{(1 - \cos\theta)}{\sin^2\theta} \tag{8.44}$$

An expansion of $(1 - \cos\theta)$ gives $(1 - \cos\theta)/\sin^2\theta \approx 1/2$ for small θ, so that

$$U_S = \frac{2}{3}\sqrt{gR} \tag{8.45}$$

Equation (8.45) is found to agree with measurements of Davies & Taylor (1950).

The radius R can be related to the volume of the bubbles by representing the bubbles by a spherical cap with an occluded angle of 100° and a relatively flat tail (Wallis, 1969, p. 249). Thus (8.45) can be written as

$$U_S = 0.79\left(gV_P^{1/3}\right)^{1/2} \tag{8.46}$$

or as

$$U_S = 1.00\sqrt{gr_P} \tag{8.47}$$

where r_P is the equivalent radius defined by the relation $V_P = (4/3)\pi r_P^3$. From (8.4), the above result indicates that the drag coefficient is approximately equal to 1.0. This is the same as found for solid discs at large Reynolds numbers (Lapple & Shepherd, 1940).

8.8 Taylor bubbles / vertical slug flow

A description of vertical slug flow is given in Section 2.7.1 repeated here: "At large gas velocities a slug flow can appear. It consists of a progression of bullet-shaped bubbles (called Taylor bubbles) that fill most of the pipe cross-section. The fronts resemble cap bubbles and the backs are approximately flat. The slugs of liquid between the Taylor bubbles can contain small bubbles. Large pressure fluctuations are observed. The liquid between the gas and the wall is moving downward. Thus, to an observer moving with the gas bubble, the liquid moves around the bubble." At the back of the bubble, the interaction of the wall layer with the liquid slug resembles a plunging jet (Lin & Donnelly, 1966). This produces the bubbles that aerate the liquid slug which follows the Taylor bubble (see Kaji *et al.*, 2009).

A discussion of the slug flow regime and of Taylor bubbles is given in Wallis (1969, Chapter 10) and Batchelor (1967, pp. 475–479). If the effects of viscosity and surface tension can be ignored, the velocity of a Taylor bubble in a stationary liquid is calculated to be

$$U_S = k_1\rho_L^{-1/2}[gd_t(\rho_L - \rho_G)]^{1/2} \tag{8.48}$$

where d_t is the pipe diameter. Wallis suggested that $k_1 = 0.345$. If ρ_G is assumed to be negligible compared with ρ_L then

$$U_S = 0.345\sqrt{2}(gr_t)^{1/2} \tag{8.49}$$

where r_t is the pipe radius. This is to be compared with the velocity of cap bubbles, equation (8.45). Wallis also develops equations that include the effects of viscosity and surface tension.

The analysis of the slug regime for vertical flows can be simplified by ignoring the influence of bubbles on liquid slugs. An equation for α is then obtained by substituting (8.29) for V_{Drift} into (8.30). A force balance gives the following equation for the pressure gradient.

$$-\frac{dp}{dz} = g[\rho_L(1-\alpha) + \rho_G\alpha] + \frac{2}{r_t}\tau_W(1-\alpha) \tag{8.50}$$

The first term on the right side gives the contribution due to hydrostatic head. The second term is the frictional pressure loss. The term $(1-\alpha)$, multiplying τ_W, appears because it is assumed that only the turbulent liquid in the slugs is making a contribution. The resisting wall stress can be approximated by the friction factor relation for single-phase flow

$$\tau_W = \frac{f}{2}\rho_L u_L^2 \tag{8.51}$$

where $u_L = u_{LS} + u_{GS}$ and f is a function of the slug Reynolds number, $\mathrm{Re} = d_t u_L \rho_L/\mu_L$.

8.9 Benjamin bubble

Consider a horizontal rectangular channel or circular tube, filled with a liquid, that is opened at one end. The liquid will drain out, as indicated in Figure 8.3. A cavity is formed at the top of the channel. A layer of liquid of height h_{L2} moves out the back of the channel. The cavity moves upstream at a velocity c_B.

View the flow in a coordinate system moving with velocity c_B. The cavity then appears stationary. Location 0 is a stagnation point where the velocity is zero. The flow at Station 1 is assumed to be moving uniformly with a velocity c_B. Thus, from this framework, fluid appears to be moving into the channel at a rate $Q = c_B H$, where Q is the volumetric flow per unit breadth. From conservation of mass, the fluid appears to be moving out of the channel at a rate $Q = c_B H$ and at a velocity of $u_{L2} = Q/h_{L2}$, where H is the height of the channel.

Benjamin (1968) provides an analysis of this flow in his paper "Gravity currents and related phenomena." Effects of surface tension are ignored. He applied the momentum theorem to a volume between upstream (Station 1) and downstream (Station 2) locations, which are in fully developed regions. Again, a coordinate system moving with a velocity of c_B is assumed. The momentum flowing into the volume is $\rho c_B^2 H$. The momentum associated with gas flow outward can be neglected because of the small density. Thus,

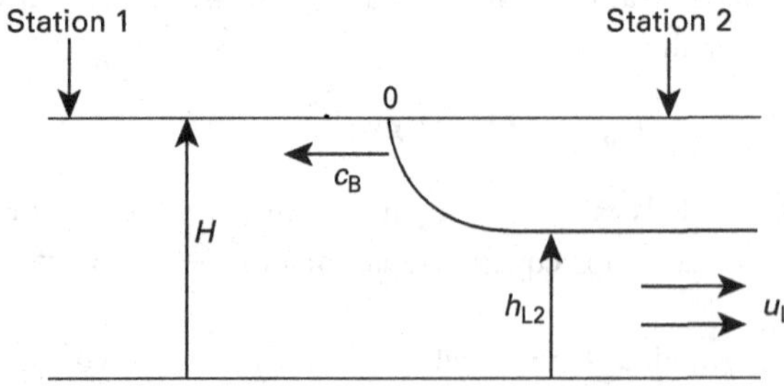

Figure 8.3 Sketch of liquid draining from a rectangular channel of height H: a Benjamin bubble moving with a velocity c_B.

the momentum flowing out is $\rho u_{L2} u_{L2} h_{L2}$. The average force on the upstream face is $(p_1 H + \rho g H^2/2)$, where the second term is the pressure force due to the gravitational head. The average force on the downstream face is $(p_2 H + \rho g h_{L1}^2/2)$. Thus, the momentum theorem gives

$$\rho u_{L2}^2 h_{L2} - \rho c_B^2 H = \left(p_1 H + \rho g \frac{H^2}{2}\right) - \left(p_2 H + \frac{\rho g h_{L2}^2}{2}\right) \tag{8.52}$$

Benjamin used (8.52), conservation of mass

$$c_B = u_{L2} \frac{h_{L2}}{H} \tag{8.53}$$

and the following two equations developed by applying the Bernoulli equation to the top wall from Station 1 and 0 and to the free streamline (0 to 2), along which the pressure is constant and equal to p_0

$$\frac{p_1}{\rho} + \frac{c_B^2}{2} = \frac{p_0}{\rho} \qquad \frac{p_0}{\rho} + 0 + gH = \frac{p_0}{\rho} + \frac{u_{L2}^2}{2} + g h_{L2} \tag{8.54}$$

He obtained

$$c_B^2 = \frac{g(H^2 - h_{L2}^2)H}{(2H - h_{L2})h_{L2}} \tag{8.55}$$

$$h_{L2} = \frac{H}{2} \tag{8.56}$$

$$u_{L2} = (2g h_L)^{1/2} \tag{8.57}$$

The shedding rate is given as

$$Q = u_{L2} h_{L2} = h_{L2}(2g h_{L2})^{1/2} \tag{8.58}$$

This analysis describes a flow caused by gravitational forces.

Benjamin also considered flow out of a circular tube of radius r_t and obtained

$$\frac{u_{L2}}{(g r_t)^{1/2}} = 1.322 \tag{8.59}$$

$$\frac{c_B}{(g r_t)^{1/2}} = 0.767 \tag{8.60}$$

The shedding rate is

$$Q = c_B \pi r_t^2 = 0.767 (g r_t)^{1/2} \pi r_t^2 \tag{8.61}$$

where Q is the volumetric flow. These results have found application in analyses of slug and plug flows, where the liquid shed by a moving slug or plug is modeled as a Benjamin bubble.

8.10 Gas lift pumps

Gas lift pumps, which have no moving parts, are widely used. See, for example, de Cachard & Delhaye (1996), Clark & Dabolt (1986). The principle of operation is that gas is introduced into an immersed vertical pipe. The gas–liquid mixture in the pipe is lighter than the surrounding liquid, so it rises.

Consider the simple example shown in Figure 8.4 for an air–water system. A pipe is submerged in water to a depth H, so the water level in the pipe is the same as in the tank. An air injector is located close to the bottom of the submerged pipe at z_0. The pressure at this location is

$$(p_0 - p_a) = \rho_L g H \tag{8.62}$$

where p_a is the ambient pressure in the splash box.

When gas is admitted at a small rate, the gas–liquid flow in the pipe looks like a bubble column (discussed in Section 8.11). As the air rate increases, the height of this column (and the void fraction) increases. Eventually it attains a sufficient height to spill over into the splash pot. The liquid carryover increases with increasing gas flow. The theoretical challenge is to predict the relation between the liquid flow and the gas flow. This is done

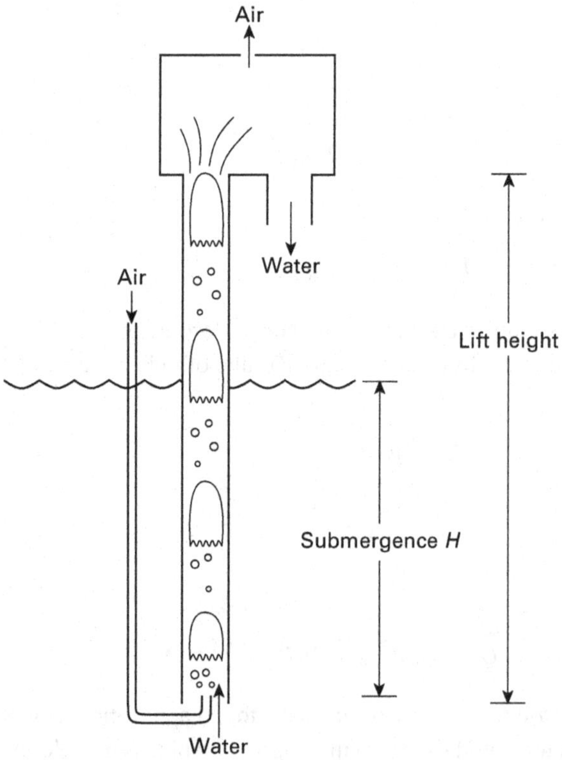

Figure 8.4 Sketch of a gas lift.

by using a momentum balance. The simplification is made that the flow may be considered to be uniform over the cross-section of the pipe. Then the one-dimensional balance is given by equation (1.73):

$$d\left[\frac{G^2Ax^2}{\alpha\rho_G} + \frac{GAx^2}{(1-\alpha)\rho_L}\right] = -Adp - g[\rho_G\alpha + (1-\alpha)\rho_L]Adz \\ -P\langle\tau_W\rangle dz + AG(u_G - u_L)dx \tag{8.63}$$

where α is the volume fraction of the gas and x is the quality, defined as the mass fraction of the flowing mixture that is a gas or a vapour and the flow is considered to be horizontal.

The flow considered in this section does not involve a phase change so x is not varying along the pipe. That is, the last term in (8.63) can be ignored. The term on the left side represents changes of the momentum flux. The second term on the right side is the change of pressure due to changes in hydrostatic head; the third term represents the frictional losses.

The pressure drop causing the flow is given by (8.62). Thus, for a given gas flow, the liquid flow increases with the submergence H/L, where L is the distance through which the liquid is lifted. Equation (8.63) can be further simplified by ignoring the influence of the momentum change. Then, a prediction of the liquid flow for a given gas flow and submergence requires information on the influences of liquid flow on the void fraction and on the frictional pressure loss.

The homogeneous model, described in Section 1.3, is not adequate. The separated flow model (Section 1.4) is needed. However, the Lockhart–Martinelli analysis (Section 1.5) is flawed since it does not, directly, recognize changes in the configuration of the phases.

The working curves for the operation of a gas lift are plots of the liquid flow as a function of the gas flow for different values of the submergence ratio, H/L. These curves show an increase in the liquid flow at small gas flows, where the pattern is a slug flow in a small-diameter pipe. Eventually, a maximum output is realized. This can occur because of a change in the flow pattern, which is associated with increases of frictional losses and of the void fraction in the column.

Clark & Dabolt (1986) give the following relation for an isothermal operation which has a 100% efficiency (no friction losses):

$$\rho_L gLV_L = V_G[p_a \ln(p_0/p_a)] \tag{8.64}$$

The term on the left side is the work to lift a volume of liquid, V_L, through a height L. The term on the right side is the work needed to compress an ideal gas with a volume, V_G, (isothermally) from a pressure p_a to a pressure p_0. The ratio V_G/V_L represents the ratio of the volumetric gas flow to the volumetric liquid flow, if the operation has a 100% efficiency.

The efficiency can be calculated along the operation curves by considering the ratio of the volumetric flows of the gas and the liquid. This increases with increasing gas velocity and reaches a maximum value before the operating curve reaches its maximum. Thus, the location of the maximum efficiency in the operating curve is at a lower gas velocity than required to reach a maximum liquid flow (see Clark & Dabolt, 1986).

The void fraction is obtained from (8.30). The momentum balance is given by (8.63). If phase changes and changes in the momentum flux are ignored, one obtains

$$A\rho_L gH = \int_0^L g[\rho_G\alpha + (1-\alpha)\rho_L)]Adz + \int_0^L P\langle\tau_W\rangle dz \tag{8.65}$$

where (8.62) has been substituted for $(p_0 - p_a)$. Slug flow is considered. The friction term is important only when liquid slugs are present. Thus τ_W is given by (8.51) and (8.53) and

$$\langle\tau_W\rangle = (1-\alpha)\tau_W \tag{8.66}$$

The friction factor is obtained from relations for single-phase turbulent flows if aeration of the liquid slugs is ignored. (Clark & Dabolt (1986) used the Martinelli equation to calculate frictional pressure losses.)

Measurements by Garich & Besserman (1986) for an air–water system in a column with a diameter of 1 inch are shown in Figure 8.5. Void fractions are plotted in Figure 8.6 Submergence ratios of 0.55, 0.68 and 0.75 were used. Slug flow is the dominant pattern at low u_{GS}. A maximum in the liquid flow occurs when a churn flow was observed. However, the location for maximum efficiency is in the region where a slug flow exists. Transition from a slug flow to a churn flow is not well defined. Garich & Besserman (and

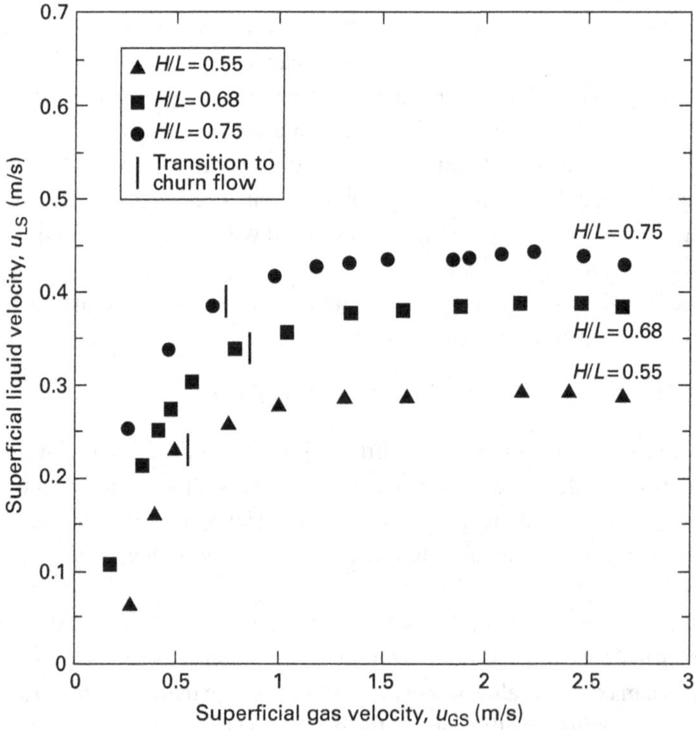

Figure 8.5 Operation curve for a gas lift in a 1.0 inch column containing water.

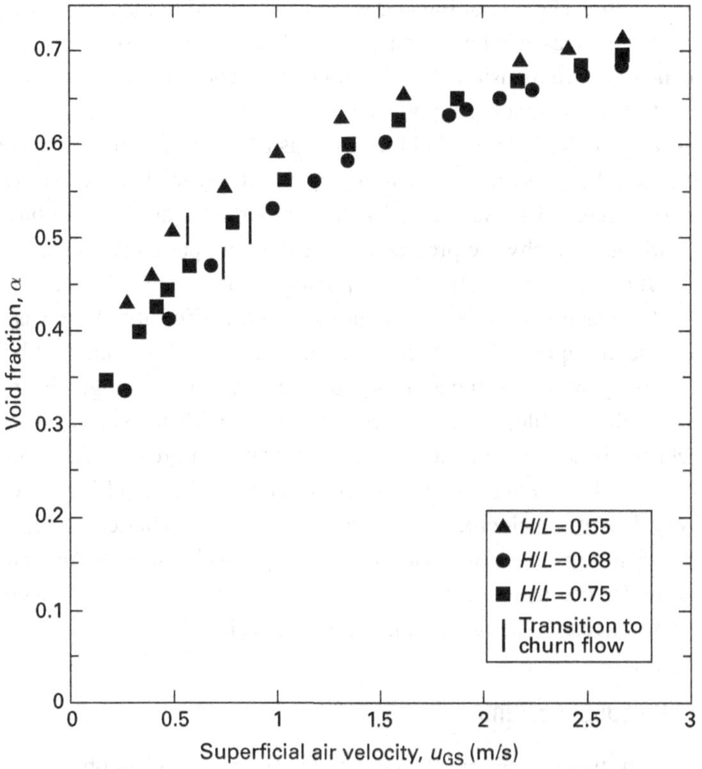

Figure 8.6 Void fractions for a gas lift in a 1.0 inch column containing water.

others) have suggested transition at a volume fraction of $\alpha \approx 0.52$. (It should be noted that Figure 2.11 indicates a transition at $u_{GS} \approx 2$ m/s.)

Gas lifts are used for pipes with a wide range of lengths and diameters. Information about the development of slug patterns is given by Kaji *et al.* (2009). If the pipe is long enough, the gas density and the void fraction can vary along the length. Then, the simplification of ignoring the change of momentum in (8.63) could be in error. Also, as pointed out in Section 2.4, slug flow is not observed in large-diameter pipes. Instead, the flow pattern is dominated by cap bubbles. Since the rise velocities of Taylor bubbles and cap bubbles are close, the equation used to predict α for slug flow might be applicable.

8.11 Bubble columns

8.11.1 Volume fraction of gas in bubble columns

Bubble columns involve the passage of gas bubbles in a stationary liquid. They find application as gas absorbers, gas strippers and reactors. All of these are mass transfer operations which involve a transfer of gaseous or vapor components between a gas (or vapor) and a liquid. The rate of transfer is strongly related to the interfacial area per unit volume, which increases with increases of volume fraction, α, and decreases in the bubble diameter. The expansion of the column, α, depends on the velocity of the bubbles, U_S, relative to the surrounding liquid.

There is a similarity between bubble columns and sedimenting or fluidizing beds of solid particles, where the solids fraction can be related to the free-fall velocity in an infinite fluid (discussed in Section 8.14). An important feature is the phenomenon of hindered settling, whereby the presence of surrounding particles decreases the free-fall velocity. Section 8.13 presents an interpretation (there are others), which explains hindered settling as due to an effective increase in the viscosity of the suspension. It is tempting to use this approach to describe bubble columns. However, bubble columns are different in many ways. Changes in liquid viscosity cause changes in the shape and orientation of the bubbles. Coalescence, associated with increased gas velocity, can create larger bubbles. Contamination of the interface can greatly affect the behavior of bubble columns (Anderson & Quinn, 1970). Hindrance of the bubble rise velocity can be seen at very low gas velocities where the bubbles are spherical. See, for example, measurements in tap water at small superficial gas velocities in Anderson & Quinn (1970, Figure 3). The usual operation involves much more complicated configurations of the gas and increases of U_S with increasing gas velocity.

8.11.2 Behavior of a bubble column

An established theoretical prediction of the behavior of bubble columns is not available. However, it is useful to give a qualitative description by citing a few examples. These rely on chemical engineering research projects performed at the University of Illinois (Mueller & Wilger, 1979; Norman & Hummel, 1980; Allen & Gillogly, 1979).

Mueller & Wilger used a sintered glass support plate of medium porosity in a glass column with a diameter of 4.45 cm. Water (1 cP, 72 dynes/cm) and pentadecane (2.81 cP, 30.2 dynes/cm) were studied. A swarm of tiny bubbles (less than 1 mm) formed right at the support. These rapidly coalesced in the vicinity of the support for low gas velocities and within 10 cm of the support at superficial gas velocities of greater than 1 cm/s.

Two major flow regimes were observed both for water and for pentadecane. These are analogous to the particulate and aggregative regimes defined for fluidized beds of solids. For want of a better choice, this terminology will be used to describe bubble beds. At low gas rates, the particulate regime exists. The bubbles remain distinct, even when in close proximity. For superficial gas velocities of about 0.3 cm/s, small spherical entrance bubbles (diameters of about 1 mm) were still present at one-third of the distance up the column. However, most of these entrance bubbles coalesced to form bubbles which are spherical for smaller bubbles and ellipsoidal for larger bubbles. They had sizes of 0.4 cm and 0.3 cm by 0.8 cm, respectively. As the flow was increased to a superficial gas velocity of about 0.97 cm/s, more collisions occurred and the entrance bubbles were present only up to 10 cm from the bottom. The flow was dominated by ellipsoidal bubbles of

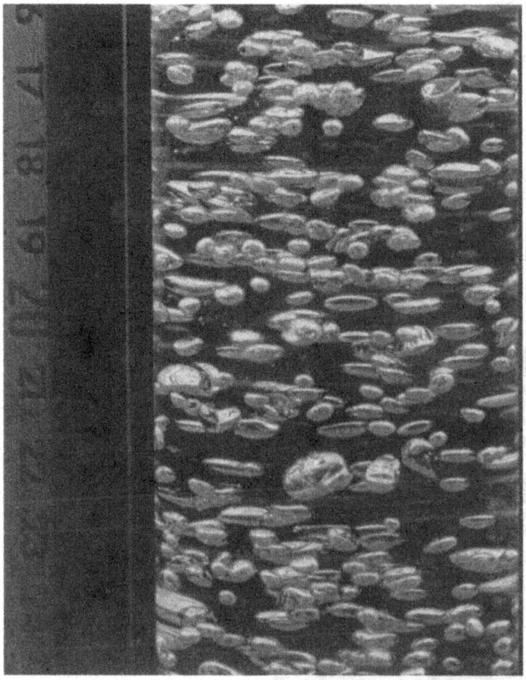

Figure 8.7 Bubbles of pentadecane near the top of the column. Superficial air velocity is 0.97 cm/s.

approximately uniform size (0.6 cm by 1.3 cm). A photograph for pentadecane is shown in Figure 8.7.

As the gas velocity increases the bubbles take on a coherent large-scale singular motion. Pockets which are devoid of bubbles, and larger bubbles which have a cap shape, appear. Figure 8.8 shows this aggregative pattern in pentadecane for a superficial air velocity of 7.7 cm/s.

The studies of Norman & Hummel (1980) in 5.15 cm and 10.45 cm columns and of Allen & Gillogly (1979) in a 5.15 cm column used plates that were punctured with 0.37 mm holes to support the liquid column. The behavior can be represented by plots of the volume fraction, α, as a function of the superficial velocity of the gas. Operation curves were found to be roughly the same for sintered glass and for sieve tray supports.

Changes of the height of the liquid in the original column of liquid had only a small effect because of the increased opportunity for coalescence. Increases in the diameter of the column were accompanied by slightly smaller void fractions.

Examples of operating curves for different liquids are shown in Figure 8.9 where D is the diameter of the tube and L_0 is the height of the liquid before aeration. These are constructed from measurements made by Mueller & Wilger (1979), Norman & Hummel (1980) and Allen & Gillogly (1979). No consistent effect of liquid properties is noted.

For example, runs with deionized water and with cyclohexane are situations for which the viscosities are the same and the surface tension decreases from 72 dynes/cm to

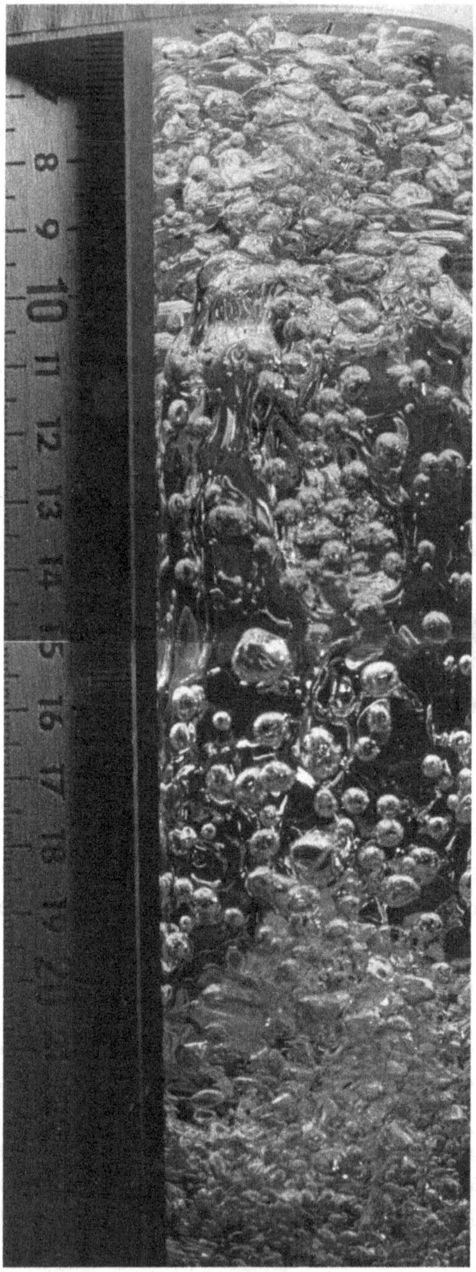

Figure 8.8 Aggregative pattern in pentadecane for a superficial air velocity of 7.7 cm/s.

25.5 dynes/cm. The decrease in surface tension is associated with an increase in void fraction, as would be expected if the bubble diameter decreases. The results from runs with deionized water and with benzene show small differences due to a large change in surface tension for liquids with viscosities of 0.652 cP and 1 cP. The remaining liquids

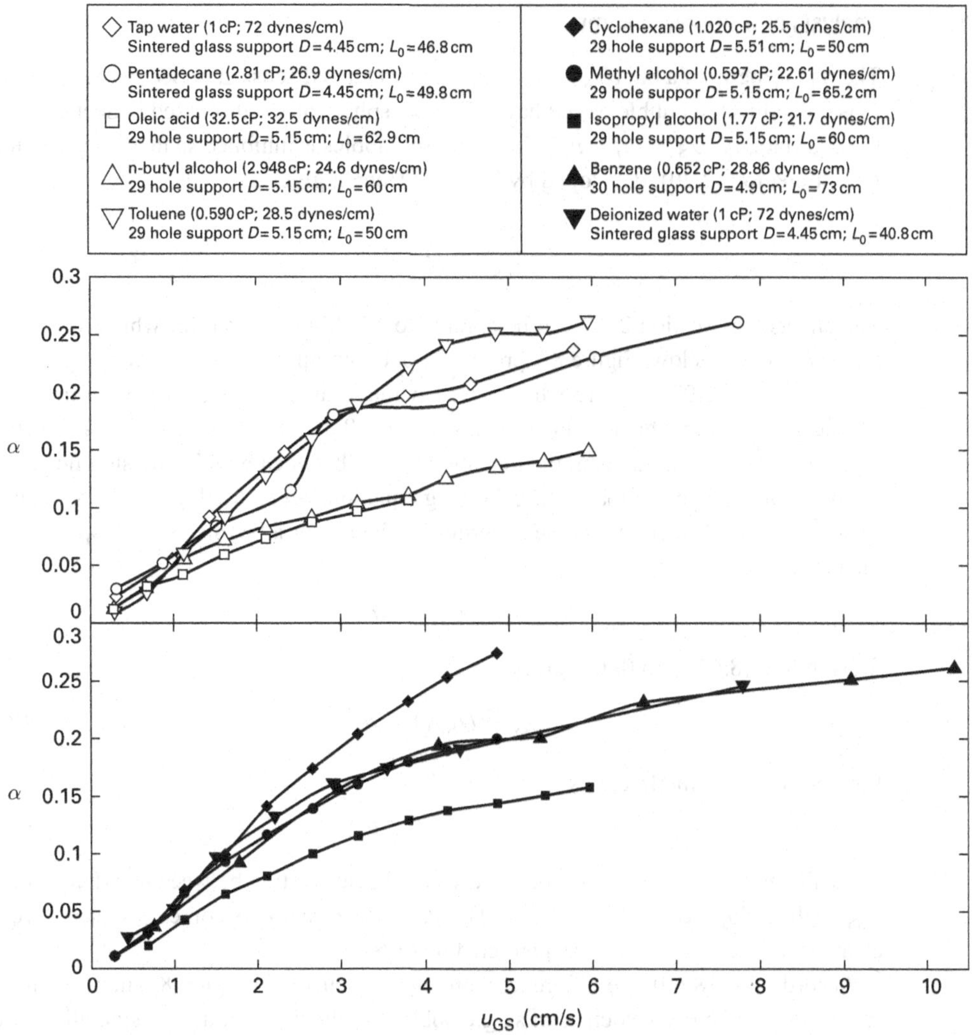

Figure 8.9 Operating curves for bubble columns.

considered in Figure 8.9 show the effect of liquid viscosity for surface tensions which are varying over a small range, 21.7 to 28.9 dynes/cm.

A consideration of the data for oleic acid (with μ = 32.5 cP) would suggest that increases in liquid viscosity are associated with large decreases in void fraction. However, the operation curve for isopropyl alcohol (μ = 1.77 cP) is close to that for oleic acid. Furthermore, a comparison of the measurements for methyl alcohol and toluene shows a large difference even though the viscosity and surface tension are close.

These experiments seem to be consistent with the findings of Anderson & Quinn (1970) that interfacial impurities are playing an important role in determining the performance of bubble columns.

8.11.3 Analysis of a bubble column

The rise velocity of a bubble, $U_{S\infty}$, in stagnant water is discussed in Section 8.5.2, where r_P is the radius the bubble would have if it were spherical. As discussed in Section 2.2.2, the rise velocity, $U_S = u_G - u_L$, in a swarm of bubbles is inhibited so that $\overline{U}_S$ is less than $U_{S\infty}$. Equation (2.20), developed by Zuber & Hench, gives

$$\frac{\overline{U}_S}{U_{S\infty}} = (1-\alpha)^{1/2} \tag{8.67}$$

The analysis in Section 2.7.2 can be applied to a bubble column, for which $u_L = 0$ and the gas velocity is low. Figure 8.7 presents a photograph of a column operating with an air velocity of 0.97 cm/s. The liquid was pentadecane. At large gas velocities, this "particulate" pattern breaks down. Thus, Figure 8.8, for a gas velocity of 7.7 cm/s, represents a situation that is in the transition region between bubbly and slug flow.

From conservation of mass, the following equation can be used to define the relation between the void fraction and the superficial velocity for a bubble column operating in the particulate regime

$$u_{GS}A = \overline{U}_S \alpha A \tag{8.68}$$

Substituting (8.67) into (8.68) gives

$$Au_{GS} = U_{S\infty}(1-\alpha)^{1/2}\alpha A \tag{8.69}$$

For very low volume fractions

$$u_{GS} = U_{S\infty}\alpha \tag{8.70}$$

Thus, the data in Figure 8.9 can be fitted with a linear relation between α and u_{GS} at low u_{GS} with a slope equal to $(U_{S\infty})^{-1}$. The data show that α becomes less sensitive to changes in u_{GS} at larger u_{GS}, as predicted by (8.69).

According to (8.69), the differences amongst the curves in Figure 8.9 indicate differences in $U_{S\infty}$, the rise velocities of single bubbles in the liquid under consideration. (See the discussion of measurements of rise velocities of single bubbles in Section 8.5.2.)

8.12 Bubbly gas–liquid flow in vertical pipes, distribution of bubbles

Figure 2.9 maps the bubbly regime for upward flow of air and water in a pipe. The results for bubble columns would be applicable in the limit of a zero value of the superficial liquid velocity. At finite liquid velocities, and small gas velocities, the bubble pattern is similar to that shown in Figure 8.7. Section 2.7.2 considers the hypothetical case of a uniform velocity field. Equation (2.18) gives a prediction of the dependency of the void fraction on the superficial gas and liquid velocities, where $\overline{U}_S$ is given by (2.20).

However, the liquid velocity profile is not uniform. This gives rise to a lift force in the r-direction, F_{Lr}, equation (10.29). For upward flows, a drift velocity of bubbles toward the wall, V_{DL}, results from this force. Measurements of void fractions, for upward bubbly flow of air and water in a 1.75 cm pipe, have been reported by Serizawa *et al.* (1975a, b). Under fully developed conditions, these show a relatively uniform distribution of void fraction over most of the pipe cross-section.

However, large maxima are observed at the wall. Turbulent mixing balances V_{DL}. A force balance in the radial direction gives an expression for V_{DL}:

$$\frac{4}{3}\pi r_{\mathrm{P}}^3 F_{\mathrm{L}r} = \frac{1}{2} C_{\mathrm{D}} \pi r_{\mathrm{P}}^2 \rho_{\mathrm{f}} (u_r - v_{\mathrm{P}r}) |\vec{u} - \vec{v}_{\mathrm{P}}| \tag{8.71}$$

where $\vec{v}_{\mathrm{P}}$ is the bubble velocity. The average liquid velocity in the r-direction, $\bar{u}_r$, is zero under fully developed conditions, so (8.71) provides an equation for the drift velocity, where $V_{\mathrm{DL}} = \bar{v}_{\mathrm{P}r}$. From (10.29)

$$F_{\mathrm{L}r} = -C_{\mathrm{L}}(\overline{u_{\mathrm{G}z}} - \overline{u_{\mathrm{L}z}})\rho_{\mathrm{L}} \frac{d\overline{u_z}}{dr} \tag{8.72}$$

where C_{L} is the lift coefficient. Substitute (8.72) into (8.71) and take an average. The substitution of zero for $\bar{u}_r$ provides a prediction of $V_{\mathrm{DL}} = \bar{v}_{\mathrm{P}r}$.

One approach to predicting the distribution of voids is to balance V_{DL} with a diffusion velocity

$$V_{\mathrm{DL}} = -\varepsilon_{\mathrm{P}}^{\mathrm{E}} \frac{d\alpha_{\mathrm{P}}}{dr} \tag{8.73}$$

Measurements of the profiles of mean velocity in the liquid phase show an approximate agreement with what would be observed for turbulent flow of a single phase. Similarly, one could assume that $\varepsilon_{\mathrm{P}}^{\mathrm{E}}$ is proportional to the eddy viscosity (Serizawa *et al.*, 1975a, b). Integration of (8.73) provides an equation for the profile of α_{P}.

The velocity difference, $\overline{u_{\mathrm{G}z}} - \overline{u_{\mathrm{L}z}}$, changes sign in downflows so that the radial lift force is toward the center of a pipe. The consequence of this is discussed in Section 2.9. The procedure outlined above could be used to calculate the spatial distribution of bubbles. However, possible coalescence could be an issue.

8.13 Performance of downcomers in tray columns

An account of the behavior of trays in a distillation column is given in Lockett (1986). A typical stage in a sieve plate contactor consists of a perforated tray supporting a liquid. A gas or vapor passing through the tray forms bubbles at the perforations. Mass transfer occurs between these bubbles and the liquid on the tray. Liquid is transferred to the tray from the tray above by a downcomer.

The level of the liquid on the tray is controlled by a weir which is part of a second downcomer that transfers liquid to the tray below it. For low flows, the liquid moves over the weir and rolls down the wall on to the liquid in the downcomer. As the liquid flow increases, it shoots over the weir and strikes the liquid with a relatively large velocity. This creates bubbles by a mechanism similar to that observed for plunging jets (Lin & Donnelly 1966; Burgess *et al.*, 1972; Van De Sande & Smith, 1973; Koga, 1982; McKeogh &

Ervine, 1981). The mixture in the downcomer then starts to resemble a bubble column. Furthermore, the impingement of the jet can cause a large-scale circulation which carries bubbles to the bottom of the downcomer and out on the lower tray. This contributes to a decrease in the efficiency of mass transfer processes. For large enough liquid flows a condition is reached for which the downcomer is completely filled with froth and foam. This could lead to flooding. A knowledge of the behavior of the gas–liquid mixture in the downcomer is thus central to understanding the behavior of the tray column.

This prompted the development of a two-dimensional transparent column to display the hydraulics of a downcomer (deZutel & Mahoney, 1992; Lee, 1993; Remus, 1994; Mclean, 1996). An example from these works is the effort of McLean to reduce bubble creation in the downcomer. The column that he used had three trays. It was 46 in high, 24 in wide and 2 in deep. Air was admitted at the bottom and liquid at the top. The downcomer on the middle tray was modified so that the impingement of a strong jet on the surface of the liquid was avoided. Thus, aeration and large-scale circulation were eliminated. A screen (50 wires per inch) at a 45° angle was located in the downcomer. This prevented the jet formed by liquid flowing over the weir from striking the surface of the liquid. Thus, a clear liquid formed under the screen. Figure 8.10

Figure 8.10 Prevention of flooding in a sieve tray column. Insertion of a screen in the downcomer prevents the creation of bubbles/foam.

shows a picture of the flow when this screen was inserted. The conditions were such that the downcomer would be completely filled with froth and foam if the insert was not used.

8.14 Sedimentation and fluidization

8.14.1 Focus

Sections 8.2 and 8.3 focus on single solid particles in a stagnant fluid. However, most applications involve a swarm of particles or bubbles for which gravity causes a relative velocity which is smaller than would be realized for single particles. This hindered settling increases with increasing volume fraction. Important parameters are the volume fraction, α_P, of the particles and the Reynolds number characterizing the flow, $\mathrm{Re_P} = d_P U_S \rho_F / \mu_F$, where U_S is the average slip velocity between the particles and the fluid, d_P is the particle diameter, ρ_F is the density of the continuous phase and μ_F is its viscosity.

Two systems are considered in this section: sedimentation and fluidization. Sedimentation involves particles settling through a stagnant fluid. Fluidization involves the flow of fluid through a bed of particles at a large enough velocity for the particles to be suspended in the fluid. Hanratty & Bandukwala (1957) provide a discussion of these processes.

8.14.2 Sedimentation

The studies of sedimentation and fluidization were carried out by Hanratty & Bandukwala in a glass pipe that had a screen on which a bed of uniformly sized spherical particles was supported. These particles were fluidized by a liquid flow. The height of the bed, and therefore the volume fraction, increases with increasing flow rate after the critical velocity needed to suspend the particles is reached.

The flow was turned off and the height of the bed decreased. Particles settled out on to the support screen. The settling velocity, U_C, was determined by measuring the rate at which the top of the bed moves downward. The volume fraction remains constant during this process. It can be changed by increasing the fluid velocity and, therefore, the height of the original expanded bed.

For single spherical particles, the settling velocity is given by equation (8.4). For small $\mathrm{Re_P}$, the drag coefficient, C_D, is given by (8.23) and the settling velocity, by (8.24). Hawksley (1950) adapted equation (8.24), for a single particle, to describe a settling suspension by considering that the medium through which a particle is settling has different properties from the fluid. Its density is that of the suspension

$$\rho_H = \alpha_P \rho_S + (1 - \alpha_P)\rho_F \tag{8.74}$$

where α_P is the volume fraction of the particles. View the bed from a frame of reference moving with the settling velocity of the bed, U_C. Because the presence of the particles constricts the area, the velocity of the fluid in the bed is

$$U_S = \frac{U_C}{(1 - \alpha_P)} \tag{8.75}$$

Hawksley represented the suspension viscosity by using an equation developed by Vand (1948), for the viscosity of a concentrated suspension of spherical particles.

$$\mu_C = \mu_F \exp\left(\frac{2.5\alpha_P}{1 - \frac{39}{64}\alpha_P}\right) \tag{8.76}$$

The following equation was derived for the sedimentation velocity at a low Reynolds number by substituting (8.74) for ρ_F, (8.76) for μ_F and (8.75) for U_S in (8.24):

$$\frac{U_C}{U_S} = (1 - \alpha_P)^2 \exp\left(\frac{-2.5\alpha_P}{1 - \frac{39}{64}\alpha_P}\right) \tag{8.77}$$

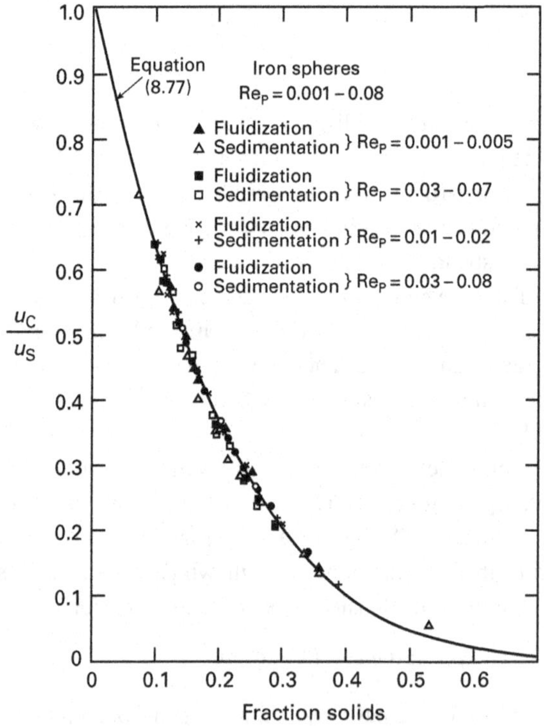

Figure 8.11 Fluidization and sedimentation data for iron spheres in glycerine–water solutions ($Re_P = 0.001$–0.08). Hanratty & Bandukwala, 1957.

Figure 8.11 presents a plot of the measured U_C versus α_P, where U_C is the settling velocity and U_S is given by (8.24). As seen in Figure 8.11, equation (8.77) provides an excellent description of measurements at $Re_P < 0.07$.

8.14.3 Fluidization

In studies of fluidization, the fluid is passed through a bed of particles resting on a porous support. At a large enough flow rate, the particles are suspended in the fluid. Increases in the flow cause the bed to expand, so that the volume fraction of the solids decreases with increasing fluid velocity. This relationship was investigated by Hanratty & Bandukwala (1957) in the same equipment in which sedimentation was studied. As seen in Figure 8.11, the Vand–Hawksley equation also describes the fluidization data at low Re_P. At high particle Reynolds numbers one should use (8.4) rather than (8.24) to calculate U_S. The relations presented in the literature for C_D for a single sphere were used by Hanratty & Bandukwala (1957). The density and viscosity of the fluid were used to calculate Re_P appearing in the relation for C_D.

Figure 8.12 shows measurements for $Re_P > 0.07$. Here, U_C is the settling velocity of a suspension or the fluidizing velocity and U_S is given by (8.4). The data for individual runs are not described so well as indicated in Figure 8.11 for low Re_P. However, rough

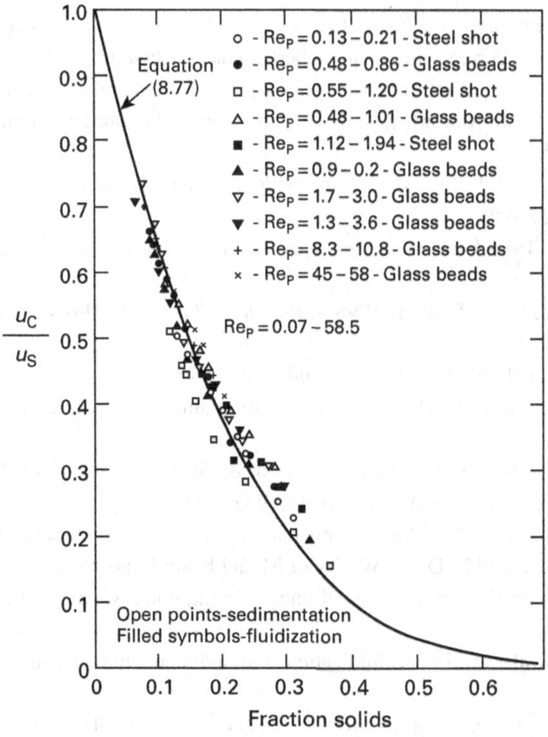

Figure 8.12 Fluidization and sedimentation data at high Reynolds numbers. ($Re_P = 0.07$–58.5). Hanratty & Bandukwala, 1957.

agreement between fluidization and sedimentation measurements is noted. Both Figure 8.11 and Figure 8.12 show the hindered settling phenomenon whereby the presence of surrounding particles causes the settling velocity or the fluidization velocity to decrease relative to U_S.

Some comments about the appearance of fluidized beds is appropriate. At particle Reynolds numbers below 0.8, the bed had a uniform appearance, with no large movements of the particles. The particles, however, were continually coming together in small groupings and then dispersing. At particle Reynolds numbers greater than 2, coherent movements of large groups of particles were observed. For the highest Reynolds numbers studied by Hanratty & Bandukwala (for both iron shot and glass spheres beads) the particle flow pattern consisted of a random eddying motion. There were large variations in the solids concentration, and the motion of the solids was quite similar to that observed in "aggregative" gas–solid systems (bubbly fluidized beds).

References

Allen, G.T. & Gillogly, D.H. 1979 The effect of viscosity on bubble coalescence. Chem. Eng. Project Report, University of Illinois.

Anderson, J.L. & Quinn, J.A. 1970 Bubble columns: transitions in the presence of trace contaminants. *Chem. Eng. Sci.* 25, 373–380.

Batchelor, G.K. 1967 *An Introduction to Fluid Dynamics*. Cambridge: Cambridge University Press.

Benjamin, T.B. 1968 Gravity currents and related phenomena. *J. Fluid Mech.* 31, 209–248.

Bird, R.B., Stewart, W.E. & Lightfoot, E.N. 1960, *Transport Phenomena*. New York: Wiley.

Burgess, J.M., Molloy, N.A. & McCarthy, M.J. 1972 A note on the plunging liquid jet reactor. *Chem. Eng. Sci* 12, 442.

de Cachard, F. & Delhaye, J.M. 1996 A slug-churn model for small-diameter airlift pumps. *Int. J. Multiphase Flow* 22, 627–649.

Clark, N.N. & Dabolt, R.J. 1986 A general design equation for air lift pumps operating in slug flow. *AIChE Jl* 32, 56–64.

Crowe, C.C., Sommerfeld, M. & Tsuji, Y. 1998 *Multiphase Flows With Droplets and Particles*. Boca Raton, FL: CRC Press.

Davies, R.M. & Taylor, G.I. 1950 *Proc. R. Soc. Lond.* A200, 375–390.

Garich, R. & Besserman, M. 1986 Design of an air lift pump. Chem. Eng. Project Report, University of Illinois.

Grace, J.R. & Weber, M.E. 1982 Hydrodynamics of drops & bubbles, 1.204–1.246. In *Handbook of Multiphase Systems*, ed. G. Hetsroni. New York: McGraw-Hill.

Haberman, W.L. & Morton, R.K. 1953 An experimental investigation of the drag and shape of air bubbles arising in various liquids. David W. Taylor Model Basin Report 802.

Hadamard, J.S. 1911 Mouvement permanent lent d'une sphere liquide et visqueux. *C. R. Acad. Sci. Paris* 152, 1735–1738.

Hanratty, T.J. & Bandukwala, A. 1957 Fluidization and sedimentation of spherical particles, *AIChE Jl* 3, 293–296.

Harmathy, T.Z. 1960 Velocity of large drops and bubbles in media of infinite and restricted extent. *AIChE Jl* 6, 281–288.

Hawksley, P.G.W. 1950 *Some Aspects of Fluid Flow*. New York: Arnold Press.

Hetsroni, G. 1982 *Handbook of Multiphase Systems*. New York: McGraw-Hill.

Kaji, R., Azzopardi, B.J. & Lucas, D. 2009 Investigation of flow development of co-current gas–liquid slug flow. *Int. J. Multiphase Flow* 35, 335–348.

Koga, M. 1982 Bubble entrainment in breaking wind waves. *Tellus* 34, 481.

Lapple, C.E. & Shepherd, C.B. 1940 Calculation of bubble trajectories. *Ind. Eng. Chem.* 32, 605–617.

Lee, D.Y. 1993 *Flow regimes in a sieve tray tower*. B.Sc. thesis, Chemical Engineering, University of Illinois.

Levich, V.G. 1962 *Physicochemical Hydrodynamics*. Englewood Cliffs, NJ: Prentice-Hall.

Lin, T.J. & Donnelly, H.G. 1966 Gas bubble entrainment by plunging laminar liquid jets, *AIChE Jl* 36, 1161.

Lockett, M.J. 1986 *Distillation Tray Fundamentals*. Cambridge: Cambridge University Press.

McKeogh, E.J. & Ervine, D.A. 1981 Air entrainment rate and diffusion Pattern of plunging liquid jets. *Chem. Eng. Sci.* 36, 1161.

McLean, R. 1996 Downcomer hydraulics in a sieve tray column. B.Sc. thesis, University of Illiois.

Mueller, M.K. & Wilger, D.M. 1979 Operation of a bubble column. Chem. Eng. Project Report, University of Illinois.

Nicklin, D.J., Wilkes, J.O. & Davidson, J.F. 1962 Two-phase flow in vertical tubes. *Trans. IChem E* 40, 61.

Norman, C.D. & Hummel, D. 1980 Operation of a bubble column. Chem. Eng. Project Report, University of Illinois.

Remus, M. 1994 Fluid mechanics of a sieve tray distillation column. B.Sc. thesis, Chemical Engineering, University of Illinois.

Schlichting, H. 1955 *Boundary Layer Theory*. New York: McGraw-Hill.

Serizawa, A., Kataoka, I. & Mishiyoshi, I. 1975a Turbulence structure of air–water bubbly flow. *Int. J. Multiphase Flow* 2, 235–246.

Serizawa, A., Kataoka, I. & Mishioshi, I. 1975b Turbulence structure of air–water bubbly flow. *Int. J. Multiphase Flow* 2, 247–259.

Soo, S.L. 1967 *Fluid Dynamics of Multiphase Systems*. Waltham, MA: Blaisdell.

Streeter, V.L. 1948 *Fluid Dynamics*. New York: McGraw-Hill.

Vand, V.J. 1948 *J. Phys. Colloid Chem.* 52, 271.

Van De Sande, E. & Smith, J.M. 1973 Surface entrainment of air by high velocity water jets. *Chem. Eng. Sci.* 28, 1161.

Wallis, G.B. 1969 *One Dimensional Two-Phase Flow*. New York: McGraw-Hill.

Zuber, N. & Findlay, J.A. 1965 Average volumetric concentration in two-phase flow systems. *J. Heat Transfer, Trans. ASME Series C*, 87, 453–468.

de Zutel, C. & Mahoney, A.W. 1992 Hydraulics of a sieve tray tower. Chemical Engineering Projects Laboratory, University of Illinois.

9 Horizontal slug flow

9.1 Prologue

Chapter 2 gives considerable attention to slug flow because of its central role in understanding the configuration of the phases in horizontal and inclined pipes. Several criteria have been identified to define the boundaries of this regime: (1) viscous large-wavelength instability of a stratified flow; (2) Kelvin–Helmholtz instability of a stratified flow; (3) stability of a slug; (4) coalescence of large-amplitude waves. Bontozoglou & Hanratty (1990) suggested that a sub-critical non-linear Kelvin-Helmholtz instability could be an effective mechanism in pipes with very large diameters, but this analysis has not been tested. A consideration of the stability of a slug emerges as being particularly important. It explains the initiation of slugs for very viscous liquids, for high-density gases, for gas velocities where wave coalescence is important and for the evolution of pseudo-slugs into slugs. Chapter 2 (Section 2.2.5) outlines an analysis of slug stability which points out the importance of understanding the rate at which slugs shed liquid. Section 9.2 continues this discussion by developing a relation for Q_{sh} and for the critical height of the liquid layer needed to support a stable slug. Section 9.3 develops a tentative model for horizontal slug flow. Section 9.4 considers the frequency of slugging.

9.2 Slug stability

9.2.1 Necessary conditions for the existence of slugs

Figure 9.1 presents simplified sketches of the front and the tail of a slug in a pipeline. The front has a velocity c_F; the back has a velocity c_B. The stratified liquid layer in front of the slug has a velocity and area designated by u_{L1}, A_{L1}. The mean velocity of the liquid in the slug is u_{L3}. The slug is usually aerated; the mean volume fraction of gas in the slug is designated by α. The gas at station 1 is moving from left to right at a velocity u_{G1}. The assumption is made that the velocity fields can be approximated as being uniform.

A first necessary condition for the existence of slugs, proposed by Ruder *et al.* (1989), is that the front of the slug may be considered to be a hydraulic jump, which is an irreversible flow with a large amount of dissipation. This model is more easily formulated for a rectangular channel. Consider a control volume surrounding the front of the slug.

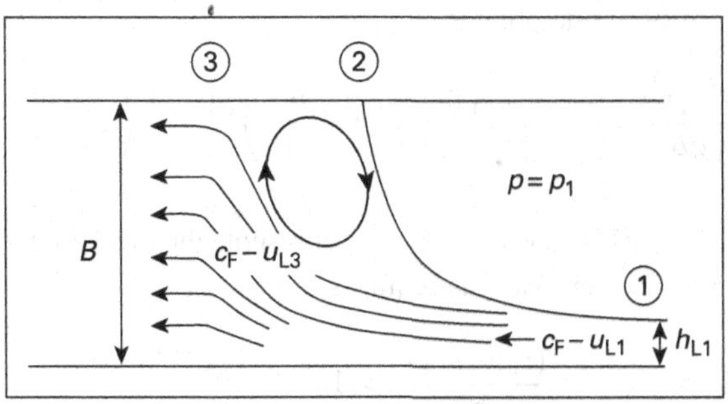

(a) Front end of a slug seen by an observer moving with a velocity c_F

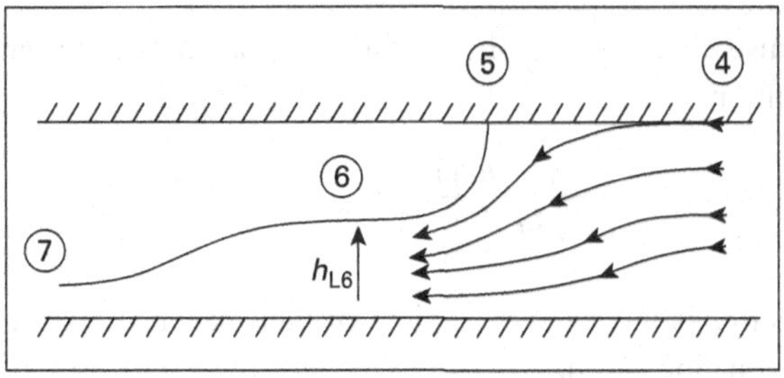

(b) Steady-state inviscid tail of a slug

Figure 9.1 Simplified sketches of the front and tail of a slug in a pipeline.

The sides are the walls of the channel. Station 1 and station 3 are sufficient distances in front of the slug and behind the slug for uniform flows to be considered. The influence of gas entrainment in the slug is ignored. Mass and momentum balances, written in a framework moving with velocity c_F, give

$$(c_F - u_{L1})h_{L1} = (c_F - u_{L3})B \tag{9.1}$$

$$\rho_L(c_F - u_{L3})^2 B - \rho_L(c_F - u_{L1})^2 h_{L1} = (p_1 - p_3)B + \frac{\rho_L g h_{L1}^2}{2} - \frac{\rho_L g B^2}{2} \tag{9.2}$$

where ρ_L is the liquid density, g is the acceleration of gravity and p_3 is the pressure at the top of the conduit at station 3. The last two terms in (9.2) are the hydrostatic forces on the front and the back of the jump; $h_{L1}/2$ and $B/2$ are the centroids of these areas. Equation (9.1) can be used to eliminate $(c_F - u_{L3})$ from (9.2). Since $p_2 = p_1$

$$\frac{(p_3 - p_2)}{\rho_L g B} = \frac{(c_F - u_{L1})^2}{gB}\left[\frac{h_{L1}}{B} - \left(\frac{h_{L1}}{B}\right)^2\right] - \frac{1}{2} + \frac{1}{2}\left(\frac{h_{L1}}{B}\right)^2 \tag{9.3}$$

If the flow changes reversibly from 1 to 3, the Bernoulli equation is applicable, so

$$\frac{(p_3 - p_2)_{\text{rev}}}{\rho_L g B} = \frac{1}{2}\frac{(c_F - u_{L1})^2}{gB}\left[1 - \frac{h_{L1}}{B} - \left(\frac{h_{L1}}{B}\right)^2\right] + \left(\frac{h_{L1}}{B} - 1\right) \tag{9.4}$$

By subtracting (9.3) from (9.4) one obtains

$$\frac{(p_1 - p_3)_{\text{rev}}}{\rho_L g B} - \frac{(p_3 - p_2)}{\rho_L g B} = \frac{(c_F - u_{L1})^2}{gB}\left[\frac{1}{2}\left(1 - \frac{h_{L1}}{B}\right)^2\right] - \frac{1}{2}\left(1 - \frac{h_{L1}}{B}\right)^2 \tag{9.5}$$

The left hand side of (9.5) must be positive since it represents the mechanical energy that is dissipated. From (9.5), it is seen that this requires

$$\frac{(c_F - u_{L1})^2}{gB} > 1 \tag{9.6}$$

Thus, for an unaerated slug to exist in a rectangular channel it is necessary that (9.6) is satisfied (see equation (2.1)).

Ruder *et al.* (1989) also considered the front of a slug in a circular pipe and derived a similar necessary condition

$$\frac{(c_F - u_{L1})}{(gd_t)^{1/2}} > N \tag{9.7}$$

where N varies with h_{L1}/d_t as indicated in Table 1 of their paper. It takes on values of 0.9459 to 1.2106 (see equation (2.2)).

The second necessary condition for a circular pipe (previously discussed in Chapter 2 depends on the behavior of the tail of the slug. An unaerated slug picks up liquid from the stratified flow in front of it at a rate given by

$$\text{Rate of pickup} = (c_F - u_{L1})(A_{L1}) \tag{9.8}$$

Liquid is shed out of the back of a slug at a rate Q_{sh}. For a growing slug, it is necessary that

$$(c_F - u_{L1})(A_{L1}) > Q_{sh} \tag{9.9}$$

For a neutrally stable slug, the stratified layer over which the slug is moving has an area, A_{L0}, given by

$$(c_F - u_{L0})(A_{L0}) = Q_{sh} \tag{9.10}$$

Thus, a slug cannot develop on a stratified layer with an area less than A_{L0}. The implementation of this necessary condition requires an understanding of Q_{sh}, that is, a physical model for the tail of a slug. An appealing approach would be to use a pseudo-steady-state solution, such as has been done for the tail of a roll wave (Sections 7.5, 7.6), for which viscous drag at the wall is an important consideration. However, photographs of the tails of slugs (Ruder *et al.*, 1989) do not seem to support such an assumption. At low throughputs, the height of the liquid behind the slug (at least, in the portion close to the top wall) is observed to decrease rapidly enough for inertia effects to dominate effects of viscous and turbulent stresses. One might, therefore, expect that, to first order, the flow may be considered inviscid.

This led Ruder *et al.* (1989) to suggest that the back of an unaerated slug can be approximated as a Benjamin bubble (see Section 8.9). Thus, for a rectangular channel

$$\frac{h_{L6}}{B} = 0.5 \tag{9.11}$$

$$(c_B - u_{L3}) = (gB)^{1/2} \tag{9.12}$$

$$Q_{sh} = 0.5B(gB)^{1/2} \tag{9.13}$$

For a circular pipe

$$\frac{h_{L6}}{d_t} = 0.563 \tag{9.14}$$

$$(c_B - u_{L3}) = 0.542(gd_t)^{1/2} \tag{9.15}$$

$$Q_{sh} = 0.542\left(\frac{\pi d_t^2}{4}\right)(gd_t)^{1/2} \tag{9.16}$$

Far enough behind a slug the liquid must return to the height of the liquid in front of the slug. This suggests that the Benjamin bubble, if it exists, would need to be matched to a region which is not inviscid.

If (9.16) is substituted into (9.10), the following equation is obtained:

$$(c_F - u_{L0})(A_{L0}) = 0.542\left(\frac{\pi d_t^2}{4}\right)(gd_t)^{1/2} \tag{9.17}$$

Height h_{L0} can be calculated from A_{L0} using the geometric relations given by (5.7) and (5.11). A_{L0} or h_{L0} provides a critical condition for a slug to be stable.

Figure 9.2, from Ruder *et al.* (1989), shows height tracings, at different u_{GS} for a constant $u_{LS} = 0.95$ m/s, downstream in a pipe. These were obtained by measuring the conductance between two wires which were strung across the pipe diameter in the vertical direction. Note that h_L/d_t does not assume a value of unity when a slug is present. This type of measurement can be used to determine the void fraction in the slugs. The solid line, in Figure 9.2, is the height needed for the formation of stable slugs, h_{L0}/d_t. Thus, the necessary condition provides a theoretical prediction of the base layer over which fully developed slugs move. The dashed line represents the h_L/d_t predicted for the initiation of intermittent flow by a viscous large-wavelength hydrodynamic instability (Lin & Hanratty, 1986). The slugs were moving too fast for the liquid height measurements to capture the detail shown in photographs. Thus, they could not be used to see whether (9.14) approximates the liquid height just behind a slug.

Figure 9.3 shows a photograph of a slug in a 2.54 cm pipe. Note that bubbles are occluded. These are caught in the tumbling action at the front of the slug; some of these escape and move out of the back of the slug. This photograph is at a low gas velocity, for which aeration is not severe. The bubble at the tail resembles a Benjamin bubble.

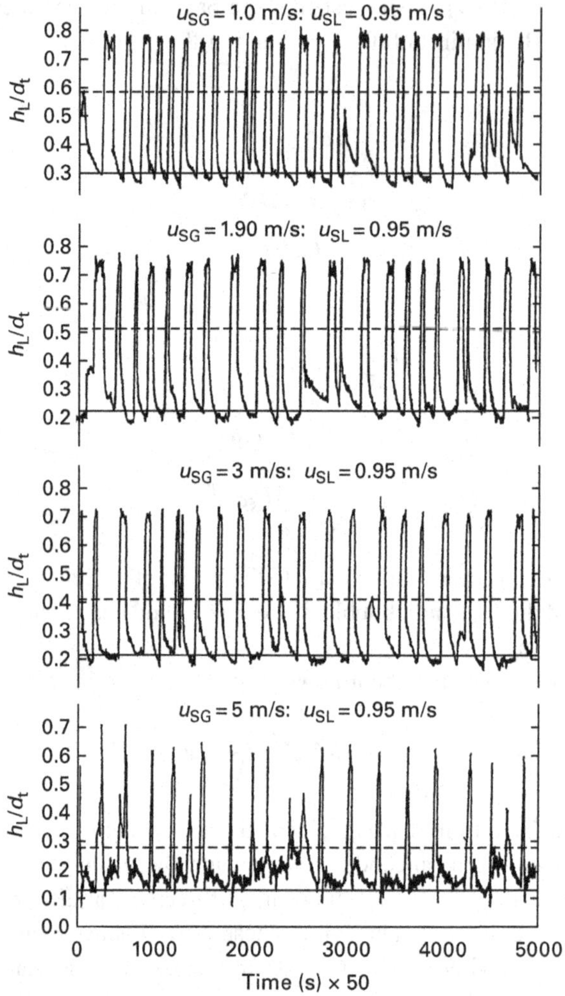

Figure 9.2 Liquid height tracings; u_{LS} = 0.95 m/s. Ruder *et al.*, 1989.

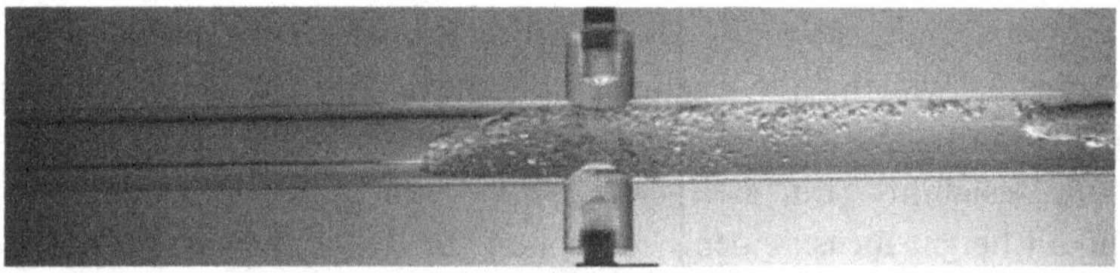

Figure 9.3 Photograph of a slug in a 2.54 cm pipe at a low gas velocity. Ruder *et al.*, 1989.

9.2.2 Transition to plug flow

As pointed out in Chapter 2, researchers have considered an intermittent regime that has two parts: slug flow and plug flow. The latter identifies a series of elongated bubbles moving along the top of the pipe.

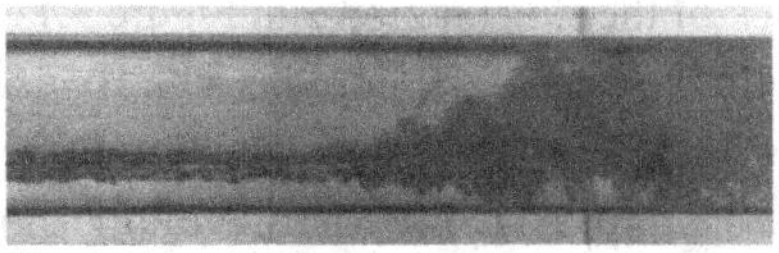

(a) Typical back of a gas cavity in the slug flow ($u_{GS} = 1.45$ m/s; $u_{LS} = 0.95$ m/s, Fr = 2.53)

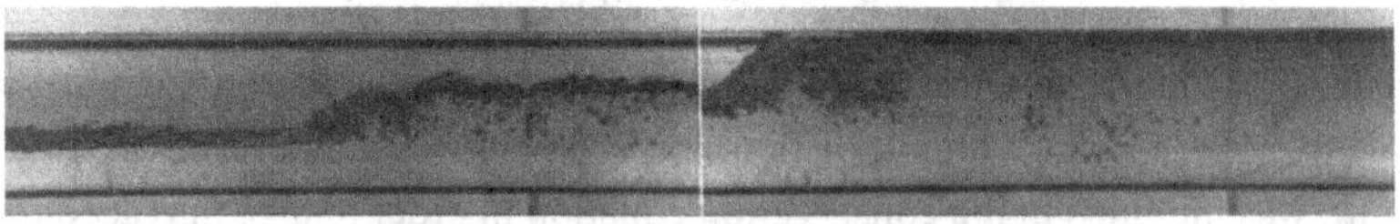

(b) Typical two-stage, staircase-like back of a gas cavity in the plug flow ($u_{GS} = 0.6$ m/s; $u_{LS} = 0.95$ m/s, Fr = 1.8)

(c) The back of a symmetrical gas cavity in a plug flow ("a Benjamin bubble") ($u_{GS} = 0.1$ m/s; $u_{LS} = 0.95$ m/s, Fr = 1.2)

Figure 9.4 Photographs of the front of a slug, or the back of an air pocket. Ruder & Hanratty, 1990.

Figure 9.4 (from Ruder & Hanratty, 1990) shows changes in the front of a slug (the back of the air cavity between slugs) for air–water flow in a horizontal 9.53 cm pipeline. The superficial liquid velocity was 0.95 m/s. The superficial air velocity decreased from 1.45 m/s to 0.1 m/s. A Froude number can be defined as $\mathrm{Fr} = (c_F - u_{L1})/(gd_t)^{0.5}$ where u_{L1} is the velocity of the liquid carpet in front of the slug. The analysis presented by Ruder *et al.* (1989) suggests that Fr > ~ 1 is required for a slug to exist (see equations (9.6) and (9.7)).

Figure 9.4a is for Fr = 2.53. It shows a highly aerated hydraulic jump. Figure 9.4b, for Fr = 1.8, shows the breakdown of the hydraulic jump into a staircase shape. Figure 9.4c, for Fr = 1.2, shows an evolution to a Benjamin bubble. For all of the situations shown in these images, the backs of the slugs/plugs are described as a Benjamin bubble. Thus, for the conditions characterizing Figure 9.4c, a symmetric elongated bubble exists between the slugs. These results are consistent with the observation by Barnea *et al.* (1980) that the plug pattern is a limiting case of slug flow when the liquid is free of entrained gas bubbles. Hurlburt & Hanratty (2002) used the above results to provide a method to predict the transition from a slug flow to a plug flow. They selected Fr = 1.2 as a criterion.

9.2.3 Measurements of the shedding rate

Measurements of the shedding rate under actual flow conditions were obtained by Woods & Hanratty (1996) for air and water flowing in a horizontal pipe at atmospheric conditions. Conductance profiles were used to determine the liquid holdup at several locations along the pipeline. These conductance measurements established the profiles of the liquid carpet and the tail of the slug, and the degree of aeration in the slug. A translating control volume was attached to the back of the slug (station 5 in Figure 9.1). This volume extends out into the stratified flow, where the area of the liquid is A_{L1}. By using the analysis for a stratified flow presented in Chapter 5, u_{L1} can be calculated from the measurements of h_{L1}.

Conservation of mass is applied to the control volume moving at velocity c_B to give the following equation for Q_{sh}:

$$Q_{sh} = (c_B - u_{L1})(A_{L1}) - \frac{dV}{dt} \tag{9.18}$$

where Q_{sh} is the volumetric flow into the volume, the first term on the right side is the net flow out and V is the volume of liquid in the control volume. The accumulation term, dV/dt, was included in the conservation equation in order to take into account changes in the liquid volume that occur in the control volume as it translates downstream. In this way, Q_{sh} was calculated from the measurements for stable, growing and decaying slugs.

9.2.4 Equations for the bubble velocity and the shedding rate

Apply conservation of mass between the back of the slug, station 5, and station 3 in the body of the slug. Choose a frame of reference moving with velocity c_B. Then, from this mass balance,

$$Q_{sh} = (c_B - u_{L3})(1 - \alpha)A \tag{9.19}$$

where Q_{sh} is the flow out of the volume. For small mixture velocities, the back of the slug may be considered to be a Benjamin bubble and α may be assumed equal to zero. Thus $(c_B - u_{L3})$ and Q_{sh} are given by (9.15) and (9.16).

For large mixture velocities, the nose of the bubble behind the slug moves downward in the pipe cross-section and the liquid in the slug is highly aerated. More general relations are needed for bubble velocity and Q_{sh}.

Common practice for vertical pipes is to relate the velocity of large bubbles to the superficial gas and liquid velocities, u_{SG} and u_{LS}. Nicklin *et al.* (1962) assumed that the bubble velocity for vertical slug flow is the sum of the centerline velocity in the liquid in front of the bubble and the drift velocity of the bubble in a stagnant liquid.

$$c_B = K_0 u_{L3} + V_{Drift} \tag{9.20}$$

where K_0 is a coefficient and V_{Drift} is the drift velocity, defined by (8.29). Nicklin *et al.* suggest that, for a turbulent liquid, K_0 is approximately 1.2, that is, the ratio of the centerline liquid velocity to the mean liquid velocity, u_{L3}, in front of the bubble.

Assuming incompressible flow, a volume balance between the inlet of the pipe and station 3 within the body of the slug gives

$$u_{GS} + u_{LS} = \alpha u_{G3} + (1 - \alpha)u_{L3} \tag{9.21}$$

Equation (9.21) is solved for u_{L3} and this is introduced into (9.20), using $Su_{L3} = u_{G3}$. The following equation for c_B is obtained:

$$c_B = C_0(u_{GS} + u_{LS}) + V_{drift} \tag{9.22}$$

where

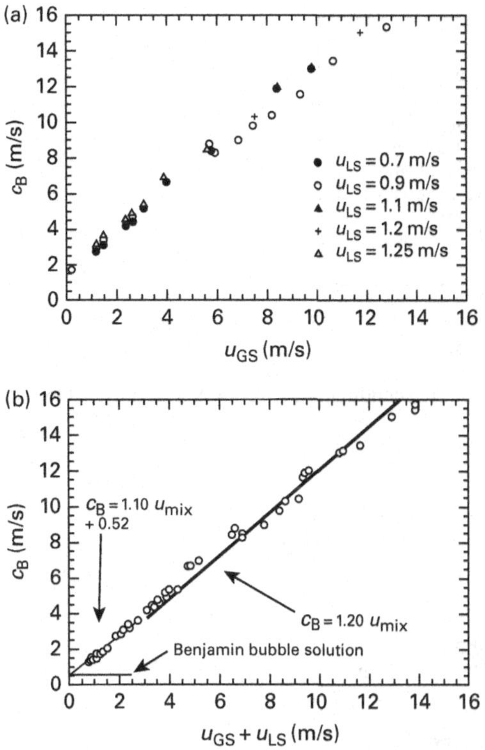

Figure 9.5 Measurements of the bubble velocity in the back of a slug. Woods & Hanratty, 1996.

$$C_0 = \frac{K_0}{1 + (S - 1)\alpha} \tag{9.23}$$

and S is the slip ratio between the liquid and gas phases within the slug. For horizontal flow, $V_{\text{Drift}} = 0.542\sqrt{gd_t}$ (see equation (9.15) with $u_{L3} = 0$). Many authors neglect the drift velocity, V_{Drift}. Therefore, expressions such as $c_B = 1.35\,(u_{LS} + u_{GS})$ (Gregory & Scott, 1969) and $c_B = 1.25{-}1.28\,(u_{LS} + u_{GS})$ (Dukler & Hubbard, 1975) are commonly used.

A plot of measurements of c_B by Woods & Hanratty (1996) for air and water flowing in a horizontal 9.53 cm pipe is given in Figure 9.5 in units of m/s. The data at low gas velocities are represented by

$$c_B = 1.10(u_{GS} + u_{LS}) + 0.52 \tag{9.24}$$

where 0.52 m/s is the velocity of a Benjamin bubble. For larger mixture velocities the influence of the drift velocity is small so that

$$c_B = 1.20(u_{LS} + u_{GS}) \tag{9.25}$$

Woods & Hanratty show that measurements in a 15 cm pipe by Kouba & Jepson (1990), in a 5 cm pipe by Theron (1989) and in a 5 cm pipe by Nydal *et al.* (1992) agree approximately with this equation.

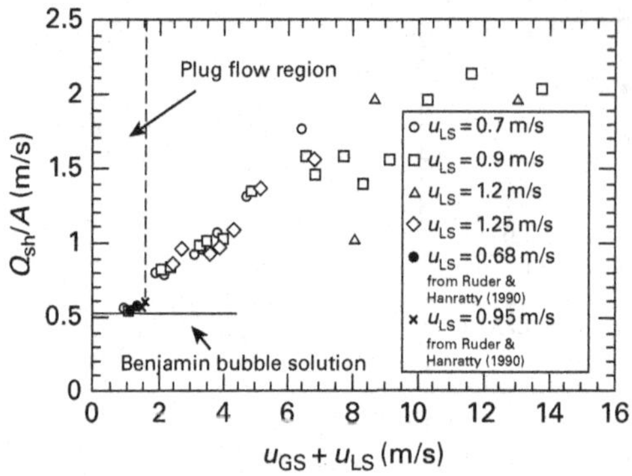

Figure 9.6 Measurements of the mean value of the shedding rate. Woods & Hanratty, 1996.

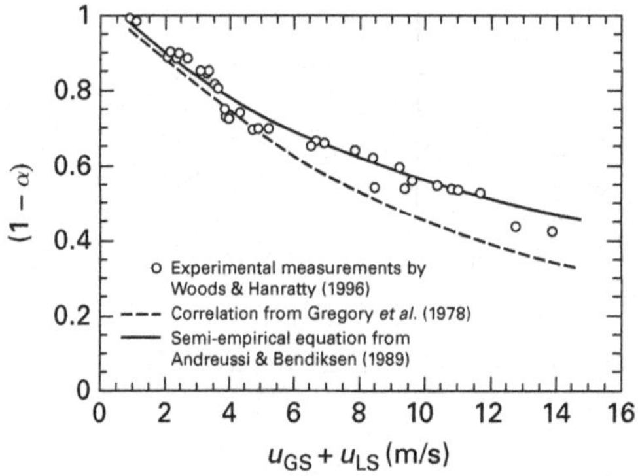

Figure 9.7 Liquid holdup measurements in a slug. Woods & Hanratty, 1996.

Measurements of Q_{sh} (Woods & Hanratty, 1996) are presented in Figure 9.6. At small $(u_{GS} + u_{LS})$, $Q_{sh}A$ is given by the inviscid solution of Benjamin. The shedding rate increases rapidly with increasing $u_{GS} + u_{LS}$ and, at $u_{mix} = 14.2$ m/s, it is approximately four times that given by the Benjamin solution. The use of (9.16) to represent Q_{sh} is valid only for very small values of $(u_{LS} + u_{GS})$.

Equation (9.19) provides a relation for Q_{sh}. From (9.19), (9.21) and (9.22)

$$\frac{Q_{sh}}{A} = \left(c_B - \frac{u_{GS} + u_{LS}}{1 + (S-1)\alpha}\right)(1-\alpha) \tag{9.26}$$

where Su_{L3} has been substituted for u_{G3}.

Figure 9.7 presents measurements of the volume fraction of gas in a slug. Values of S were calculated with (9.26) using measurements of Q_{sh} (Figure 9.6) and of α (Figure 9.7). A slip ratio equal to 1 is a rough approximation at low gas velocities. At high mixture velocities $(u_{GS} + u_{LS}) > 7\,\mathrm{m/s}$, a slip ratio of 1.5 is indicated (see Figure 9.8).

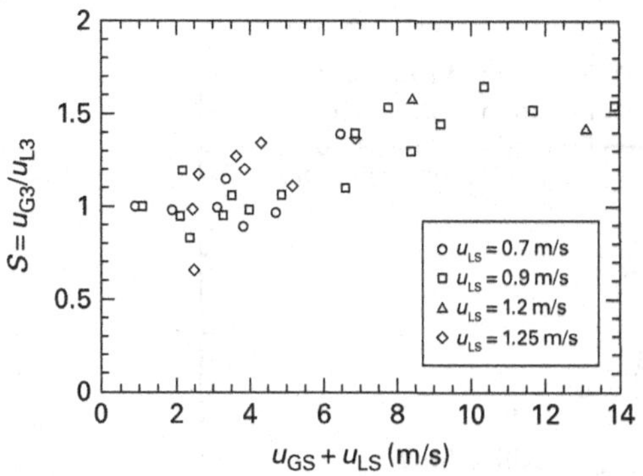

Figure 9.8 Measurements of the velocity ratio in slugs. Woods & Hanratty, 1996.

Bendiksen (1984) suggested that Froude number

$$\mathrm{Fr}_{\mathrm{mix}} = \frac{(u_{\mathrm{GS}} + u_{\mathrm{LS}})}{(gd_{\mathrm{t}})^{1/2}} \tag{9.27}$$

is a measure of the relative roles of inertia and gravity on the behavior of the tail of the bubble behind a slug. At small Froude numbers, gravitation effects are large enough for the nose of the bubble to be at the top of the pipe. As the Froude number increases, inertia becomes more important and, eventually, becomes dominant; the nose of the bubble moves from the top of the pipe toward the centerline. This asymptotic behavior is shown in Figures 9.6 and 9.8 at mixture velocities larger than 8 m/s. For mixture velocities greater than 14 m/s, the slug flow changes to an annular flow.

Of interest is the finding from Figure 9.8 that $(S - 1)$ is positive at large mixture velocities. This suggests that the gas flow in the slug changes from a situation (at low mixture velocities) for which gas is entrained in the front and rises to the top of the pipe because of gravitational effects, to one for which gas enters at the back of the slug. One could then speculate that a transition to annular flow occurs when the gas blows through the back of the slug. More attention needs to be given to this suggestion.

9.2.5 Calculation of the stability of a slug

Slug stability is usually discussed by defining the critical height (or area) needed to sustain slugs. This is defined by (9.10). Thus, (9.26) can be substituted into (9.10) to produce the following expression:

$$\left(\frac{A_{\mathrm{L0}}}{A}\right) = \frac{\left[\left(C_0 - \frac{1}{1+(S-1)\alpha}\right)u_{\mathrm{mix}} + U_{\mathrm{S}}\right](1-\alpha)}{C_0 u_{\mathrm{mix}} + U_{\mathrm{S}} - u_{\mathrm{L1}}} \tag{9.28}$$

For large mixture velocities, U_{S} and u_{L1} can be neglected, so

$$\left(\frac{A_{\mathrm{L0}}}{A}\right)_{\mathrm{critical}} = \frac{C_0 - \frac{1}{1+(S-1)\alpha}}{C_0}(1-\alpha) \tag{9.29}$$

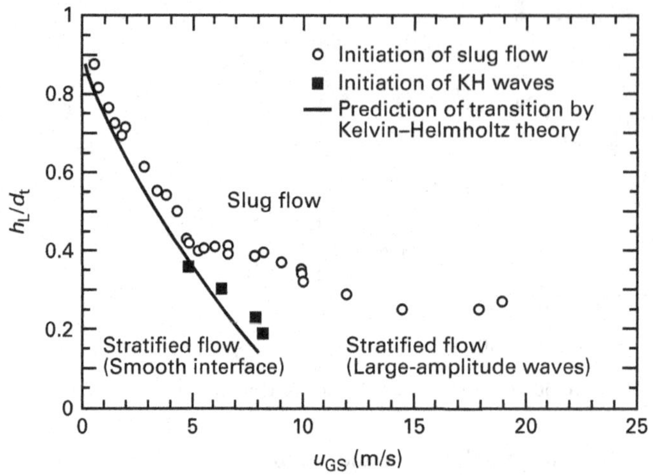

Figure 9.9 Initiation of slugs and Kelvin–Helmholtz waves for air and a 100 cP liquid flowing in a 0.0953 m pipe. Woods & Hanratty, 1996.

From Figures 9.8 and 9.7, $S \approx 1.5$ and $(1-\alpha) \approx 0.5$ at very large mixture velocities. From (9.22) and Figure 9.5, $C_0 \approx 1.2u_{\text{mix}}$. Substitute into (9.29) to get $(A_{L0}/A)_{\text{critical}} \approx 0.16$. From Figure 5.1 and Figure 10.38 in Govier & Aziz (1972), the critical height ratio is then given, approximately, as $h_L/d_t = 0.2$.

Figure 9.9 presents measurements made by Andritsos *et al.* (1989) for air and a 100 cP liquid flowing in a 9.53 cm pipe. (Note that the transition data are plotted as h_L/d_t versus u_{GS}, rather than as u_{SL} versus u_{SG}. For given values of h_L/d_t and u_{SG}, u_{SL} is fixed.) The open points represent observations of the conditions needed for the initiation of slugs. The solid points designate the initiation of Kelvin–Helmholtz waves. The curve is the prediction of instability based on KH theory. The above estimation of a critical $h_L/d_t \approx 0.2$ at very large superficial gas velocity is close enough to the data to support the notion that the initiation of slugs is dictated by the stability of slugs (see Fan *et al.*, 1992).

At low gas velocities, the interface is smooth. The initiation of waves is predicted by KH theory. At $u_{GS} < 5$ m/s the dimensionless height is larger than the critical height so slugs are predicted to be stable; KH waves rapidly evolve into slugs. At $u_{GS} > 5$ m/s the KH waves do not immediately evolve into slugs because the slugs would be unstable. The initiation of slugging is then predicted from a consideration of the conditions needed for stable slugs to exist.

9.3 Modeling of slug flow in a horizontal pipe

Dukler & Hubbard (1975) developed a model for slug flow that is characterized by the intermittent appearance of slugs. A slug is accompanied by a sudden increase in both the liquid level and the pressure. This implies that there would be a discontinuous increase in pressure over each of the slugs in the pipeline. This section uses this picture to describe a flow which is approximately fully developed.

From discussions presented in this chapter, the stratified flow between slugs can be pictured as having a height, h_{Lo}, which is defined from the necessary condition for the existence of a slug. This portion of the flow could be described by the methods used in Chapter 5, as mentioned earlier in this chapter. A critical issue is the description of the behavior of a slug and, in particular, the pressure drops over individual slugs. A second issue is the prediction of the distribution of slug lengths. This is addressed in Section 9.4.

Figure 9.1 presents an idealized drawing of a slug. The pressure drop is considered as being composed of three parts: the change associated with the hydraulic jump in front of the slug, from station 1 to 3, $\Delta\tilde{P}_{\mathrm{h}}$; a frictional loss in the body of the slug, from station 3 to 4, $\Delta\tilde{P}_{\mathrm{f}}$; a pressure variation associated with the velocity change in the rear of the slug from station 4 to 5, $\Delta\tilde{P}_{\mathrm{r}}$. Thus, for a horizontal pipe, the total pressure change over a slug is given as

$$\Delta\tilde{P}_{\mathrm{T}} = \Delta\tilde{P}_{\mathrm{h}} + \Delta\tilde{P}_{\mathrm{f}} + \Delta\tilde{P}_{\mathrm{r}}. \tag{9.30}$$

by Fan *et al.* (1993), where $\tilde{P}$ is the pressure made dimensionless with $\rho_{\mathrm{L}} g d_{\mathrm{t}}$. This is similar to the treatment by Singh & Griffith (1970), who used only the first two terms on the right side of (9.30), but did consider gravitational effects needed to analyze slugging in inclined pipes.

The front is moving downstream with a velocity c_{F}. The liquid and gas velocities at locations 1, 3 and 4 are assumed to be uniform. The liquid layer in front of the slug has a velocity u_{L1} in a laboratory framework. The flows of the liquid and gas in the slug body are uniform after a short distance behind the slug front. For simplification, the gas and liquid are assumed to have the same velocity in the slug ($S = 1$). The pressure drop over a hydraulic jump in a rectangular channel is given by equation (9.3). For small $\rho_{\mathrm{G}}/\rho_{\mathrm{L}}$ Fan *et al.* (1993) give the following relation for a horizontal circular pipe:

$$\Delta\tilde{P}_{\mathrm{h}} = \rho_{\mathrm{L}}\left[\frac{A_{\mathrm{L1}}}{A_{\mathrm{L3}}} - \left(\frac{A_{\mathrm{L1}}}{A_{\mathrm{L3}}}\right)^2\left(\frac{1}{1-\alpha_3}\right)\right](c - u_{\mathrm{L1}})^2 + \rho_{\mathrm{L}} g\left[\frac{A_{\mathrm{L1}}}{A_{\mathrm{L3}}} h_{\mathrm{L1}}^{\mathrm{C}} - (1-\alpha_3) h_{\mathrm{L3}}^{\mathrm{C}}\right] \tag{9.31}$$

where the terms with a superscript of C are centroid heights for hydrostatic pressure in the liquid layer and the slug body. They are length scales that give the average hydrostatic pressure over the area being considered.

The pressure drop in the body of the slug may be pictured as resulting from the frictional drag at the wall, so that

$$\Delta\tilde{P}_{\mathrm{f}} = \tilde{P}_4 - \tilde{P}_2 = \frac{\tau_{\mathrm{W}} \pi d_{\mathrm{t}} L_{\mathrm{S}}}{A} \tag{9.32}$$

where τ_{W} is the average resisting stress at the wall and L_{S} is the length of the slug. Because of a lack of better information, the wall stress is approximated by the Blausius equation

$$\tau_{\mathrm{W}} = 0.046\left(\frac{d_{\mathrm{t}}}{\upsilon_{\mathrm{L}}}\right)^{-0.20}\left(\frac{\rho_{\mathrm{L}} u_{\mathrm{L3}}{}^{1.8}}{2}\right) \tag{9.33}$$

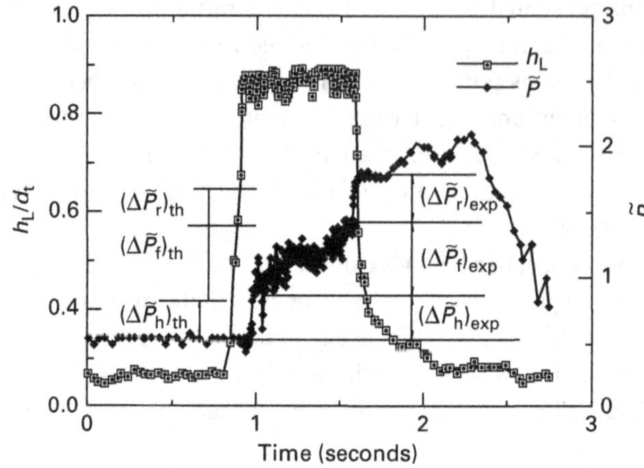

Figure 9.10 Pressure distribution for a stable slug at $u_{GS} = 1.10$ m/s, $u_{LS} = 0.6$ m/s, where $\tilde{P}$ is the pressure made dimensionless with $\rho_L g d_t$. Fan *et al.*, 1993.

where υ_L is the kinematic viscosity of the liquid. Dukler & Hubbard (1975) used

$$\rho_L = \rho_L(1 - \alpha) + \rho_G \alpha \tag{9.34}$$

in (9.33). Fan *et al.* (1993b) simply use the density of the liquid. Their approach is followed in the calculations presented in this section.

Fan *et al.* (1993b) made measurements of the pressure variation over individual slugs, with a piezoresistive transducer, for air and water flowing in a 0.095 m pipe. The time-varying liquid holdup was determined from measurements of the electric resistance between two parallel wires strung across the pipe diameter. The results from one of the experiments, at $u_{GS} = 1.10$ m/s and $u_{LS} = 0.6$ m/s, are plotted in Figure 9.10. Both dimensionless holdup, h_L/d_t, and the dimensionless pressure $\tilde{P} = P/\rho_L g d_t$, are plotted. For $h_L/d_t = 1$ the liquid is not aerated so the value of h_L/d_t provides a measurement of the void fraction. The experiment was conducted at a low gas velocity, for which the resolution of the pressure profile is good.

Note that the pressure experiences a sudden increase when a slug appears. This is defined in Figure 9.10 by $(\Delta\tilde{P}_h)_{exp}$. This is followed by a more gradual increase, $(\Delta\tilde{P}_f)_{exp}$. Another sudden change occurs in the back of the slug, $(\Delta\tilde{P}_r)_{exp}$. Calculated values of $(\Delta\tilde{P}_h)_{th}$ and $(\Delta\tilde{P}_f)_{th}$ were obtained from (9.31) and (9.32). These show good agreement with values suggested from the measurements. The calculation of $(\Delta\tilde{P}_f)_{th}$ requires the specification of the length of the slug. This was obtained as the product of the slug velocity and the time interval over which the slug was observed.

Fan *et al.* (1993b) calculated the pressure change in the rear by picturing it as a Benjamin bubble. This requires that the top of the slug tail is a stagnation point in a frame of reference moving with the velocity c_F and the pressure change along the top wall can be calculated with the Bernoulli equation. In this framework, the top of the tail, station 5 in Figure 9.1, is a stagnation point. Furthermore, the pressure at 5 equals the pressure at 6.

Table 9.1 Pressure drop over stable slugs

u_{SG}	u_{SL}	c_F	$\Delta\tilde{P}_h$	$(\Delta\tilde{P}_f)_{th}$	$(\Delta\tilde{P}_r)_{th}$	$(\Delta\tilde{P}_T)_{th}$	$(\Delta\tilde{P}_T)_{exp}$
1.109	0.60	2.07	0.29	0.64	0.19	1.12	1.23
1.994	0.50	4.09	1.59	1.53	0.27	3.39	3.23
1.994	0.60	3.73	1.59	1.55	0.38	3.52	4.05
1.994	0.60	3.73	1.73	1.48	0.50	3.71	4.53
1.994	0.70	3.80	1.76	1.53	0.43	3.72	3.87
2.965	0.60	4.35	2.19	2.02	0.64	4.85	5.21
2.965	0.60	4.94	2.43	2.45	0.25	5.13	5.56
2.965	0.90	5.96	3.20	2.39	0.51	6.10	6.43
4.060	0.50	6.33	4.36	2.55	0.81	7.72	8.07
4.060	0.50	6.53	5.22	3.22	1.38	9.81	9.26
4.060	1.00	6.60	5.18	4.20	1.22	10.60	11.21
4.985	0.60	8.17	7.23	3.46	1.29	11.98	11.19
5.976	0.70	7.60	4.30	4.51	0.83	9.64	9.16
7.036	0.60	9.02	8.72	1.52	4.03	14.27	13.90
7.036	0.60	9.74	5.66	6.60	0.85	13.10	15.45
7.036	0.90	10.92	8.60	9.49	1.45	19.53	19.97
9.087	0.60	13.19	10.74	6.09	1.70	18.53	19.01
9.087	0.90	15.88	14.06	8.98	1.90	24.93	23.38
13.18	0.80	17.27	15.09	12.80	0.58	28.47	28.55
13.18	0.90	18.45	18.29	11.97	2.45	32.72	30.67
15.93	0.80	17.27	12.02	17.44	1.22	30.67	34.50
15.93	1.00	18.36	12.16	23.79	0.95	36.90	40.57

In this table $\Delta\tilde{P}$ stands for $\Delta P/\rho_L g d_t$ and velocities are given in units of m/s. Subscript th refers to theoretical calculation; subscript exp refers to a measurement. Fan *et al.*, 1993

$$\Delta P_r = P_6 - P_4 = P_5 - P_4 \tag{9.35}$$

Applying the Bernoulli equation between 4 and 5

$$\Delta P_r = \frac{1}{2}\rho_L(c_F - u_{L4})^2 \tag{9.36}$$

Equation (9.36) was used to calculate $(\Delta P_r)_{th}$ in Figure 9.10. At higher gas velocities, the delineation of the different contributions to the pressure drop becomes more difficult. Therefore, Table 9.1, which summarizes all of the measurements with stable slugs by Fan *et al.* (1993b), compares calculated and measured overall pressure drops, made dimensionless with $\rho_L g d_t$.

The calculated pressure drops over slugs roughly predict measurements. This agreement is surprising considering that the following simplifying assumptions were made: In analyzing the hydraulic jump, the bubbles in the slug body were assumed to be uniformly distributed and to be moving with the liquid velocity. Friction factor relations for single-phase flow are used to calculate the frictional pressure losses in the body of the slug. The effect of voidage is taken into account only by considering its role in increasing the liquid

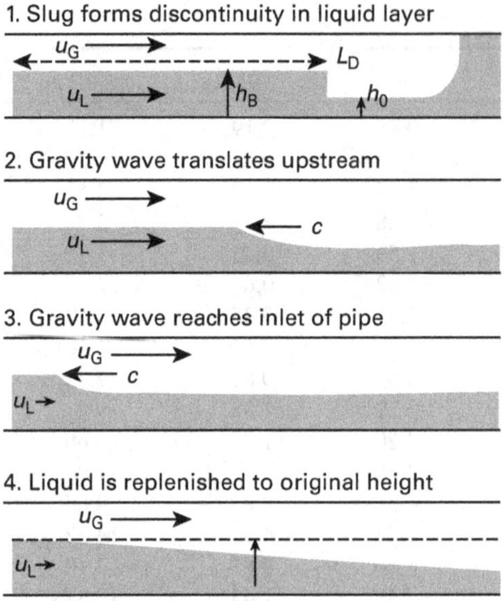
1. Slug forms discontinuity in liquid layer
u_G
L_D
u_L
h_B
h_0
2. Gravity wave translates upstream
u_G
c
u_L
3. Gravity wave reaches inlet of pipe
u_G
c
u_L
4. Liquid is replenished to original height
u_G
u_L

Figure 9.11 Schematic of a gravity wave translating upstream after the formation of a slug. Woods & Hanratty, 1999.

velocity. The Benjamin bubble model, which was used to calculate the pressure drop in the rear of the slug, could be too simple at high gas velocities. Clearly, more work on this problem is needed.

9.4 Frequency of slugging

9.4.1 Shallow liquid inviscid theory

The formation of a slug on a stratified flow is accompanied by a depletion of the liquid. Replenishment occurs rapidly enough for inertia to dominate. Therefore, the effects of viscous shear stresses, turbulent shear stresses, and air flow can be ignored. At the instant a slug appears, the liquid behind the slug is approximated as a discontinuity (see Figure 9.11). Two means exist for rebuilding the liquid level: the flow of new liquid into the pipe and the upstream propagation of a depression (gravity) wave from the location at which the slug formed. This treatment is the same as the breaking dam problem discussed by Stoker (1957). The relative importance of these two mechanisms can be determined by considering the propagation velocity of a large-wavelength inviscid gravity wave in a stationary shallow liquid (see Section 4.4):

$$c = \sqrt{gh_L} \tag{9.37}$$

In a flowing liquid with a velocity u_L, a gravity wave moves upstream with a velocity

$$c = -\sqrt{gh_L} + u_L \tag{9.38}$$

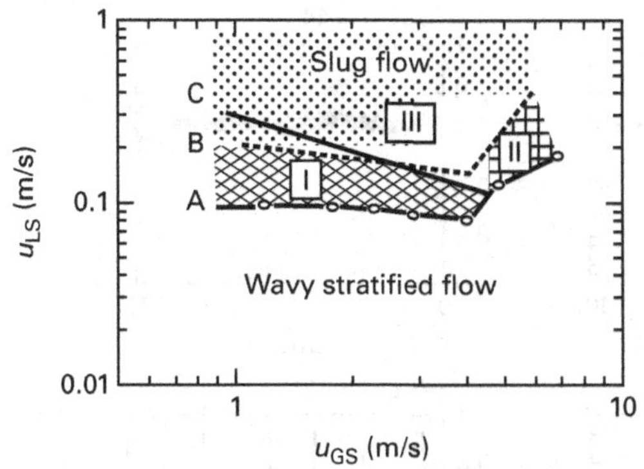

Figure 9.12 Flow regime for air–water flow in a horizontal 0.0763 m pipe. Curve A indicates the transition to slug flow. Between curves A and B, slugs form downstream of $L/d_t \approx 40$. Along curve C, Fr = 1 at the inlet. Woods & Hanratty, 1999.

A consideration of (9.38) shows it is not possible for a depression wave to move upstream if the Froude number, $\mathrm{Fr} = u_L/\sqrt{gh_L}$, is greater than unity. In this case, the refilling will occur by a bore (Stoker, 1957, p. 320) moving downstream.

The inviscid shallow liquid equations can be used to describe the refilling for Fr < 1:

$$\frac{\partial h_L}{\partial t} + u_L\frac{\partial h_L}{\partial x} + h_L\frac{\partial u_L}{\partial x} = 0 \tag{9.39}$$

$$\frac{\partial u_L}{\partial t} + u_L\frac{\partial u_L}{\partial x} + g\frac{\partial h_L}{\partial x} = 0 \tag{9.40}$$

where (9.39) is conservation of mass and (9.40) is conservation of momentum.

Figure 9.12 presents a flow regime map for air and water in a horizontal 7.63 cm pipe (Woods & Hanratty, 1999). Curve A represents conditions for the initiation of slugs. Slugs are observed to form far downstream for conditions close to the transition curve. As u_{LS} increases for a fixed u_{GS}, slugs form closer to the inlet. For conditions below B, slugs appear at $L/d_t <$ ~40 so that the formation could be dependent on the design of the inlet.

Curve C describes conditions for which Fr = 1 at the inlet. Thus refilling can occur with a depression wave at conditions below this curve. Zone I in Figure 9.12 is the region in which Fr < 1 and slugs form downstream of $L/d_t <$ ~40. The region of the flow map in which Fr > 1 and slugs form at $L/d_t <$ ~40 is defined as zone II. For high u_{LS} slugs form upstream of $L/d_t <$~40. This is denoted as zone III.

9.4.2 Regular slugging

Three situations can be cited for which the slugging is regular, that is, periodic. These are horizontal flows at low gas and low liquid velocities, inclined pipes, declined pipes.

The first of these involves situations for which Fr < 1. It is illustrated in Figure 9.13, which describes measurements for air–water flow at Fr = 0.36, $u_L = 25$ m/s by Woods & Hanratty (1999). The phases were combined in a tee section with water in the run and air

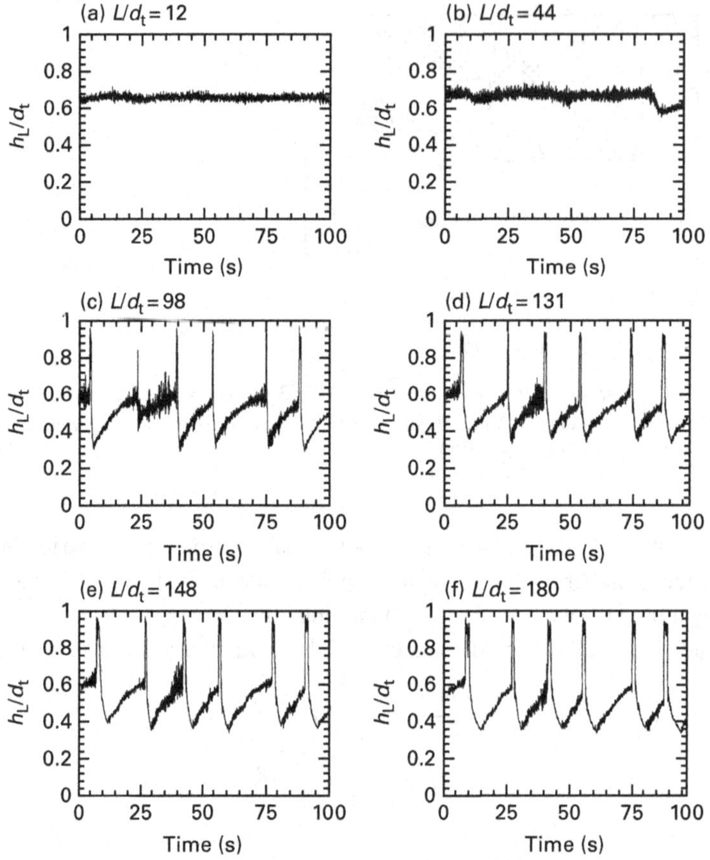

Figure 9.13 Liquid holdup measurements for u_{GS} =1.2 m/s, u_{LS}=0.16 m/s. Woods & Hanratty, 1999.

entering from the top. The air flow was forced through an orifice located five diameters upstream of the tee section. The velocity of the air through this orifice approaches the local velocity of sound. Consequently, downstream variations in the gas-phase pressure caused by the formation of slugs do not strongly affect the inlet air flow.

Measurements of liquid holdup, presented in Figure 9.13 were obtained with six conductance probes located along the pipe. No slugs form for $L/d_t < 40$.The presence of slugs at $L/d_t > 98$ is indicated by peaks in h_L/d_t.

At the transition from a stratified flow to slug flow, h_L/d_t was approximately 0.6. Measurements at $L/d_t = 12$ show that the stratified flow at the inlet is maintained at $h_L/d_t = 0.6$, where h_{LS} is the height at transition. If no slugs were present in the pipe, the stratified flow would assume an equilibrium height, h_{Le}, which is larger than h_{LS}. Thus, the slugs feed information to the inlet, through backward moving waves, to regulate the inlet height. The height of the stratified flow required for a stable slug to exist was $h_{L0}/d_t = 0.3$. One notes that this is a lower limit for the height of the layer behind the slug.

The time interval over which a slug is detected at $L/d_t = 180$ is larger than at $L/d_t = 98$. This suggests that the slug is growing in length since its velocity is approximately

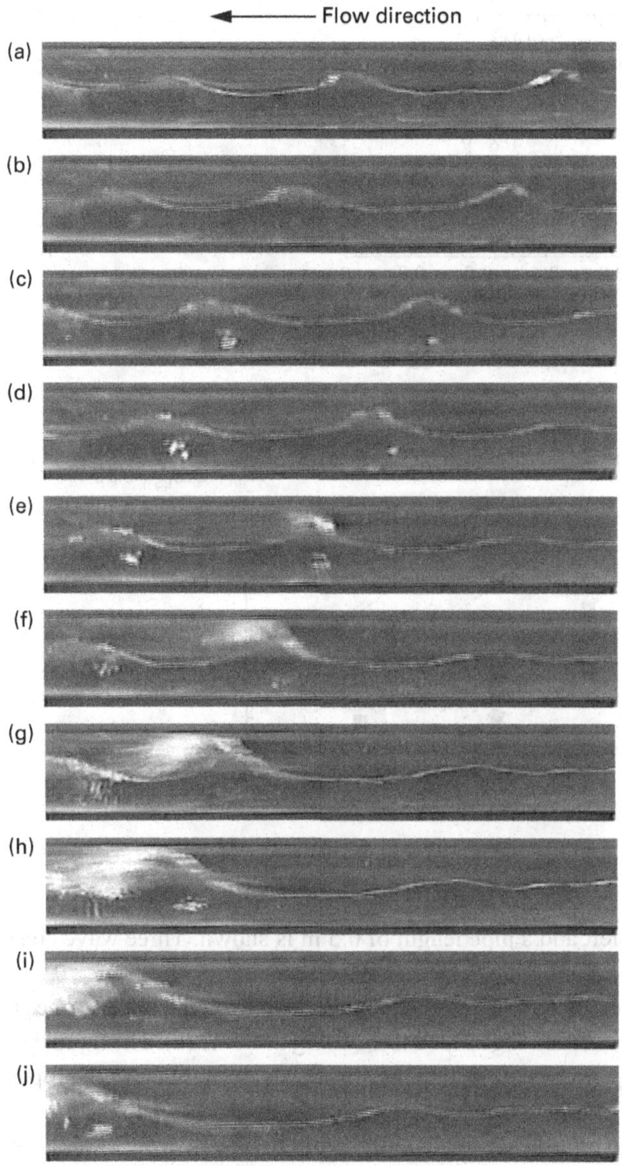

Figure 9.14 Video images illustrating the formation of a slug within zone I. Woods & Hanratty, 1999.

constant as it moves downstream. The height of the liquid in front of the slug is greater than was needed to stabilize it, $h_L > h_{L0}$. The slugs grow by consuming liquid in the stratified layer in front of them. In a much longer pipe than was used in this experiment, one could presume that slugs stop growing (unless they coalesce) and that the stratified layer between slugs would have a height slightly greater than h_{L0}.

Figure 9.14 shows the formation of a slug in zone I. It gives 10 sequential frames obtained with a high-speed video camera. The time between frames is 1/30 s. The flow

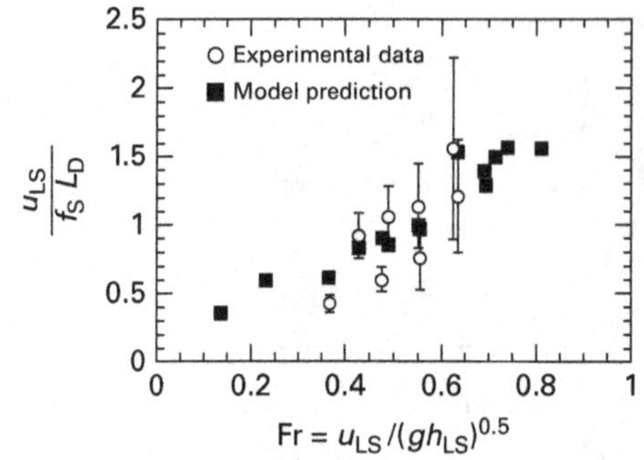

Figure 9.15 Measurements of the frequency of slugging. Woods & Hanratty, 1999.

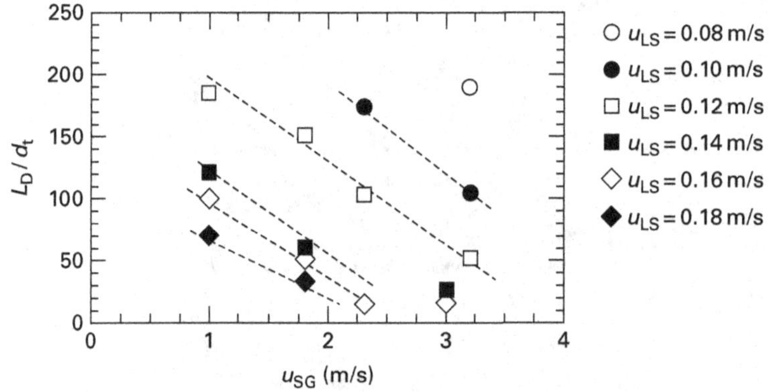

Figure 9.16 Variation of L_D/d_t with u_{GS} and u_{LS}. Woods & Hanratty, 1999.

is from right to left and a pipe length of 0.5 m is shown. Three wave crests, 16–20 cm apart are observed in Figure 9.14a. They correspond to a frequency 0.5 Hz. The wave velocity is approximately 0.9 m/s. The third wave from the left becomes unstable as it propagates downstream. It grows to form a slug in frame (h). The liquid behind the slug decreases to h_{L0}. The waves behind the slug decay because they are propagating over a thinner layer.

Figure 9.11 gives a sketch of the refilling of the pipe by a backward moving wave in the pipe. Woods & Hanratty (1999) used equations (9.39) and (9.40) to calculate the time interval between slugs. Figure 9.15 gives the results of this calculation plotted as $u_{LS}/f_S L_D$ versus Fr $= u_{LS}/(gh_{LS})^{0.5}$. Here, $t_S = 1/f_S$ is the time interval between slugs and u_{LS}, h_{LS} are the liquid velocity and the height corresponding to the instability conditions. The length of pipe needed to develop an unstable wave train is designated by L_D. The bars in Figure 9.15 reflect the errors in estimating L_D. Figure 9.16 shows the variation of the mean values of L_D with u_{GS} and u_{LS}. This parameter was estimated by measuring the liquid holdup at various locations in the pipeline for several different flow conditions and by estimating the probe at which a slug was first observed. Each mean value in Figure 9.16 is based on an ensemble of at least 100 slugs.

The influence of the declination of a pipe on the initiation of slug flow is discussed in a paper by Woods *et al.* (2000) and in Section 5.5.3. As seen in Figure 5.8, transition to slugging occurs at larger u_{LS} when the pipe is declined. The difference between horizontal and declined pipes is that the liquid velocity is larger in declined pipes, when comparison is made for the same height of the liquid layer. Both horizontal and declined pipes show regular slugging (see Figure 5.9). A declined pipe, however, will be characterized by $Fr > 1$ at the inlet. Thus, refilling by backward-moving gravity waves does not occur.

In a horizontal pipe, slugs evolve as a result of a viscous large-wavelength instability, so the frequency of slugging equals the frequency of these waves. (The actual transition appears to be associated with a Kelvin–Helmholtz instability at the crest of a large-wavelength wave, as shown in Figure 5.9.)

Section 5.5.2 describes the behavior in inclined pipes, for which reversed flows can exist. In this regime an intermittent behavior is experienced. Slugs are formed at the inlet by a Kelvin–Helmholtz instability. They move downstream and are followed by a backward moving stratified flow. The liquid level at the inlet drops after the formation of a slug. It rebuilds by liquid flowing into the pipe and by the backward-moving stratified flow. When the height at the inlet reaches a critical value, a slug forms. Slugs appear periodically with a frequency defined by the time required to rebuild the liquid height to a critical value needed for a slug to form.

Thus, at low gas velocities regular slugging can occur for horizontal, inclined and declined pipes. However, the mechanisms governing slugging for these three cases are different.

9.4.3 Stochastic slugging

Stochastic slugging is observed in cases for which the Froude number is greater than unity (regions II and III in Figure 9.12). For these situations, a bore formed behind the slug replenishes the liquid picked up by the slug. Since the slug moves downstream at a faster velocity, the stratified layer between the bore and the slug lengthens. The stratified flow has a height, h_{L0}, which is equal to that required for a stable or growing slug to exist.

This process is illustrated in Figure 9.17, which displays holdup measurements (Woods *et al.*, 2006) obtained for air and water flowing through a horizontal 0.0763 m pipe. The liquid and gas flows were $u_{LS} = 1.0$ m/s, $u_{GS} = 1.8$ m/s (zone III). A pressure transducer at $L/d_t = 22$ reflects the passage of a slug (see Figure 9.17c). A highly disturbed interface is observed at $L/d_t = 4.0$. The frequency of the disturbances decreases dramatically downstream until at $L/d_t = 200$ an approximately fully developed slug pattern is observed.

A sharp drop in the level behind a disturbance, to a value of h_{L0}, is the signature for the presence of an incipient slug at $L/d_t = 4$. At $L/d_t = 37$, both slugs and large-amplitude waves are present. The slower-moving large-amplitude waves are consumed by the slugs so the number of peaks decreases between $L/d_t = 37$ and $L/d_t = 200$.

Measurements such as those presented in Figure 9.17 suggest the following picture of the behavior of interfacial disturbances: Incipient slugs develop close to the entry. These grow if $h_L > h_{L0}$ and decay if $h_L < h_{L0}$. For h_L close to h_{L0}, they will collapse if the slug is

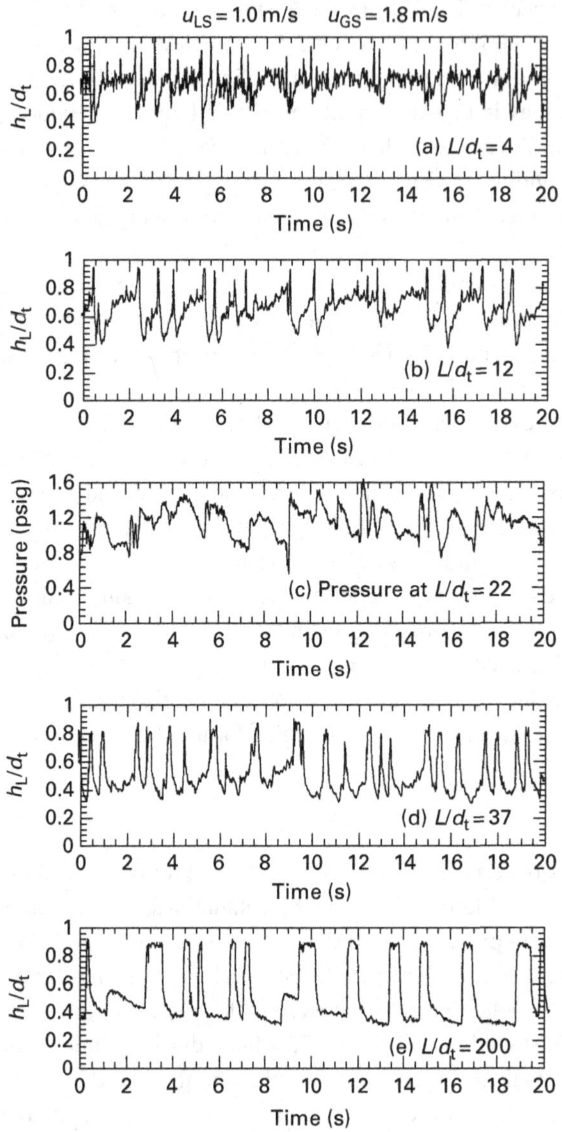

Figure 9.17 Liquid holdup at four different locations and pressure pulsations at $L/d_t = 22$, $u_{GS} = 1.8$ m/s, $d_t = 0.0763$ m. Woods *et al.*, 2006.

not long enough. Collapsed slugs form large-amplitude roll waves which propagate downstream. They are incorporated into faster-moving slugs.

Measurements of slug frequency at a pipe exit have been reported by a number of investigators (Gregory & Scott,1969; Heywood & Richardson, 1979; Crowley *et al.*, 1986; Woods *et al.*, 2006). These are summarized in Figure 9.18, where $f_S d_t/u_{LS}$ is plotted against $u_{LS}/(u_{LS} + u_{GS})$. Note that this plot captures the effects of liquid flow and pipe diameter and that $f_S d_t/u_{LS} = 0.05$ is a rough approximation of the data.

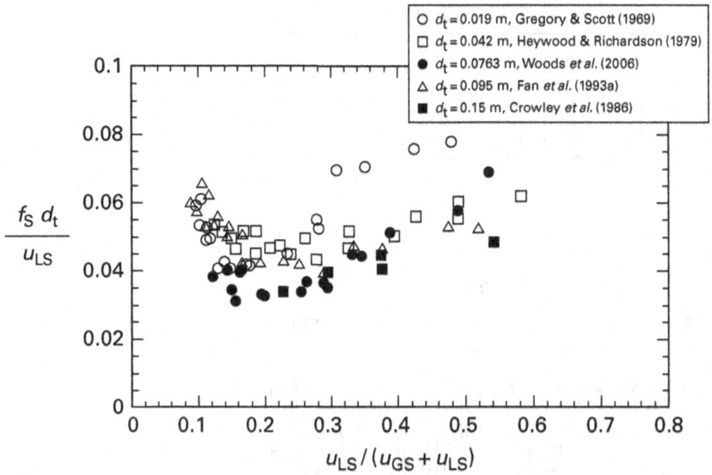

Figure 9.18 Measurements of slug frequencies for different pipe diameters. Woods *et al.*, 2006.

The slug frequency can be related to the distribution of slug lengths, l_S, for a flow consisting of slugs embedded in a stratified layer with height h_{L0} and velocity u_{L0}. An intermittency, I, is defined as the fraction of the time a slug is observed by a stationary observer

$$I = \frac{f_S \overline{L}_S}{c} \tag{9.41}$$

where c is the slug velocity. It can be roughly approximated as

$$I \approx \frac{u_{LS}}{u_{LS} + u_{GS}} \tag{9.42}$$

Furthermore (from Section 9.2.4 and equation (9.25).

$$c = 1.2(u_{GS} + u_{LS}) \tag{9.43}$$

If (9.42) and (9.43) are substituted into (9.41), a first approximation for the relation between f_S and $I = f_S\overline{L}_S/c$ under fully developed conditions is obtained:

$$\frac{f_S d_t}{u_{LS}} = 1.2\left(\frac{\overline{L}_S}{d_t}\right)^{-1} \tag{9.44}$$

In addition to the use of a highly simplified model for slug flow, equation (9.44) ignores the liquid carried in the stratified flow, and the gas carried by the slugs. A derivation that takes these factors into account is given by Woods (1998).

Information on slug length was obtained by Woods *et al.* (2006) from measurements of the holdup for air–water flow at the pipe outlet, such as shown in Figure 9.19. Values of the void fraction in the slugs were calculated from peaks in h_L/d_t. The arrows in Figure 9.19 that are pointing downward indicate large-amplitude waves, which are

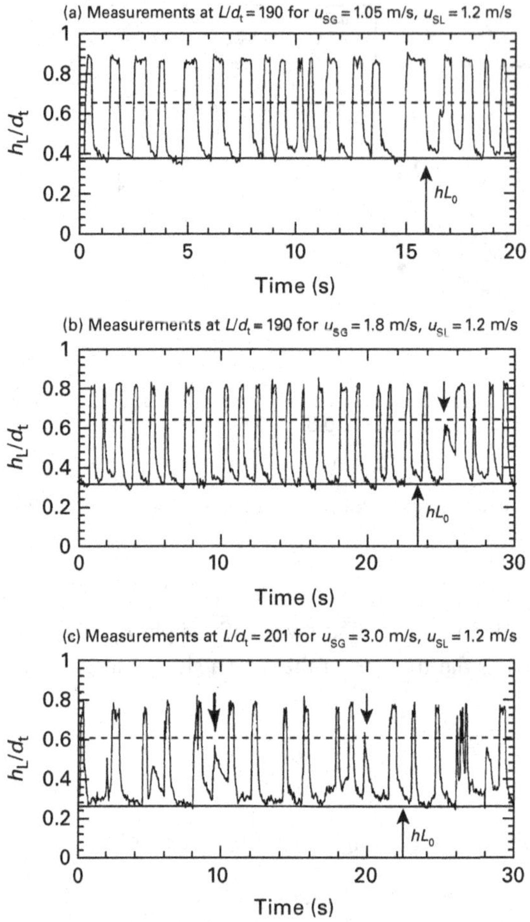

Figure 9.19 Holdup measurements at $u_{LS} = 1.2$ m/s, $d_t = 0.0763$ m. Woods *et al.*, 2006.

moving at a lower velocity than the slugs. These are overtaken by slugs, thus causing an increase of the mass of liquid in the slugs.

Measurements of the average slug length were determined by setting an h_L/d_t at which to measure the time interval for the slug passage. The triggers (the dashed lines in Figure 9.19) were selected to give the correct slug frequency. For large gas velocities, the measurements are less accurate because the slugs are highly aerated and the h_L/d_t are smaller. This can make it difficult to differentiate slugs from large-amplitude waves. In these cases, measurements of pressure pulses, characteristic of slugs, can be used to differentiate fast-moving slugs from slow-moving large-amplitude waves at the outlet of the pipe.

Average lengths measured close to the pipe outlet are not sensitive to changes in u_{LS} and weakly sensitive to u_{GS}. A value of $\overline{L}_S/d_t = 18$ was obtained by Woods & Hanratty (2006) for $u_{GS} = 2$–4 m/s, $L/d_t = 200$, where L is the length of the pipe. Grenier *et al.* (1997) carried out a study in a 0.053 m pipe, which had a length of 90 m ($L/d_t = 1700$), and found $\overline{L}_S/d_t = 20$ for mixture velocities greater than 1.2 m/s. A similar result was obtained by Ferre (1979) and by Bernicott & Drouffe (1991) in a pipe with $L/d_t = 9500$. It is of interest that this roughly equals the value of 18 obtained by Woods *et al.* (2006) for $L/d_t = 200$. The experiments of Nydal *et al.* (1992) are of particular interest since a large range of gas (0.5–20 m/s) and liquid (0.6–3.5 m/s) superficial velocities were considered. Two pipes with diameters of 0.053 m and 0.09 m, and lengths of 17 m were used. They measured $\overline{L}_S/d_t = 15–20$ in the 0.053 m pipe and $\overline{L}_S/d_t = 12–16$ in the 0.09 m pipe. Experiments by Saether *et al.* (1990) in a 0.032 m pipe with $L/d_t = 560$ gave $\overline{L}_S/d_t = 27$ for $(u_{GS} + u_{LS}) = 7.58\,\text{m/s}$.

A striking feature of these measurements is that the reported average values of L_S/d_t are roughly the same for a wide range of pipe lengths. Woods (1998) obtained measurements of the standard deviation for the distribution of slug lengths of about 0.55 m for u_{GS} = 2–4 m/s and 0.5 m for $u_{GS} > 4$ m/s. The ratio $\sigma_{LS}/\overline{L}_S$ was measured as 0.35 for $u_{GS} = 1$ m/s and as 0.5 for $u_{GS} > 4$ m/s. Nydal *et al.* (1992) obtained $\sigma_{LS}/\overline{L}_S = 0.37$ for $(u_{GS} + u_{LS}) \leq 5\,\text{m/s}$. They measured a drop of $\sigma_{LS}/\overline{L}_S$ at $(u_{GS} + u_{LS}) \leq 5\,\text{m/s}$. A value of $\sigma_{LS}/\overline{L}_S = 0.31$ was obtained for $(u_{GS} + u_{LS}) = 10\,\text{m/s}$. The study of Saether *et al.* (1990) at $(u_{GS} + u_{LS}) = 7.58\,\text{m/s}$ gave $\sigma_{LS}/\overline{L}_S = 0.26$.

A number of researchers have suggested that there is a minimum length below which slugs are unstable (Taitel *et al.* 1980). However, there are differences in its suggested value (Barnea & Brauner, 1985; Dukler *et al.*, 1985). Measurements by Woods *et al.* (2006) of $\overline{L}_S$ and $d\overline{L}_S/dt$ suggest that $(\overline{L}_S/d_t)_{min} \approx 5$. The cumulative probability density function of slug lengths obtained at u_{GS} = 5 m/s by Nydal *et al.* (1992) gives $(\overline{L}_S/d_t)_{min} \approx 8$. Measurements by Grenier *et al.* (1997) give a 100% chance for a slug with $\overline{L}_S/d_t = 9$ to be stable.

The probability density function for slug length can be approximated by a log-normal function (Brill *et al.*, 1981; Nydal *et al.*, 1992; Woods *et al.*, 2006)

$$p(L_S) = \frac{1}{L_S C_2 \sqrt{2\pi}} \exp\left[\frac{-(\ln L_S - C_1)^2}{2C_2^2}\right] \tag{9.45}$$

where constants C_1 and C_2 can be obtained from measurements of $\overline{L}_S$ and σ^2_{LS}.

Scott *et al.* (1987) have suggested two stages of slug growth from field measurements in very long petroleum pipelines, at conditions close to those needed to initiate slugging. The first involves an accumulation of waves by slugs which would culminate in the "fully developed" pattern discussed above. The second involves the coalescence of slugs. This second stage has been discussed by Barnea and Taitel (1993).

Several mechanisms for the growth of slugs in the second stage have been discussed. The focus has been on differences in the velocities of the gas bubbles behind two successive slugs with different lengths. (This approach presumes that the bubble velocity behind a slug increases with decreasing slug length.)

Measurements of the bubble velocity at the back of a slug, c_B, have been made by Woods *et al.* (2006). These are shown in Figure 9.20. It is noted that c_B can have a small, but significant, range of velocities for given u_{LS} and u_{GS}. This indicates that adjacent

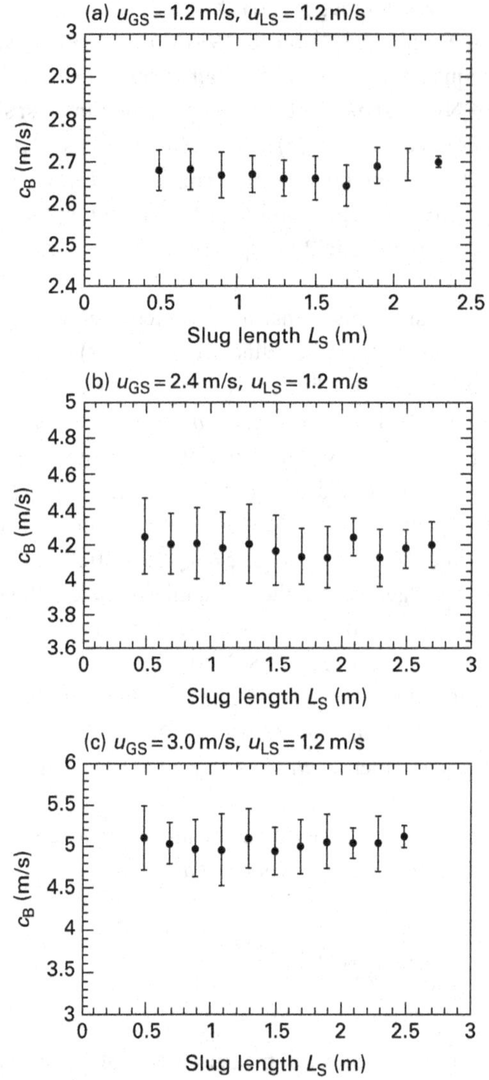

Figure 9.20 Effect of slug length on the velocity of a bubble behind the slug. Woods *et al.*, 2006.

slugs can overtake one another given sufficient time. However, note from Figure 9.20 that the average value of c_B is independent of slug length. This seems to overrule the mechanism involving coalescence outlined above.

These results, as well as the L_S/d_t measurements discussed earlier, suggest that the occurrence of a "second stage" would require an extremely long pipe, such as used in field operations. The coalescence of two adjacent slugs in this mode of operation should be a stochastic event.

The holdup tracings shown in Figure 9.19 could be misleading in that spatial variations in the flow direction are compressed. Slug velocities are given by (9.43). Thus, one second in Figure 9.19b represents 36 m or 47 pipe diameters. Photographs taken with a

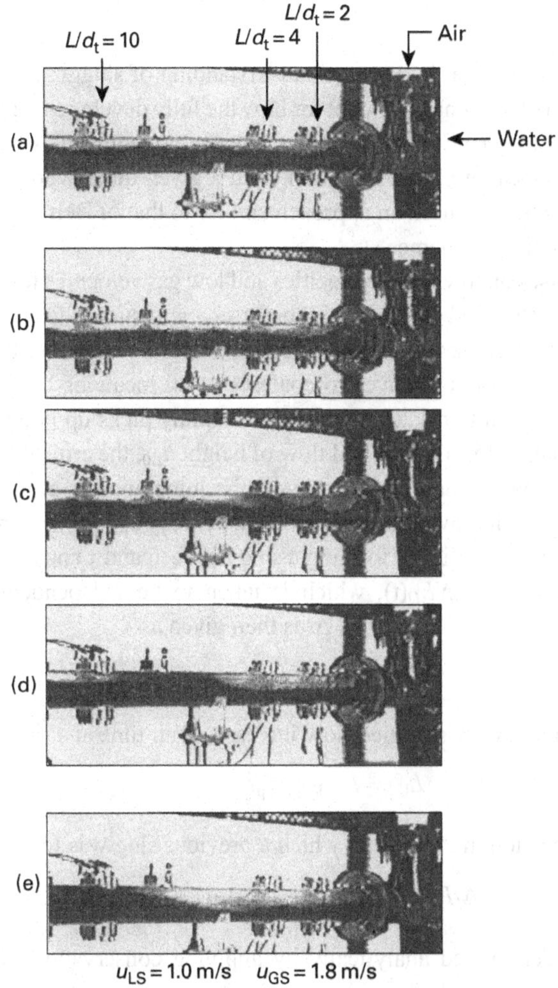

Figure 9.21 Video images at the entrance of a pipe for u_{GS} = 1.8 m/s, u_{LS} = 1 m/s, d_t = 0.0763 m. Woods *et al.*, 2006.

high-speed video camera (Woods *et al.*, 2006) give a better representation of what is happening. Figure 9.21 is an example. The conditions are u_{GS} = 1.8 m/s, u_{LS} = 1 m/s, d_t = 0.073 m (the same as for Figure 9.17). The liquid was dyed to produce contrast. The tee section in which air and water are mixed is at the right. The probes at $L/d_t = 2$, 4 and 10 are shown in (a). The dimensionless length $L/d_t = 10$ corresponds to a length of 0.76 m. The velocity of a slug under these conditions is 3.4 m/s so the photos would correspond to a time interval of 0.22 s. The formation of an incipient slug is shown in (a)–(c). It appears as a growing slug in photo (d), where $L_S/d_t = 5$. A developing tail is shown in photos (d) and (e). The height of the liquid in front of the slug is larger than h_{L0} so it is anticipated that this slug reaches a large enough length to be stable. In photo (e), it can be seen that the depleted liquid behind the slug is being replaced by an incoming bore.

9.4.4 Stochastic models

A critical issue in establishing a fundamental understanding of slugging at large liquid flows is the formulation of a stochastic model for how the fully developed region evolves. Bernicott & Drouffe (1991) proposed a model whose prerequisite is a probability function for a slug to appear at a given location, x, at a given time. Woods & Hanratty (1999) and Woods *et al.* (2006) used an approach similar to that of Bernicott & Drouffe and developed computational schemes.

The 1999 paper focuses on low liquid velocities and low gas velocities for which slugs appear at $L/d_t > 40$. A stratified flow with a height h_{LS} is assumed to exist for L/L_D. Beyond L_D, an unstable irregular wave pattern appears over a length L_U. It is assumed that it is equally probable for roll waves to coalesce at all locations between L_D and L_D+L_U. After a slug forms, it translates downstream, rapidly picks up liquid at its front and sheds liquid at its tail to form a stratified flow of height h_{L0}, the critical height above which stable slugs can exist. Since Fr > 1, a hydraulic jump appears at the location at which the slug was formed; it moves downstream with velocity c_{jump} to restore the height of the liquid to h_{LS}. The probability of forming a slug between and x and $x + dx$ during a time interval Δt is defined as $NL_U(t)$, which is taken to be independent of x. The probability of forming a slug in the interval Δt is then given as

$$\Delta t P(t) = \Delta t N L_U(t) \tag{9.46}$$

The length of the unstable wavy stratified flow increases with time at a rate given by

$$L_U = L_f + c_{jump} t \tag{9.47}$$

where $L_f + L_D$ is the location in the pipe at which a previous slug was formed. Thus

$$\Delta t P(t) = \Delta t N \left(L_f + c_{jump} t\right) \tag{9.48}$$

The velocity c_{jump} is determined analytically by applying conservation of mass in a reference plane moving with the hydraulic jump

$$\left(c_{jump} - u_{slug}\right) A_{LS} = \left(c_{jump} - u_{L0}\right) A_{L0} \tag{9.49}$$

where A_{LS} is the area of the liquid in a slug and A_{L0} is the critical area needed to sustain a stable or growing slug. The randomness of the slugging process was modeled through the use of random number generators. At $t + \Delta t$, a random number, R_i, between 0 and 1 is selected and compared with the value of $\Delta tP(t)$. If $R_i > \Delta tP(t)$, a slug does not form. If $R_i < \Delta tP(t)$, a slug has formed somewhere in L_U. A second random number generator was then used to determine where, along L_U, this occurred. This becomes a new value of L_f. The algorithm is repeated to provide an ensemble of time intervals between slug formation and an ensemble of slug lengths, L_S.

In the second paper, Woods *et al.* (2006) used this computational scheme to represent stochastic slugging for large superficial liquid velocities. For this case, slugs form close to the inlet so L_D was taken to be zero. Furthermore, the 2006 paper employed the concept of minimum slug length for a slug to be stable. A value of $(L_S/d_t)_{min} = 4$ was used. The

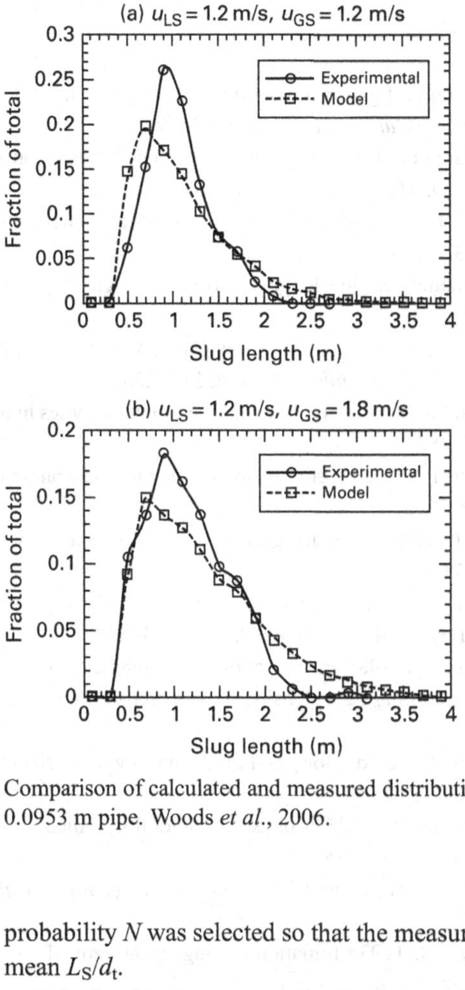

Figure 9.22 Comparison of calculated and measured distributions of slug lengths for air–water flow in a 0.0953 m pipe. Woods *et al.*, 2006.

probability N was selected so that the measured mean L_S/d_t is the same as the calculated mean L_S/d_t.

Calculated and measured distributions of L_S for air–water are compared in Figure 9.22 (Woods *et al.*, 2006) under conditions that the flow is fully developed or close to fully developed. Rough agreement is noted. Both show skewed distributions. However, the model predicts the existence of slug lengths larger than observed in experiments. Clearly, there is room for improvement.

9.5 Triggering of slug formation

Two modes of operation can be considered. The data obtained by Woods & Hanratty (1999) and by Woods *et al.* (2006) were obtained while the gas flow rate was kept constant by using a critical orifice.

Another possibility is to keep the pressure drop constant. This type of operation is discussed by Scott *et al.* (1987). A slug leaving the pipe gives rise to a sudden increase in the flow rate. This, in turn, can trigger the formation of a new slug in the pipe.

References

Andreussi, P. & Bendikson, K. 1989 Investigation of void fraction in liquid slugs for horizontal and inclined gas–liquid pipe flow. *Int. J. Multiphase Flow* 15, 937–946.

Andritsos, N., Williams, L. & Hanratty, T. J. 1989 Effect of liquid viscosity on stratified-slug transition in horizontal pipes. *Int. J. Multiphase Flow* 18, 877–892.

Barnea, D. & Brauner, N. 1985 Holdup of the liquid in two-phase intermittent flow. *Int. J. Multiphase Flow* 10, 467–483.

Barnea, D. & Taitel, Y. 1993 A model for slug length distribution in gas–liquid slug flow. *Int. J. Multiphase Flow* 10, 467–483.

Barnea, D., Shoham, O., Taitel, Y. & Dukler, A. E. 1980 Flow pattern transition for gas–liquid flow in horizontal and inclined pipes. *Int. J. Multiphase Flow* 6, 217–225.

Bendiksen, K. H. 1984 Experimental investigation of the motion of long bubbles in inclined tubes. *Int. J. Multiphase Flow* 10, 467–482.

Bernicott, M. F. & Drouffe, J. M. 1991 A slug length distribution law for multiphase transportation systems. *SPE Prod. Eng.* 19, 829–838.

Bontozoglou, V. & Hanratty, T. J. 1990 Capillary-gravity Kelvin–Helmholtz waves close to resonance. *J. Fluid Mech.* 217, 71–91.

Brill, J. P., Schmidt, Z., Coberly, W. A. & Moore, D. W. 1981 Analysis of two-phase tests in large diameter flow lines in Prudhoe Bay Field. *Soc. Petrol. Eng. J.* 363–377.

Crowley, C. J., Sam, R. G. & Rothe, P. H. 1986 Investigation of two-phase flow in horizontal and inclined pipes at large pipe sizes and high gas density. Report prepared for the American Gas Association by Creare Inc.

Dukler, A. E. & Hubbard, M. G. 1975 A model for gas–liquid slug flow in horizontal tubes. *Ind. Eng. Chem. Fund.* 14, 337–347.

Dukler, A. E., Maron, D. M. & Brauner, N. 1985 A physical model for predicting minimum slug length. *Int. J. Multiphase Flow* 40, 1379–1385.

Fan, Z., Jepson, W. P. & Hanratty, T. J. 1992 A model for stationary slugs. *Int. J. Multiphase Flow* 18, 477–494.

Fan, Z., Lusseyran, F. & Hanratty, T.J. 1993a Initiation of slugs in horizontal gas–liquid flows. *AIChE Jl* 39, 1741–1753.

Fan, Z., Ruder, Z. & Hanratty, T. J. 1993b Pressure profiles for slugs in horizontal pipelines. *Int. J. Multiphase Flow* 19, 3421–3437.

Ferre, D. 1979 Ecoulements diphasiques aporches en conduite horizontale. *Rev. Ind. Fr. Pet.* 34, 113–142.

Govier, G. W. & Aziz, K. 1972 *The Flow of Complex Mixtures in Pipes*. New York: Van Nostrand Rheinhold.

Gregory, G. A. & Scott, D. S. 1969 Correlation of liquid slug velocity and frequency in concurrent gas–liquid slug flow. *AIChE Jl* 15, 833–835.

Gregory, G. A., Nicholson, M.K. & Aziz, K. 1978 Correlation for liquid volume fraction in the slug for gas–liquid flow. *Int. J. Multiphase Flow* 4, 33–39.

Grenier, P., Fabre, J. & Fagundes Netto, J. R. 1997 *Slug Flow in Pipelines: Recent Advances and Future Developments*. Bedford: BHR Group, pp. 107–121.

Heywood, N. I. & Richardson, J. F. 1979 Slug flow in air–water mixtures in a horizontal pipe; determination of liquid holdup by gamma ray absorption. *Chem. Eng. Sci.* 28, 17–30.

Hurlburt, E.T. & Hanratty, T.J. 2002 Prediction of the transition from stratified to slug and plug flow for long pipes. *Int. J. Multiphase Flow* 20, 707–729.

Kouba, G.K. & Jepson, W.P. 1990 The flow of slugs in horizontal two phase pipelines. *Trans. ASME* 112, 20–25.

Lin, P.Y. & Hanratty, T.J. 1986 Prediction of the initiation of slugs with linear stability theory. *Int. J. Multiphase Flow* 12, 79–98.

Nicklin, D.J., Wilkes, J.O. & Davidson, J.F. 1962 Two phase flow in vertical pipes. *Trans. Inst. Chem. Engs*. 102, 61–68.

Nydal, O.J., Pintus, S. & Andreussi, P. 1992 Statistical characterization of slug flow in horizontal pipes. *Int. J. Multiphase Flow* 18, 439–452.

Ruder, Z. & Hanratty, T.J. 1990 A definition of gas–liquid plug flow in horizontal pipes. *Int. J. Multiphase Flow* 16, 233–242.

Ruder, Z., Hanratty, P.J. & Hanratty, T.J. 1989 Necessary conditions for the existence of stable slugs. *Int. J. Multiphase Flow* 15, 209–226.

Saether, G., Bendiksen, K., Muller, J. & Froland, E. 1990 The fractal statistics of liquid slug lengths. *Int. J. Multiphase Flow*, 16, 1117–1126.

Scott, S.L., Shoham, O. & Brill, J.P. 1987 Modeling slug growth in large diameter pipes. Paper presented at the Third International Joint Conference on Multiphase Flow, The Hague, Netherlands, May 1987, paper B2.

Singh, G. & Griffith, P. 1970 Determination of the pressure optimum pipe size for two-phase flow in an inclined pipe. *TASME J. Eng. Ind*. 92, 717–726.

Stoker, J.J. 1957 *Water Waves*. New York: Interscience, pp. 313–333.

Taitel, Y., Barnea, D. & Dukler, A. E. 1980 Modelling of flow pattern transitions for steady upward gas–liquid flow in vertical tubes. *AIChE Jl* 3, 345–354.

Theron, B. 1989 Ecoulements diphasiques instatationaires en conduite horizontale. Ph.D. thesis, Institut National Polytechnique de Toulouse.

Woods, B.D. 1998 Slug formation and frequency of slugging in gas–liquid flows. Ph.D. thesis, University of Illinois.

Woods, B.D. & Hanratty, T.J. 1996 Relation of slug stability to shedding rate. *Int. J. Multiphase Flow* 22, 809–828.

Woods, B.D. & Hanratty, T.J. 1999 Influence of Froude number on physical processes determining frequency of slugging in horizontal gas–liquid flows. *Int. J. Multiphase Flow* 25, 1195–1223.

Woods, B.D., Hurlburt, E.T. & Hanratty, T.J. 2000 Mechanism of slug formation in downwardly inclined pipes. *Int. J. Multiphase Flow* 26, 977–998.

Woods, B.D., Fan, Z. & Hanratty, T.J. 2006 Frequency and development of slugs in a horizontal pipe at large liquid flows. *Int. J. Multiphase Flow* 32, 902–925.

10 Particle dispersion and deposition

10.1 Prologue

Particles entrained in a turbulent fluid are dispersed by velocity fluctuations; they assume a motion that is related to the fluid turbulence. If the suspension flows through a conduit, deposition on a wall depends on the particle turbulence. An understanding of these processes is needed to describe the annular flow regime for which liquid flows along the walls and as drops in the gas flow. The fraction of liquid that is entrained by the gas depends on the rate at which the film is atomized and the rate at which drops deposit on the film.

Equations for trajectories of spherical drops and bubbles in a turbulent flow field are developed. These are used to relate the turbulence properties and the dispersion of particles to the turbulence properties of the fluid in which they are entrained. Of particular interest is the development of relations for the influence of drop size on drop turbulence and on drop dispersion.

The deposition of particles is strongly related to their size. A wide range of behaviors is observed. Brownian motion is realized for very small particles. Particle motion becomes independent of fluid turbulence for very large particles. It is hard to capture all of these behaviors in a single laboratory experiment. Therefore, numerical studies have been carried out in which a modified Langevin equation is used to provide a stochastic representation of the fluid turbulence.

Results from laboratory and numerical experiments are used to develop correlations for the rate of deposition and for the velocities with which particles strike a wall.

10.2 Dispersion of fluid particles in homogeneous isotropic turbulence

A starting point for this chapter is the description of the dispersion of marked fluid particles from a point source, provided by Taylor (1921). The problem is simplified by assuming a homogeneous, isotropic turbulence and a zero mean velocity ($\bar{u}_i = 0$).

The velocity of a diffusing fluid particle is designated by three components in the x_1, x_2, x_3 directions, v_1, v_2, v_3. The location is designated by y_1, y_2, y_3. The magnitude of the particle turbulence is represented by the mean-square values of the velocity fluctuations, $\overline{v_1^2}$, $\overline{v_2^2}$, $\overline{v_3^2}$. For a homogeneous field, the components $\overline{v_i^2}$ do not vary with location.

Consider a marked fluid particle located at $y_i = 0$, at time zero. The distance traveled by the particle in the y_i-direction by time t is given as

$$y_i(t) = \int_0^t v_i(t')dt' \tag{10.1}$$

If this measurement is repeated a large number of times, the ensemble average at time t is given by $\overline{y_i}(t) = 0$ since v_i is just as likely to be plus or minus. However, $\overline{y_i^2}(t) \neq 0$; it is a measure of the average spread determined from a number of trials.

Einstein (1905) has shown that the molecular diffusion coefficient, D, can be related to the change of mean-square dispersion over a time interval dt by the equation

$$\frac{1}{2}\frac{d\overline{y_i^2}}{dt} = D \tag{10.2}$$

From kinetic theory,

$$D \propto \lambda_{\mathrm{L}} c \tag{10.3}$$

where c is the molecular velocity and λ_{L} is the molecular mean-free path.

Taylor (1921) derived an analogous equation for turbulent dispersion. He pointed out that

$$\frac{dy_i^2}{dt} = 2y_i\frac{dy_i}{dt} = 2v_i\int_0^t v_i(t')dt'$$

$$\frac{d\overline{y_i^2}}{dt} = 2\int_0^t \overline{v_i(t)v_i(t')}dt' \tag{10.4}$$

Thus, a turbulent diffusivity can be defined as

$$\varepsilon = \frac{1}{2}\frac{d\overline{y_i^2}}{dt} = \int_0^t \overline{v_i(t'+s)v_i(t')}dt' \tag{10.5}$$

where $t = t' + s$ has been substituted into (10.4). The turbulence is steady in time so $\overline{v_i(t'+s)v_i(t')} = \overline{v_i(0)v_i(s)}$. Thus, the turbulent diffusivity, ε, is time-dependent

$$\varepsilon(t) = \overline{v_i^2}\int_0^t R_i^{\mathrm{L}}(s)ds \tag{10.6}$$

where

$$R_i^{\mathrm{L}}(s) = \frac{\overline{v_i(0)v_i(s)}}{\overline{v_i^2}} \tag{10.7}$$

is the Lagrangian correlation coefficient. For $s \to 0$, $v_i(0)$ and $v_i(s)$ are perfectly correlated so $R_i^{\mathrm{L}}(t \to 0) \approx 1$ and

$$\varepsilon(t \to 0) = \overline{v_i^2}t \tag{10.8}$$

As s increases, $v_i(0)$ and $v_i(s)$ eventually become completely uncorrelated; that is, $v_i(s)$ is just as likely to be plus or minus if $v_i(0)$ is plus, so $R_i^{\mathrm{L}}(\infty) = 0$. Define a Lagrangian time-scale as

$$\tau^{\mathrm{L}} = \int_0^{\infty} R_i^{\mathrm{L}}(s)ds \tag{10.9}$$

From (10.6)

$$\varepsilon(\infty) = \overline{v_i^2}\tau^{\mathrm{L}} \tag{10.10}$$

Define a Lagrangian length scale as

$$\Lambda^{\mathrm{L}} = \left(\overline{v_i^2}\right)^{1/2}\tau^{\mathrm{L}} \tag{10.11}$$

Thus,

$$\varepsilon(\infty) = \left(\overline{v_i^2}\right)^{1/2}\Lambda^{\mathrm{L}} \tag{10.12}$$

where the two terms on the right side are analogous to the molecular velocity and the molecular mean-free path.

An integration of (10.5) gives

$$\overline{y_i^2} = 2\overline{v_i^2}\int_0^t dt'\int_0^{t'} R_i^{\mathrm{L}}(s)ds \tag{10.13}$$

As shown by Hinze (1959), the integration of (10.13) gives the following relation:

$$\overline{y_i^2} = 2\overline{v_i^2}\int_0^t (t-s)R_i^{\mathrm{L}}(s)ds \tag{10.14}$$

For s close to zero, $R_i^{\mathrm{L}}(s) \approx 1$. Equation (10.14) then gives

$$\overline{y_i^2} = \overline{v_i^2}t^2 \tag{10.15}$$

for small t. For large t,

$$\overline{y_i^2}(t) = 2\overline{v_i^2}t\int_0^t R_i^{\mathrm{L}}(s)ds - 2\overline{v_i^2}\int_0^t sR_i^{\mathrm{L}}(s)ds \tag{10.16}$$

In order to calculate $\overline{y_i^2}(t)$, one needs to specify $R_i^{\mathrm{L}}(s)$. Hinze defines a Lagrangian micro-time-scale as

$$\frac{1}{(\lambda^{\mathrm{T}})^2} = \frac{1}{2}\left(\frac{\partial^2 R_i^{\mathrm{L}}}{\partial s^2}\right)_{s\to 0} \tag{10.17}$$

Thus, as $s \to 0$

$$R_i^{\mathrm{L}}(s) = 1 - \frac{s^2}{(\lambda^{\mathrm{T}})^2} \tag{10.18}$$

A commonly used functionality for $R_i^{\mathrm{L}}(s)$ is

$$R_i^{\mathrm{L}}(s) = \exp\left(-s/\tau^{\mathrm{L}}\right) \tag{10.19}$$

Substitution of (10.19) into (10.16) yields

$$\overline{y_i^2}(t) = 2\overline{v_i^2}\tau^{\mathrm{L}}t - 2\overline{v_i^2}\left(\tau^{\mathrm{L}}\right)^2\left[1 - \exp\left(-t/\tau^{\mathrm{L}}\right)\right] \tag{10.20}$$

Equation (10.19) is not accurate as $t \to 0$ since it gives

$$R_i^{\mathrm{L}}(s) = 1 - \frac{s}{\tau^{\mathrm{L}}} \tag{10.21}$$

in disagreement with (10.18). However, at large Reynolds numbers, this is not a serious problem since (10.18) is valid only over a small range of diffusion times.

Measurements of the dispersion of marked fluid elements originating from the center of a pipe have been made by Towle & Sherwood (1939), Boothroyd (1967), Groenhof (1970) and Taylor & Middleman (1974). Concentration profiles downstream of the origin are represented by Gaussian distributions with standard deviations of $\overline{y_i^2}$. A turbulent diffusivity is calculated from (10.5). Data from the above references for large times are represented by the equation

$$\frac{\varepsilon}{v^* 2r_{\mathrm{t}}} = 0.037 \tag{10.22}$$

where r_{t} is the pipe radius and v^* is the friction velocity, $\left(\tau_{\mathrm{W}}/\rho_{\mathrm{f}}\right)^{1/2}$. The shear stress at the wall, τ_{W}, is calculated from measurements of the pressure drop under fully developed conditions.

$$\tau_{\mathrm{W}} = \frac{r_{\mathrm{t}}}{2}\left|\frac{\Delta P}{\Delta z}\right| \tag{10.23}$$

where vertical bars indicate an absolute value.

10.3 Equation of motion of a particle or a bubble

The equation of motion of a spherical particle, entrained in turbulence, at relative velocities in the range where the force of the fluid on the particle is small enough for Stokes law to be applicable has been given by Maxey and Riley (1983) as

$$\begin{aligned}\frac{4}{3}\pi r_{\mathrm{P}}^3\frac{d\vec{v}_{\mathrm{P}}}{dt} &= \frac{4}{3}\pi r_{\mathrm{P}}^3(\rho_{\mathrm{P}} - \rho_{\mathrm{f}})\vec{g} + 6\pi\mu_{\mathrm{f}}r_{\mathrm{P}}\left(\vec{u} - \vec{v}_{\mathrm{P}}\right)\\ &+ 6\pi\mu_{\mathrm{f}}r_{\mathrm{P}}\int_0^t \frac{d\left(\vec{u} - \vec{v}_{\mathrm{P}}\right)}{d\tau}\frac{d\tau}{\left[\pi\upsilon_{\mathrm{f}}(t-\tau)\right]^{1/2}}\\ &+ \frac{2}{3}\pi r_{\mathrm{P}}^3\rho_{\mathrm{f}}\frac{d}{dt}\left(\vec{u} - \vec{v}_{\mathrm{P}}\right) + \frac{4}{3}\pi r_{\mathrm{P}}^3\rho_{\mathrm{f}}\frac{D\vec{u}}{Dt}\end{aligned} \tag{10.24}$$

where $\vec{u}$ is the fluid velocity, $\vec{v}_P$ is the particle velocity and $(\vec{u} - \vec{v}_P)$ is the relative velocity which dictates the force of the fluid on the particle. Lift forces have been ignored. The term on the left side is the mass of the particle times its acceleration. The first term on the right side is the force of gravity minus the buoyancy force on the particle. The second term is Stokes drag force for a steady flow. The third term is the Basset force (Basset, 1888), which is a correction to Stokes law because the flow is unsteady. The fourth term is the added mass, which accounts for the fluid in the continuum which is accelerated when the particle is accelerated. The last term is the force due to pressure gradients in the fluid. This can be understood by recognizing that $D\vec{u}/Dt$ is the substantial derivative of the fluid. According to inviscid theory $\rho_f D\vec{u}/Dt$ equals the local pressure gradient accompanying the fluid acceleration.

Mei *et al.* (1991) show that the Basset force predicts too long a memory in the time domain. Thus, its use in (10.24) is questionable. For the present, it is best to ignore this term.

In cases for which $\rho_P >> \rho_f$, such as a solid or liquid sphere in a gas, (10.24) can be simplified to

$$\begin{aligned}\frac{4}{3}\pi r_P^3 \rho_P \frac{d\vec{v}_P}{dt} = {} & \frac{4}{3}\pi r_P^3 \vec{g}\rho_P + \vec{F}_L \frac{4}{3}\pi r_P^3 \\ & + \frac{1}{2} C_D \pi r_P^3 \rho_f (\vec{u} - \vec{v}_P)|\vec{u} - \vec{v}_P| \\ & + \frac{2}{3}\pi r_P^3 \rho_f \frac{d}{dt}(\vec{u} - \vec{v}_P) + \frac{4}{3}\pi r_P^3 \rho_f \frac{D\vec{u}}{Dt}\end{aligned} \tag{10.25}$$

where a lift force per unit volume, $\vec{F}_L$, is included, $\vec{u}$ is the fluid velocity seen by the particle and non-Stokesian behavior is considered. For cases in which $\rho_f >> \rho_P$, such as a spherical gas bubble in a liquid, (10.25) can be written as

$$\begin{aligned}0 = & -\frac{4}{3}\pi r_P^3 \rho_f \vec{g} + \frac{4}{3}\pi r_P^3 F_L + C_D \frac{1}{2}\pi r_P^2 \rho_f (\vec{u} - \vec{v})|\vec{u} - \vec{v}_P| \\ & + \frac{2}{3}\pi r_P^3 \rho_f \frac{d(\vec{u} - \vec{v}_P)}{dt} + \frac{4}{3}\pi r_P^3 \rho_f \frac{D\vec{u}}{Dt}\end{aligned} \tag{10.26}$$

Equation (10.25) describes the time change of the velocity of a particle, $\vec{v}_P$, as it moves around in a fluid. The path is given by solving the equation

$$\frac{d\vec{y}}{dt} = \vec{v}_P \tag{10.27}$$

Note that the solution of (10.25) requires a knowledge of the fluid velocity along the path of the particle, which is not known *a priori*. This presents a major problem which has been solved with an iterative method by Reeks (1977).

Equations (10.26) and (10.27) can be used to describe the change of bubble velocity and location with time. Equation (10.26) differs from (10.25) in that the time change of the momentum of the particle is not an important consideration for bubble motion, because of its small density. Instead, the acceleration of the added mass introduces the temporal change of the particle velocity.

10.4 Lift forces and lateral drift

10.4.1 Inviscid lift forces

Several lift forces have been defined. One is the classical inviscid result for a rotating sphere (Magnus effect) described by Schlichting (1979). For this situation,

$$\vec{F}_{\mathrm{L}} = C_{\mathrm{L}}\rho_{\mathrm{L}}\vec{U}_{\mathrm{S}} \times \vec{\omega}_{\mathrm{P}} \tag{10.28}$$

where $\vec{U}_{\mathrm{S}}$ is the slip velocity of the particle or bubble (velocity of the particle minus the velocity of the surrounding fluid) and $\vec{\omega}_{\mathrm{P}}$ is the angular velocity of the particle (which is positive for a counterclockwise rotation). For a bubble in upward flow, $\vec{U}_{\mathrm{S}}$ is positive and the lift force is toward the wall if $\vec{\omega}_{\mathrm{P}}$ is positive.

As already discussed in Section 2.9 in the chapter on flow regimes, bubbles tend to accumulate close to the wall for upward flows in a pipe and away from the wall for downward flows. This supports (10.28) if $\omega_{\mathrm{P}} = -\vec{i}_r\, d\overline{u}_z/dr$, so that

$$F_{\mathrm{L}r} = -C_{\mathrm{L}}(\overline{u}_{\mathrm{G2}} - \overline{u}_{\mathrm{L2}})\rho_{\mathrm{L}}\frac{d\overline{u}_z}{dr} \tag{10.29}$$

However, this agreement does not prove that the bubbles are rotating. As shown by Beyerlein *et al.* (1985), a bubble embedded in a flow which has gradients in the mean velocity would experience a difference in fluid velocity on the two sides of the bubble. Because of a Bernoulli effect this results in a pressure difference. This, in turn, provides a lateral net force on the bubble.

Furthermore, the bubbles are not spherical in gas–liquid flows. The force on these bubbles could have a component in the lateral direction which is related to their shape and orientation. Thus, the theoretical justification for using (10.29) is not established.

These lift forces on bubbles, associated with a non-uniform mean velocity, lead to drift velocities in the radial direction. A drift velocity can also be realized if the fluid turbulence varies spatially. The particles will then tend to move from regions of high turbulence to regions of low turbulence. Section 10.9 develops an analytical expression for this effect.

10.4.2 Saffman viscous lift force

Saffman (1965, 1968) showed that, for low particle Reynolds numbers in a shear flow, the lift force due to particle rotation can be neglected. He developed the following relation for a particle in a steady laminar flow due to uniform simple shear

$$\vec{F}_{\mathrm{L}} = 6.46\mu_{\mathrm{f}}r_{\mathrm{P}}^2\left(\frac{1}{\upsilon_{\mathrm{f}}}\frac{d\overline{u_z}}{dy}\right)^{1/2}(\overline{u_z} - \overline{v_{\mathrm{P}z}}) \tag{10.30}$$

where y is the distance from the wall. If the particle leads the fluid, F_{L} is negative, that is, toward the wall. If the particle lags the fluid, F_{L} is positive. Since the fluid drag is described by Stokes law, the particle Reynolds number needs to be small. Furthermore, the dimensionless shear rate needs to be large and the dimensionless particle diameter needs to be small.

$$\frac{\left(\frac{d\overline{u_z}}{dy}\right)^{1/2} \upsilon_{\mathrm{f}}^{1/2}}{|\overline{u_z} - \overline{v_{\mathrm{P}z}}|} >> 1 \qquad \left(\frac{d\overline{u_z}}{dy}\right)\frac{(d_{\mathrm{P}})^2}{\upsilon_{\mathrm{f}}} << 1 \tag{10.31}$$

These restrictions suggest that Saffman's equation should not be used for $y^+ > 70$ in a turbulent flow (see McLaughlin, 1991).

10.5 Characterization of fluid turbulence

Particles in the presence of a turbulent velocity field will assume a turbulent motion, because of the fluctuating drag associated with the fluctuating fluid velocity. Because of the inertia of the particles and the gravitational force, the particles do not follow the velocity fluctuations exactly. The goal of theory is to relate the particle turbulence and the particle diffusivity to the fluid turbulence and the fluid turbulent diffusivity. Homogeneous, isotropic turbulence with zero mean velocity is assumed.

Fluid turbulence is characterized by the magnitudes of the three components of the fluid velocity fluctuations, $\overline{u_i^2}$, by correlation coefficients and by spectra. Eulerian spatial and temporal correlation coefficients can be defined analogous to the Lagrangian correlation (10.7). Thus, for a fixed location, the temporal correlation is defined as

$$R_i^{\mathrm{E}}(x, s) = \frac{\overline{u_i(t)u_i(t+s)}}{\overline{u_i^2}} \tag{10.32}$$

where $u_i(t)\, u_i\,(t + s)$ is the product of u_i at two different times and at a fixed location, x_i. A spatial Eulerian correlation can be defined for a fixed time as

$$R_i^{\mathrm{E}}(t, \xi) = \frac{\overline{u_i(x_i)u(x_i + \xi)}}{\overline{u_i^2}} \tag{10.33}$$

where $u_i(x_i)u_i(x_i{+}\xi_i)$ is the product of u_i at two different locations at a fixed time. The overbars are averages for a large number of realizations of the field. For $\xi = 0$ or $s = 0$, these correlation coefficients are equal to unity. For $\xi = \infty$, $R_i^{\mathrm{E}} = 0$, since for a fixed time the velocities at locations $x_i + \xi$ and x_i are uncorrelated for large ξ. Thus, from (10.33),

$$\Lambda_i = \int_0^\infty R_i^{\mathrm{E}} d\xi \tag{10.34}$$

is a measure of the extent of the turbulent velocity fluctuations. An Eulerian time-scale is obtained from (10.32):

$$\tau_i^{\mathrm{E}} = \int_0^\infty R_i^{\mathrm{E}}(s) ds \tag{10.35}$$

which is a measure of the duration of the turbulent fluctuations at a given location.

An alternative way to describe turbulence is through frequency spectra, whereby the intensity $\overline{u_i^2}$ can be viewed as containing contributions from a number of frequencies. Thus $E_i(n)dn$ represents contributions to $\overline{u_i^2}$ from frequencies between n and $n + dn$ and

$$\overline{u_i^2} = \int_0^\infty E_i(n)(dn) \tag{10.36}$$

Hinze (1959) derived the following relations between the spectral function $E_i(n)$ and the Eulerian temporal correlation:

$$R_i^{\mathrm{E}}(t) = \frac{1}{\overline{u_i^2}} \int_0^\infty dnE_i(n) \cos 2\pi nt \tag{10.37}$$

$$E_i(n) = 4\overline{u_i^2} \int_0^\infty dt R_i^{\mathrm{E}}(t) \cos 2\pi nt \tag{10.38}$$

Note that

$$E_i(0) = 4\overline{u_i^2} \int_0^\infty dt R_i^{\mathrm{E}}(t) = 4\overline{u_i^2}\tau_i^{\mathrm{E}} \tag{10.39}$$

For a homogeneous isotropic field, $\overline{u_1^2} = \overline{u_2^2} = \overline{u_3^2}$, two fundamental spatial correlations can be defined. The longitudinal correlation considers two components in the same direction, say x_1, separated by a distance r. Thus,

$$f(r) = \overline{u_1(x_1)u_1(x_1 + r)} \tag{10.40}$$

The lateral correlation relates lateral components, u_2 or u_3, of the velocity

$$g(r) = \overline{u_2(x_1)u_2(x_1 + r)} \tag{10.41}$$

For isotropic turbulence, conservation of mass gives

$$g(r) = f(r) + \frac{1}{2} r \frac{df}{dr} \tag{10.42}$$

See Hinze (1959).

Consider the case

$$f(r) = e^{-r/\Lambda} \tag{10.43}$$

The lateral correlation is given as

$$g(r) = \left(1 - \frac{r}{2\Lambda}\right) e^{-r/\Lambda} \tag{10.44}$$

This shows interesting differences between $f(r)$ and $g(r)$. The longitudinal correlation $f(r)$ is positive for all values of r. However, $g(r)$ takes on negative values. The positive values of $g(r)$ correspond to a region in which the lateral velocity has the same sign. The correlation must have negative values at large r in order that conservation of mass is satisfied.

If a homogeneous isotropic turbulence is being transported with a uniform velocity in the x_1-direction, $\overline{U_1}$, the Taylor hypothesis can be used whereby the time variation of the turbulence can be related to the spatial variation by assuming

$$\frac{\partial}{\partial t} = -\overline{U_1}\frac{\partial}{\partial x_1} \tag{10.45}$$

Thus, the temporal correlation can be related to the spatial correlation as follows

$$f(x_1) = R_1^{\mathrm{E}}\left(t = x_1/\overline{U_1}\right) \tag{10.46}$$

(See Hinze (1959, pp. 40–41) for a discussion of the Taylor hypothesis.)

From (10.37) and (10.38), the spectral function and the correlations can be defined as follows:

$$f(x_1) = \frac{1}{\overline{u_1^2}}\int_0^\infty dnE_1(n)\cos\frac{2\pi n x_1}{\overline{U_1}} \tag{10.47}$$

$$E_1(n) = 4\overline{u_1^2}\int_0^\infty \frac{dx_1}{\overline{U_1}} f(x_1)\frac{\cos 2\pi n x_1}{\overline{U_1}} \tag{10.48}$$

Suppose $f(x_1)$ is approximated by an exponential function, $f(x_1) = \exp(-x_1/\Lambda_1)$, except for very small x_1 (which is representative of high frequencies). Then from (10.48)

$$\frac{\overline{U_1}E_1(n)}{4\overline{u_i^2}\Lambda_1} = \frac{1}{1 + \left(4\pi^2 n^2/\overline{U_1^2}(\Lambda_1)^2\right)} \tag{10.49}$$

A plot of data, constructed by Favre and colleagues is presented by Hinze (1959, p. 61). The measurements were made behind a square grid, located in a wind tunnel, so as to approximate a homogeneous, isotropic turbulence convected with a velocity $\overline{U_1}$. The solid line in this plot represents (10.49). The filled points are measurements. The open points are calculated from measurements of $f(x_1)$. Hinze points out that (10.49) "appears to approximate measured spectrum curves satisfactorily, except, of course, for that part of the $E_1(n)$ -curve that pertains to high values of n. When n approaches zero, the value of $\overline{U_1}E_1(n)/\overline{u_1^2}\Lambda_1 = 4$ agrees very satisfactorily with experimental data."

10.6 Relation of particle turbulence to fluid turbulence

An inertial time constant, $1/\beta$, can be defined as

$$\frac{1}{\beta} = \frac{3C_{\mathrm{D}}\rho_{\mathrm{f}}}{4d_{\mathrm{p}}\rho_{\mathrm{p}}}\left|\vec{u} - \vec{v}_p\right| \tag{10.50}$$

where, for Stokes flow,

$$\beta = \frac{18\mu_f}{d_P^2\rho_P} \tag{10.51}$$

Equation (10.25) for $\rho_P >> \rho_f$ can then be written as follows if the effect of gravity and the last two terms of (10.25) are ignored, and β is defined by (10.50)

$$\frac{d\vec{v}_P}{dt} = \beta(\vec{u} - \vec{v}_P) \tag{10.52}$$

or

$$\frac{d\vec{v}_P}{dt} + \beta\vec{v}_P = \beta\vec{u} \tag{10.53}$$

Thus, $1/\beta$ is a time constant which characterizes the ability of a particle to follow the fluid turbulence. It plays a central role in understanding the behavior of particles in a turbulent field. (Consider a particle released in a stagnant fluid with a velocity $\vec{v}_0$. The product of $\vec{v}_0$ and $1/\beta$ represents the stopping distance.)

The integration of (10.53) yields

$$v_{Pi} = v_{Pi}(0)\exp(-\beta t) + \beta\exp(-\beta t)\int_0^t \exp(\beta T)u_i dT \tag{10.54}$$

where $v_{Pi}(0)$ is a component of the particle velocity vector and u_i is a component of the fluid velocity vector that the entrained particle sees at time T. The first term on the right side of (10.54) is the solution to the homogeneous part of (10.53) and the second term is the particular solution. Squaring (10.54) and, then, taking the mean value, over many particles, of each of the terms, gives

$$\begin{aligned} \overline{v_{Pi}^2} &= \overline{v_{Pi}^2}(0)\exp(-2\beta t) + 2\beta\exp(-2\beta t)\int_0^t \exp(\beta T)\overline{v_P(0)u(T)}dT \\ &+ \overline{u_i^2}\beta\exp(-2\beta t)\int_0^t R_i^{\text{path}}(\theta)[\exp(-\beta\theta + 2\beta t) - \exp(\beta\theta)]d\theta \end{aligned} \tag{10.55}$$

where

$$R_i^{\text{path}}(\theta) = \frac{\overline{u_i(t)u_i(t+\theta)}}{\overline{u_i^2}} \tag{10.56}$$

The correlation $R_i^{\text{path}}(\theta)$ is not an Eulerian or a Lagrangian correlation in that it considers velocities of the fluid seen by an entrained particle as it moves around the field. In the limit of large times, (10.55) gives

$$\overline{v_{Pi}^2} = \overline{u_i^2}\beta\int_0^\infty \exp(-\beta\theta)R_i^{\text{path}}(\theta)d\theta \tag{10.57}$$

Suppose

$$R_i^{\mathrm{path}}(\theta) = \exp(-\alpha\theta) = \exp\left(-\theta/\tau^{\mathrm{path}}\right) \qquad (10.58)$$

where $\tau_i^{\mathrm{path}} = \alpha^{-1}$ is the time-scale for this correlation. Then (10.57) gives the limiting value of $\overline{v_{\mathrm{P}i}^2}$

$$\overline{v_{\mathrm{P}i}^2} = \overline{u_i^2}\left(\frac{\beta}{\alpha+\beta}\right) \qquad (10.59)$$

Since β is the reciprocal of the inertial time constant, equation (10.59) predicts that particles with large inertial time constants (small β) will have much smaller turbulent velocity fluctuations than the fluid. This is because the entrained particles do not respond to high frequencies and thus "see" only a fraction of the fluid velocity fluctuations.

Equation (10.55) is plotted in Figure 10.1 for the case of $\overline{v_{\mathrm{P}i}^2}(0) = 0$. The abscissa, βt, is the ratio of the time to the inertial time constant of the particles. Note that there is a transient region in which the particle becomes entrained in the fluid turbulence. Its duration is given as βt = 2–3. Thus, particles with large inertial time constants require a long time to come to a stationary state.

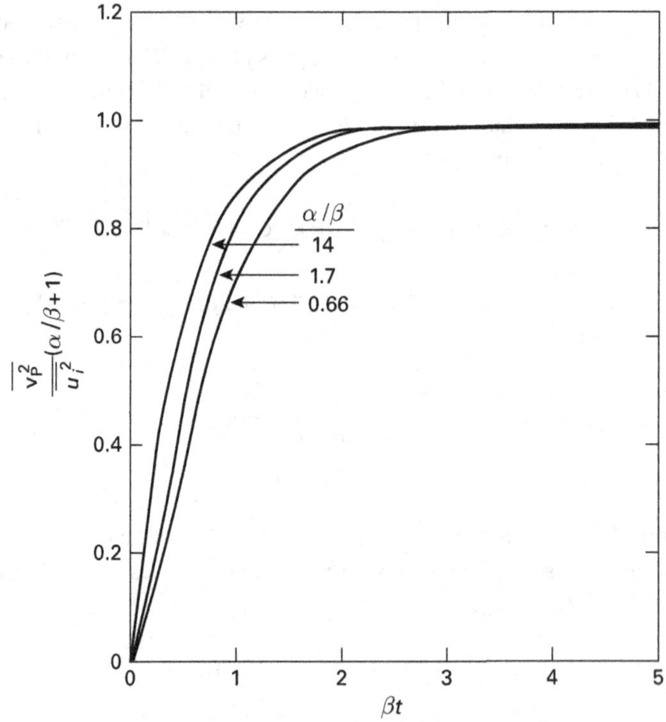

Figure 10.1 Particle turbulence as a function of the dimensionless time for the case of $\overline{v_{\mathrm{P}}^2}(0) = 0$.

10.7 Dispersion of particles in a turbulent gas which is homogeneous and isotropic

10.7.1 No effect of gravity

Friedlander & Johnstone (1957) integrated (10.53) twice, using Stokes law, to obtain the following equation for the dispersion of entrained particles, from a point source, at large times:

$$\varepsilon_{\mathrm{P}} = \frac{\overline{X_{\mathrm{P}}^2}}{2t} = \overline{u_i^2}\int_0^t R_i^{\mathrm{path}}\,d\theta \quad \text{as} \quad t \to \infty \tag{10.60}$$

where $\overline{u_i^2}$ represents the fluid turbulence and R_i^{path} is the correlation coefficient of fluid velocity fluctuations seen by the entrained particle. If the correlation coefficient is represented by (10.58), equation (10.60) yields

$$\varepsilon_{\mathrm{P}\infty} = \frac{\overline{u_i^2}}{\alpha} = \frac{\overline{u_i^2}\tau^{\mathrm{L}}}{\alpha\tau^{\mathrm{L}}} = \frac{\varepsilon(\infty)}{\alpha\tau^{\mathrm{L}}} \tag{10.61}$$

where τ^{L} is the Lagrangian time-scale for the fluid turbulence, defined by (10.9). For homogeneous turbulence, the Eulerian average of the fluid velocity fluctuations at a given location, $\overline{u_i^2}$, is usually taken to be equal to the Lagrangian average over particle paths. Thus, in the above equation, the fluid diffusivity $\varepsilon(\infty)$ is defined by (10.10).

If $\beta\tau^{\mathrm{L}}$ is large, the particle closely follows the fluid ($\alpha\tau^{\mathrm{L}} = 1$). Equations (10.59) and (10.61) yield $\overline{v_{\mathrm{P}i}^2} = \overline{u_i^2}$ and $\varepsilon_{\mathrm{P}}(\infty) = \varepsilon(\infty)$. If $\beta\tau^{\mathrm{L}}$ is small, the particles will not follow the turbulence. In this limit, the particles will be almost stationary; they are, effectively, sampling the turbulence at a fixed location; that is, $\alpha^{-1} = \tau^{\mathrm{E}}$, where τ^{E} is the Eulerian time-scale, defined by (10.35). Theoretical work by Reeks (1977) gives $\tau^{\mathrm{E}} = (1 - 1.3)\tau^{\mathrm{L}}$ as $\beta\tau^{\mathrm{L}}$ varies from 2 to 0. Reeks, therefore, suggests $\varepsilon_{\mathrm{P}\infty} \approx \varepsilon_{\infty}$.

This is consistent with the analysis of Hinze (1959, p. 466, equation 5.207), which uses spectral representations of $R_{\mathrm{P}}^{\mathrm{L}}(t)$ and of $R^{\mathrm{L}}(t)$ to suggest that

$$\varepsilon_{\mathrm{P}\infty} = \varepsilon_{\infty} \tag{10.62}$$

The above result is remarkable since (10.59) predicts that the intensity of the particle turbulence, $\overline{v_{\mathrm{P}i}^2}$, can be much smaller than the intensity of the fluid turbulence, $\overline{u_i^2}$, if the inertial time-scale of the particles is large.

From Section 10.2, one can describe the diffusivity of particles at large times as the product of a velocity and a length scale.

$$\varepsilon_{\mathrm{P}}(\infty) = \left(\overline{v_{\mathrm{P}i}^2}\right)^{1/2} \Lambda_{\mathrm{P}i} \tag{10.63}$$

where the length scale is defined as

$$\Lambda_{\mathrm{P}i} = \left(\overline{v_{\mathrm{P}i}^2}\right)^{1/2} \tau_{\mathrm{P}}^{\mathrm{L}} \tag{10.64}$$

For large inertial time constants, the velocity scale decreases with increasing $1/\beta$. Hinze's suggestion that $\varepsilon_{\rm P} \approx \varepsilon$ implies that $\Lambda_{\rm P}$ must increase with increasing $1/\beta$ just enough for $\varepsilon_{\rm P}(\infty)$ to remain roughly constant. This can be explained because inertia of entrained particles inhibits their ability to respond to high-frequency velocity fluctuations. Thus, the entrained particles have larger time constants and, therefore, larger length scales.

The prediction of $\overline{v_{{\rm P}i}^2}$ in equation (10.59) depends on the representation of $R_i^{\rm path}$ by (10.58) and the definition of an empirical reciprocal time constant α. As mentioned above, some success in relating this correlation to Eulerian properties of the fluid turbulence has been realized by using the iterative approach of Reeks (1977).

[Equation (10.62) is strictly applicable to a homogeneous turbulence, which is approximated in the central region of a pipe or channel. In fact, computational results exist which suggest that $\varepsilon_{\rm P} \neq \varepsilon_{\rm f}$ close to a wall where the turbulence is non-homogeneous. See, for example, Brooke *et al.* (1994).]

10.7.2 Effect of gravity; heavy particles

The preceding section points out that heavy particles do not follow the fluid turbulence exactly. Their inertia causes the root-mean-square of the turbulent velocity fluctuations to be smaller than the root-mean-square of the velocity fluctuations of the fluid. Furthermore, high-frequency fluid velocity fluctuations become less effective in dispersing particles, so that the length scale characterizing their dispersion increases.

Yudine (1959) pointed out an additional effect due to the influence of gravity, which he called a "crossing of trajectories." Turbulent dispersion of fluid particles is strongly related to turbulent structures in which fluid motion is highly correlated. Heavy particles entrained in the fluid possess a free-fall velocity. Yudine recognized that this velocity could cause the particles to move out of eddies responsible for the dispersion of fluid particles. Thus, the velocity correlation of heavy particles decreases more rapidly with time than does the correlation for fluid particles. This "crossing of trajectories" effect can have a strong influence on the dispersion of heavy particles.

However, the expectation is that the "crossing of trajectories" should have little, if any, effect on the measured intensity of turbulent velocity fluctuations of the particles since it only affects the method of sampling.

For the case of very large ratios of the free-fall velocity to the fluid turbulence, $\overline{U_{\rm S}}/\left(\overline{u^2}\right)^{1/2}$, the turbulence seen by a particle may be considered as frozen. Thus, the fluid velocity fluctuations seen by a particle are characterized by the Eulerian spatial correlation for the fluid turbulence:

$$R_{ii}^{\rm L}(s) = R_{ii}^{\rm E}\left(x = \overline{U_{\rm S}s}\right) \tag{10.65}$$

where x is the coordinate in the direction of free fall. The particle diffusivity for large diffusion times is then given by

$$\varepsilon_{\mathrm{P}} = \overline{v_{\mathrm{P}i}^2} \int_0^\infty R_{ii}^{\mathrm{E}}(x = U_{\mathrm{S}} s) dx \tag{10.66}$$

For very small $\overline{U_{\mathrm{S}}}/\left(\overline{u_i^2}\right)^{1/2}$ and large diffusion times, the turbulent diffusivity of the particles is given by (10.60). For intermediate $\overline{U_{\mathrm{S}}}/\left(\overline{u_i^2}\right)^{1/2}$, interpolation formulas (Csanady, 1963; Meek & Jones, 1973) can be used.

For example, Csanady (1963) suggests that the correlation coefficient can be given by

$$R(s) = \exp\left[-\frac{s}{\Lambda_i}\left(\overline{U_{\mathrm{S}}}^2 + \overline{v_{\mathrm{P}}^2}\right)^{1/2}\right] \tag{10.67}$$

Thus, for large $\overline{U_{\mathrm{S}}}/\left(\overline{v_{\mathrm{P}}^2}\right)^{1/2}$

$$R(s) = \exp\left(-\frac{s\overline{U_{\mathrm{S}}}}{\Lambda_i}\right) \tag{10.68}$$

For small $\overline{U_{\mathrm{S}}}/\left(\overline{v_{\mathrm{P}}^2}\right)^{1/2}$

$$R(s) = \exp\left(-\frac{s\left(\overline{v_{\mathrm{P}}^2}\right)^{1/2}}{\Lambda_i}\right) \tag{10.69}$$

The analysis of Meek & Jones (1973) uses the spectral function. They argue that the average wave number is unchanged by the crossing of trajectories. However, the convection velocity changes from c_0 to $c_0 + \overline{U_{\mathrm{S}}}$. Similarly, the frequency changes from $\omega_0 = kc_0$ to $\omega = k\left(\overline{U_{\mathrm{S}}} + c_0\right)$. Meek & Jones suggest the following interpolation formula for ω:

$$\omega = \omega_0 \left[1 + \left(\frac{\overline{U_{\mathrm{S}}}}{c_0}\right)\right]^{1/2} \tag{10.70}$$

Thus, the crossing of trajectories can have a much greater effect on the spatial scale of the turbulence seen by the particles than does particle inertia. The calculations performed by Reeks (1977) indicate that the effect of crossing of trajectories starts to become important when the free-fall velocity is roughly equal to the magnitude of the turbulent fluctuations $\left(\overline{U_{\mathrm{S}}}/q \approx 1\right)$.

10.8 Measurements of particle turbulence for suspension flow in a pipe

10.8.1 The experiments of Lee *et al.*

Photographic measurements of particle turbulence have been made by Lee *et al.* (1989) and by Young & Hanratty (1991). These were carried out for downward flow in a 5.08 cm pipe.

The experiments of Lee *et al.* were done by injecting drops of water, with diameters of 50–150 μm, into an air stream at the center of a pipe. Light sheets were formed at three

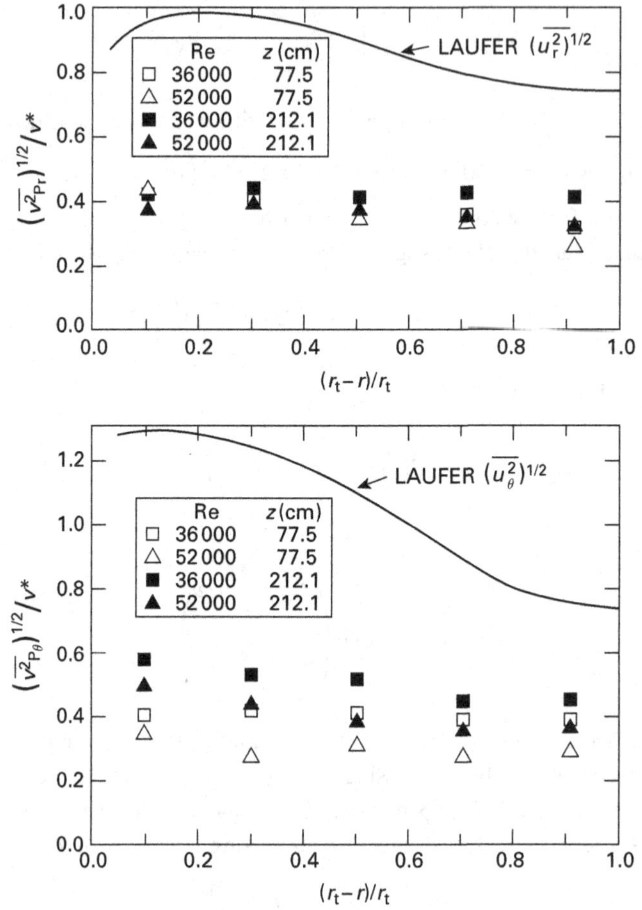

Figure 10.2 Measurements of particle turbulence for 50 μm drops in air. Lee *et al.*, 1989.

locations along the pipe by electronic flash units. A camera, located at the bottom of the pipe, obtained axial-view photographs of the particles at illuminated cross-sections. The timing between flashes provided three photographs of particles as they moved down the pipe. The distances traveled by the particles and the time interval between flashes can be used to calculate a velocity. In this way, the motion of the droplets in a pipe cross-section could be measured. This technique has two interesting aspects: (1) The use of axial-viewing photography filters out the much larger axial flow component so that accurate measurements of turbulent motion in planes perpendicular to the mean flow direction are possible. (2) Direct (rather than the usual indirect) measurements were made of local mass fluxes in the radial direction, needed to define Eulerian eddy diffusion coefficients.

Figure 10.2 shows measurements of the dimensionless root-mean-square of the radial and tangential components of the velocity fluctuations of 50 μm drops, $\left(\overline{v_{Pr}^2}\right)^{1/2}$

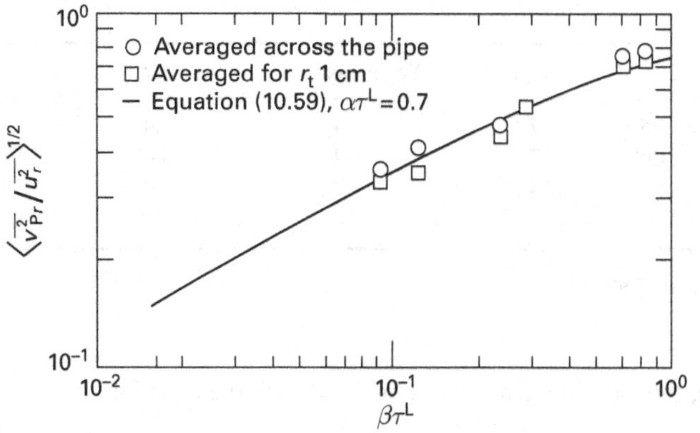

Figure 10.3 Spatially averaged ratio of the particle to fluid turbulence as a function of $\beta\tau^L$

and $\left(\overline{v_{P\theta}^2}\right)^{1/2}$ over a pipe cross-section. The curved line represents measurements of the fluid velocity fluctuations, $\left(\overline{u_i^2}\right)^{1/2}/v^*$, obtained by Laufer (1954) in a pipe flow. It is noted that the turbulence of entrained particles is significantly lower than the fluid turbulence. Figure 10.3 shows a plot of the spatially averaged ratio of particle turbulence to fluid turbulence as a function of $\beta\tau^L$, where τ^L is the Lagrangian time-scale of the fluid. The fluid time-scale is obtained from (10.22) since $\varepsilon = \left(\overline{v_r^2}\right)\tau^L$. The solid curve is calculated with (10.59), using $\alpha = 0.7/\tau^L$. This fitting parameter is reasonable since it indicates that the time-scale of the fluid velocity fluctuations seen by the particles (in the absence of gravitational effects) is approximately 30 % larger than the Lagrangian fluid time-scale. From Figure 10.3, it is seen that particle inertia affects the intensity of the particle velocity fluctuations for $\beta\tau^L$ less than 1. No evidence of an effect of a crossing of trajectories on $\overline{v_r^2}$ was observed, since $\left(\overline{U}_S/q\right) < 1$.

Values of the average velocity in the r-direction, $\overline{V_r}$, were also obtained. These give a direct measurement of the flux of particles since

$$\overline{N} = \overline{V_r}\overline{C} \tag{10.71}$$

The Eulerian diffusion coefficient could be calculated from measurements of $\overline{V_r}$ and $\overline{C}(r)$ since

$$\varepsilon_P^E(r) = \frac{\overline{V_r}\overline{C}}{\left(-\partial\overline{C}/\partial r\right)} \tag{10.72}$$

The experiments of Lee *et al.* show that concentration profiles of particles downstream of the injector could be approximated by a Gaussian function. Then (10.72) gives

$$\varepsilon_P = \frac{\overline{V_r}\overline{X_P^2}}{r} \tag{10.73}$$

Table 10.1 Estimates of turbulent diffusion coefficients

d_P (μm)	Re	z_P (cm)	$\overline{X_P^2}$ (measured)	ε_P (cm^2/s)	ε_f (cm^2/s)	$\varepsilon_P/\varepsilon_f$
90	36 000	77.5	1.4	8.4	9.5	.088
90	52 000	77.5	1.4	13.4	13.1	1.02
150	36 000	114.0	1.32	5.9	9.5	0.62
150	52 000	114.0	1.0	8.9	13.1	.068

Source: Lee *et al.* (1989)

Furthermore, the measurements could be fitted with the equation

$$\overline{V_r} = kr \tag{10.74}$$

where r is the radial location and k is an empirical constant.

Values of ε_P calculated in this way are presented in Table 10.1, taken from Lee *et al.* (1989). These measurements show that ε_P is close to ε_f provided $\overline{U_S}/\left(\overline{v_{Pr}^2}\right)^{1/2}$ is less than 0.85. The values of $\varepsilon_P/\varepsilon_f = 0.62$, 0.68 for 150 μm particles can be interpreted as due to a "crossing of trajectories" effect.

10.8.2 The experiments of Young & Hanratty

The experiments of Young & Hanratty (1991) used a water flow. Glass or steel spheres with a diameter of 100 μm were injected at the center of a 5.08 cm pipe. In one set of experiments the cross-section of the pipe was illuminated over 3–5 mm lengths by flash units. The positions of the particles were photographed at different distances from the injector by a camera located at the bottom of the pipe. These experiments provided measurements of $\overline{X_P^2}$ versus time $= z/\overline{v}_{Pz}$. The Lagrangian turbulent diffusivity at large times was obtained from the relation $\varepsilon_P = (1/2)\left(d\overline{X_P^2}/dt\right)$. In a second set of experiments the velocity components of the particles in a plane perpendicular to the flow were determined by using five colored flashes located immediately downstream of one another. This allowed measurements of the particle acceleration, as well as the particle velocity.

The equation of motion for a solid particle in a liquid is a situation for which the assumption that $\rho_P >> \rho_f$ might not be satisfied. Based on the work of Tchen, discussed in Hinze (1959, 1975), Young & Hanratty (1991) recommended the following approximation for (10.25) and (10.52) for liquid–particle flows:

$$\frac{d\vec{v}_P}{dt} = \tilde{\beta}(\vec{u} - \overline{v_P}) + \frac{\left(\rho_P - \rho_f\right)}{\rho_P}\vec{g} \tag{10.75}$$

where

$$\tilde{\beta} = \frac{3C_{\mathrm{D}}\rho_{\mathrm{f}}|\vec{u} - \vec{v}_{\mathrm{P}}|}{2d_{\mathrm{P}}(2\rho_{\mathrm{P}} + \rho_{\mathrm{f}})} \tag{10.76}$$

Note that (10.76) is the same as (10.50) for small ρ_{f}. The relative velocity appearing in (10.76) is given as

$$|\vec{u} - \vec{v}_{\mathrm{P}}|^2 = \overline{(\bar{u}_z - \bar{v}_{\mathrm{P}z})^2} + \sum_{i=1}^{i=3} \overline{(u_i^t - v_{\mathrm{P}i}^t)^2} \tag{10.77}$$

The second term (due to particle turbulence) on the right side was ignored so that $|\vec{u} - \vec{v}_{\mathrm{P}}| = |\bar{u}_z - \bar{v}_{\mathrm{P}z}|$. Young & Hanratty (1991) used their measurements of particle turbulence to justify this assumption.

The experiments are characterized by $\tilde{\beta}\tau^{\mathrm{L}}$ much greater than unity so the inertial time constant was small enough for the particles to follow the fluid velocity fluctuations closely. Thus, $\overline{v_{\mathrm{P}r}^2} = \overline{u_r^2}$. One of the experiments, using stainless steel particles, was done with $\overline{U_{\mathrm{S}}^+} = 1.77$ and $\tilde{\beta}\tau^{\mathrm{L}} = 53.8$. This produced a value for the Eulerian particle diffusivity of $\varepsilon_{\mathrm{P}}/\varepsilon_{\mathrm{f}} = 0.57$, where $\overline{U_{\mathrm{S}}^+}$ is the average slip velocity between the solid particles and the fluid (see Young & Hanratty, 1991, Table 4). This result, for large $\overline{U_{\mathrm{S}}^+}$, is consistent with the "crossing of trajectories" theory of Csanady (1963). Young & Hanratty support the suggestion that this effect becomes important for $\overline{U_{\mathrm{S}}}/\left(\overline{v_{\mathrm{P}r}^2}\right)^{1/2} \geq 0.85$.

10.9 Turbophoresis

10.9.1 The concept of turbophoresis

The results presented above consider the turbulence to be homogeneous. However, for flow in a pipe or channel, the turbulence is changing in space. This can lead to a force that causes particles to move from regions of high turbulence to regions of low turbulence (Caporaloni *et al.*, 1975; Reeks, 1983). This process is called turbophoresis. Caporaloni *et al.* and Reeks derived the following equation for the turbophoretic velocity:

$$V_{\mathrm{tp}} = -\frac{1}{\tilde{\beta}}\frac{d\overline{v_{\mathrm{P}r}^{t2}}}{dr} \tag{10.78}$$

where $v_{\mathrm{P}r}^t$ is the r-component of the turbulent velocity of the particle.

A straightforward derivation of this relation is presented by Young & Hanratty (1991). From (10.75), if gravitational effects are ignored,

$$A_{\mathrm{P}i} = \tilde{\beta}(u_i - v_{\mathrm{P}i}) \tag{10.79}$$

where $A_{\mathrm{P}i}$ is the acceleration of the particle in the i-direction and $\tilde{\beta}$ is given by (10.76). For very large $\rho_{\mathrm{P}}/\rho_{\mathrm{f}}$ (particles in a gas)

$$\tilde{\beta} = \frac{3C_{\mathrm{D}}\rho_{\mathrm{f}}|\vec{u} - \vec{v}_{\mathrm{P}}|}{4d_{\mathrm{P}}\rho_{\mathrm{P}}} \tag{10.80}$$

Take an average of (10.79) at a given location and let $i = r$,

$$\overline{\frac{1}{\tilde{\beta}}A_{\mathrm{P}r}} = (\overline{u}_r - \overline{v_{\mathrm{P}r}}) \tag{10.81}$$

Since $\overline{u_r} = 0$

$$V_{\mathrm{tp}} = \overline{v_{\mathrm{P}r}} = -\overline{\frac{1}{\tilde{\beta}}A_{\mathrm{P}r}} \approx -\frac{1}{\overline{\tilde{\beta}}}\overline{A_{\mathrm{P}r}} \tag{10.82}$$

Thus, if $\overline{A_{\mathrm{P}r}}$ is not equal to zero, a turbophoretic velocity, $V_{\mathrm{tP}} = \overline{v_{\mathrm{P}r}}$, exists.

10.9.2 Formulation in cylindrical polar coordinates

Consider the Eulerian representation of $A_{\mathrm{P}r}$ in cylindrical polar coordinates

$$\overline{A_{\mathrm{P}r}} = \overline{\frac{dv_{\mathrm{P}r}}{dt}} - \overline{\frac{v_{\mathrm{P}\theta}^2}{r}} \tag{10.83}$$

where $\overline{v_{\mathrm{P}\theta}^2/r}$ is the centripetal acceleration and $\overline{dv_{\mathrm{P}r}/dt}$ is the average of the substantial derivative, $Dv_{\mathrm{P}r}/Dt$.

$$\frac{dv_{\mathrm{P}r}}{dt} = \frac{\partial v_{\mathrm{P}r}}{\partial t} + v_{\mathrm{P}r}\frac{\partial v_{\mathrm{P}r}}{\partial r} + \frac{v_{\mathrm{P}\theta}}{r}\frac{\partial v_{\mathrm{P}r}}{\partial \theta} + v_{\mathrm{P}z}\frac{\partial v_{\mathrm{P}r}}{\partial z} \tag{10.84}$$

If changes of the particle concentration are ignored (or if the concentration is very small), conservation of mass dictates that

$$0 = v_{\mathrm{P}r}\left(\frac{\partial v_{\mathrm{P}r}}{\partial r} + \frac{v_{\mathrm{P}r}}{r} + \frac{1}{r}\frac{\partial v_{\mathrm{P}\theta}}{\partial \theta} + \frac{\partial v_{\mathrm{P}z}}{\partial z}\right) \tag{10.85}$$

Add (10.85) to (10.84) and take a time average

$$\overline{\frac{dv_{\mathrm{P}r}}{dt}} = \frac{1}{r}\frac{\partial\left(r\overline{v_{\mathrm{P}r}^2}\right)}{\partial r} + \frac{1}{r}\frac{\partial(\overline{v_{\mathrm{P}r}v_{\mathrm{P}\theta}})}{\partial \theta} + \frac{\partial(\overline{v_{\mathrm{P}r}v_{\mathrm{P}z}})}{\partial z} \tag{10.86}$$

From (10.83) and (10.86)

$$\overline{A_{\mathrm{P}r}} = -\frac{\overline{v_{\mathrm{P}\theta}^{\mathrm{t}2}}}{r} + \frac{1}{r}\left(\frac{\partial\left(r\overline{v_{\mathrm{P}r}^2}\right)}{\partial r}\right) + \frac{1}{r}\frac{\partial(\overline{v_{\mathrm{P}r}v_{\mathrm{P}\theta}})}{\partial \theta} + \frac{\partial(\overline{v_{\mathrm{P}r}v_{\mathrm{P}z}})}{\partial z} \tag{10.87}$$

Substitute $\overline{v_{\mathrm{P}r}^2} = \overline{v_{\mathrm{P}r}}^2 + \overline{v_{\mathrm{P}r}^{\mathrm{t}2}}$, where $v_{\mathrm{P}}^{\mathrm{t}}$ is a turbulent velocity fluctuation. If the flow is fully developed, z-derivatives can be ignored. Because of symmetry, θ-derivatives can also be ignored. Then (10.87) simplifies to

$$\overline{A_{Pr}} = -\frac{\overline{v_{P\theta}^{t2}}}{r} + \frac{1}{r}\frac{\partial(r\bar{v}_{Pr}^2)}{\partial r} + \frac{1}{r}\frac{\partial\left(r\overline{v_{Pr}^{t2}}\right)}{\partial r} \tag{10.88}$$

Since $\bar{v}_r^2 << \overline{v_r^{t2}}$

$$\overline{A_{Pr}} = -\frac{\overline{v_{P\theta}^{t2}}}{r} + \frac{\overline{v_{Pr}^{t2}}}{r} + \frac{\partial\overline{v_{Pr}^{t2}}}{\partial r} \tag{10.89}$$

From (10.82)

$$V_{tp} = \frac{1}{\bar{\bar{\beta}}}\left(-\frac{\partial\overline{v_{Pr}^{t2}}}{\partial r} + \frac{\overline{v_{P\theta}^{t2}}}{r} - \frac{\overline{v_{Pr}^{t2}}}{r}\right) \tag{10.90}$$

Assume $\overline{v_{P\theta}^{t2}} \approx \overline{v_{Pr}^{t2}}$ (see Figure 10.2). Then (10.78) is obtained.

The measurements of $\overline{dv_P/dt}$ presented by Young & Hanratty (1991) are plotted in Figure 10.4, where runs 1 & 2 represent glass spheres at Re = 70 800 and at Re = 16 400.

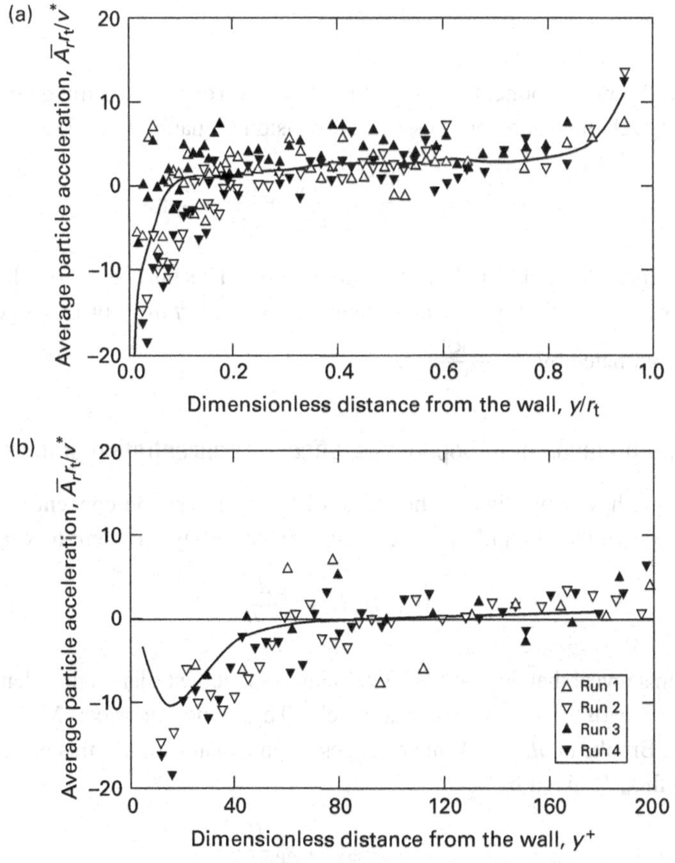

Figure 10.4 (a) The temporal change of V_R measured by Young & Hanratty. The solid curve is calculated with (10, 86). (b) Measurements close to the wall. Young & Hanratty, 1991.

Runs 3 and 4 are for steel spheres at Re = 72 900 and Re = 15 700. The curves in Figure 10.4a and b represent theoretical calculations. The interesting feature of these measurements is that the sign of $\overline{dv_P/dt}$ changes. For low flow rates, this change occurs at $y/r_t \approx 0.2$ and for high flow rates, at $y/r_t \approx 0.05$, where y is the distance from the wall. This can be understood better when measurements are plotted against y, made dimensionless with wall parameters, as shown in Figure 10.4b. Here, it is seen that the change in sign occurs in the range $y^+ = 80$–120. This corresponds to the location of a maximum in the root-mean-square of the radial velocity fluctuations.

Thus, turbophoresis exists if $\overline{dv_P/dt}$ has a finite value. Then a force given by the right side of (10.75) will be acting on the particle. This requires that $\overline{v}_P \neq 0$.

10.9.3 Formulation in Cartesian coordinates

Several researchers have formulated the turbophoretic velocity in Cartesian coordinates. These works are summarized by Mito & Hanratty (2004b). From (10.82), The turbophoretic velocity is exactly defined as

$$V_{tp} = -\overline{\frac{1}{\overline{\beta}}\frac{dv_{2P}}{dt}} \tag{10.91}$$

where v_{2P} is the x_2-component of the particle velocity.Young & Leeming (1997) derived the following relation for a compressible suspension (equation (10.85) is not valid):

$$\overline{\frac{dv_{2P}}{dt}} = \overline{v_{2P}}\frac{\partial \overline{v_{2P}}}{\partial x_2} + \frac{\partial \overline{v_{2P}^{t2}}}{\partial x_2} + \overline{\frac{\partial \ln C}{\partial x_i} v_{2P}^{t} v_{iP}} \tag{10.92}$$

where the repeated i-index indicates a summation. The third term on the right side accounts for compressibility of the suspension. Cerbelli *et al.* (2001) suggested that it can be approximated by $\overline{v_{2P}^{t2}}\frac{\partial(\ln\overline{C})}{\partial x_2}$.

10.9.4 Physical interpretation of turbophoresis; effect on concentration profiles

Several people have investigated the effect of turbophoresis on concentration profiles. They have argued that the turbophoretic flux is balanced by turbulent mixing.

$$0 = V_{tp}C - \varepsilon_P^E \frac{dC}{dy} \tag{10.93}$$

Measurements reveal that the use of (10.93) requires that the Eulerian turbulent diffusivity is negative in parts of the concentration field. This is non-physical. Mito & Hanratty (2004b) and Brooke *et al.* (1994) have suggested an explanation for this paradox.

Define a flux, F, given by

$$F = V_{tp}C - \varepsilon_P^E \frac{dC}{dy} \tag{10.94}$$

This need not be zero, as suggested by (10.93). Mito & Hanratty (2004b) carried out numerical experiments in a channel flow for which the fluid turbulence is represented by the Langevin equation (see Section 10.11). One aspect of the investigation by Mito & Hanratty was an extensive study of turbophoresis. This work and the study by Brooke *et al.* (1994) speculate that turbophoretic motions resemble "free-flight" deposition described in Section 10.10.

Particles entrained in a turbulent field tend to follow the turbulence. However, if the particles experience a sudden decrease in the fluid turbulence, they will become detached from the turbulence and undergo a free-flight. Thus, there will be two populations of particles. One is entrained in the fluid turbulence and a second, which is smaller, is undergoing free-flights. Particles can end their free-flights by again becoming entrained in the turbulence or by depositing in the wall region. The velocity of these free-flight particles is associated with the turbophoretic velocity.

At a given y, particles can be leaving or entering the entrained population at rates of S^0 and S^i (in the units of mass per unit time per unit area). Thus, the equation describing the variation of the flux with y is given, for a stationary state, as

$$\frac{dF}{dy} = S^i - S^0 \tag{10.95}$$

where F is given by (10.94) and a unit length in the spanwise direction is assumed. Mito & Hanratty (2004b) suggest that S^i and S^0 represent contributions from a large-scale velocity field similar to a secondary flow.

10.10 Deposition of particles

The design equation that is used to describe the rate of deposition on a unit wall area, R_D, is

$$R_D = k_D C_B \tag{10.96}$$

where C_B is the bulk composition and k_D is the deposition coefficient, with units of velocity. The most important parameter characterizing the deposition is the inertial time constant $\tau = \beta^{-1}$, defined by (10.50).

For small $\tau^+ = \tau v^{*2}/\upsilon$ particles closely follow the fluid velocity fluctuations. Under these circumstances, deposition occurs by Brownian motion, which is characterized by a very large Schmidt number, $\mathrm{Sc} = \upsilon/D$. The equation for the deposition coefficient is

$$\frac{k_D}{v^*} = c\mathrm{Sc}^{-n} \tag{10.97}$$

where c is a constant. The classical approach used to explain the effect of Schmidt number on the average mass transfer rate is to assume an analogy between momentum and mass transfer. This leads to the assumption that the eddy diffusivity varies with the cube of the distance from the wall throughout the concentration boundary layer and a

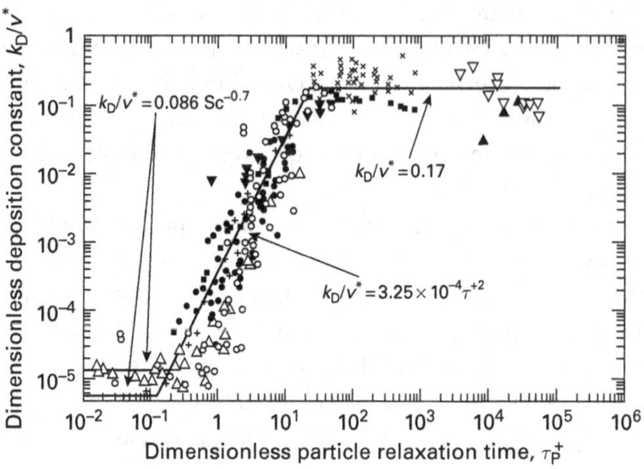

Figure 10.5 Summary of deposition data in vertical flows by McCoy & Hanratty. McCoy & Hanratty, 1977.

prediction that $n = -2/3$. However, experiments by Shaw & Hanratty (1977) and by Na *et al.* (1999) show that $n = -0.704$. For Sc = 631−37 200,

$$\frac{k_D}{v^*} = 0.0889\mathrm{Sc}^{-0.704} \tag{10.98}$$

The argument that the eddy diffusion coefficient varies with the cube of the distance from the wall is correct since $u_y^t \propto y^2$ and $u_x^t \propto y$, where u_y^t is the turbulent velocity component perpendicular to the wall and u_x^t is the turbulent velocity component parallel to the wall. However, the assumption that this holds throughout the concentration boundary layer is not correct. This prompted theoretical work by Vasiliadou & Hanratty (1988), which shows, for extremely large Sc, that $n = -3/4$. Since Brownian motion is characterized by a much larger Sc than studied by Shaw & Hanratty (1977), it is expected that $n = -3/4$ would be more appropriate for very small aerosol particles.

When $\tau_P^+ > \sim 0.2$, particles do not follow the fluid turbulence exactly over the whole field; they impinge on the wall by an inertial mechanism, called a "free-flight." For $0.2 < \tau_P^+ < 20$, particles moving toward the wall disengage from the turbulence in the viscous wall region where the magnitudes of the turbulent fluid velocities are decreasing rapidly as the distance from the wall decreases. Thus, the rate of deposition increases rapidly with increasing τ_P^+.

Figure 10.5 was created by McCoy & Hanratty (1977). It summarizes measurements of k_D made in vertical flows by Farmer (1969), Forney & Spielman (1974), Friedlander & Johnstone (1957), Llori (1971), Liu & Agrawal (1974), Schwendiman & Postma (1961), Sehmel (1973), Wells & Chamberlain (1967), Jagota *et al.* (1973), Cousins & Hewitt (1968). At very small τ_P^+, the dimensionless deposition coefficient, k_D/v^*, is seen to be independent of τ_P^+. (The particles mix by Brownian motion.) For $0.2 < \tau_P^+ < 22.9$, k_D/v^* increases with τ_P^{+2}, as suggested by Kneen & Straus (1969) and by Liu & Agrawal (1974). It is noted that, at a given τ_P^+, there can be an order of magnitude variation of the measurements made in different laboratories. McCoy & Hanratty (1977) suggested that the following equation, very roughly, captures the trend:

$$\frac{k_D}{v^*} = 3.25 \times 10^{-4} \tau_P^{+2} \tag{10.99}$$

In the range where (10.99) is applicable, there is a remarkably strong increase of k_D with gas velocity and with particle diameter, $k_D \propto v^{*5}$ and $k_D \propto d_P^4$.

For a given particle size, k_D/v^* becomes independent of τ_P^+ at very small τ_P^+. The data shown in Figure 10.5 are in this range for particles with diameters of 0.65 to 1.1 µm, which correspond to Sc of 3.3×10^5 to 1.2×10^6.

At $\tau^+ > \sim 22.9$, the value of k_D/v^* appears to be relatively insensitive to particle diameter or fluid velocity. McCoy & Hanratty suggest that an average fit to the data at large τ_P^+ is

$$\frac{k_D}{v^*} = 0.17 \tag{10.100}$$

Friedlander & Johnstone (1957) were the first to model deposition in the range $0.2 < \tau_P^+ < 20$ as due to a free-flight mechanism. They pictured the particles as arriving at a certain distance from the wall, y_{ff}, by turbulent diffusion and then moving to the wall by a free-flight. Here, y_{ff}, the location at which the particles start a free-flight to the wall, is defined as $y_{ff} = v_{ff}\tau_P$. Friedlander & Johnstone assumed that v_{ff} is the root-mean-square of the fluid velocity fluctuations at $y^+ = 30$. Theoretical work had mainly involved modifications of this theory. For example, Davies (1966) argued that v_{ff} is representative of the fluid velocity fluctuations at y_{ff}, rather than $y^+ = 30$, but predicted deposition rates which are too small.

Studies of particle motion in direct numerical simulations (DNS) of the Navier–Stokes equation for turbulent flow in a channel have provided an opportunity to study free-flights (Chen & McLaughlin, 1995; McLaughlin, 1991; Brooke *et al.*, 1992, 1994). These studies involved the seeding of the flow with spherical particles of uniform size, and removing them from the field when they are located at one particle radius from the wall. Their paths were calculated with (10.25) and (10.27).

Over the outer flow field the particles follow, but lag, the fluid velocity fluctuations. Eventually, they move toward the wall where the fluid velocity decreases rapidly and the particles disengage from the fluid turbulence as they move toward the wall with a higher velocity than the surrounding fluid. The change from a situation where the particles lag the fluid velocity fluctuations to one in which they are disengaged from the fluid velocity fluctuations defines the initiation of a "free-flight."

These free-flight particles can move directly to the wall or they can stop in a layer closer to the wall where velocity fluctuations in a direction perpendicular to the wall are very small. The trapping of particles at the wall has been noted by several investigators (Kallio & Reeks, 1989; McLaughlin, 1989; Sun & Lin, 1986; Brooke *et al.*, 1992). Observations by Young & Hanratty (1991) are displayed in Figure 10.6. These authors showed that the particles are trapped in a necklace formation at a distance of less than one particle diameter from the wall. The location is dictated by a balance between the Saffman lift force and a wall-induced force (Brenner, 1961; Maude, 1961; Goldman *et al.*, 1967).

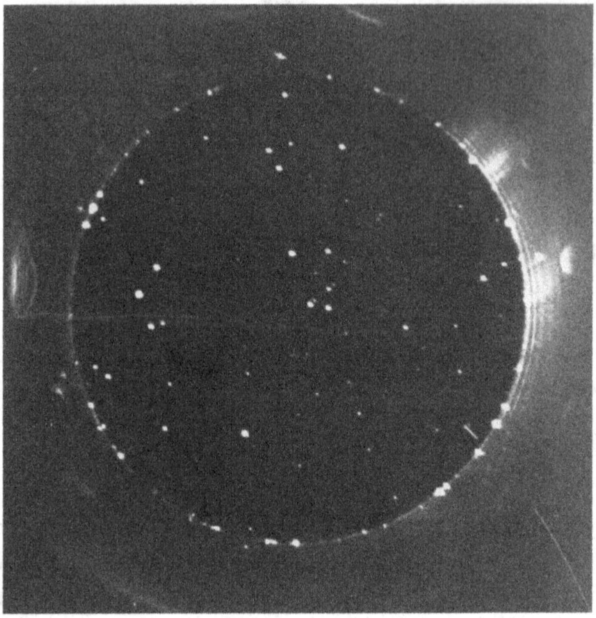

Figure 10.6 Particles trapped at a wall. Young & Hanratty, 1991.

Several new or revised concepts arose from computer experiments (Brooke *et al.*, 1992, 1994):

(1) At a given distance from the wall, a small fraction of the particles, with the highest velocities perpendicular to the wall, start a free-flight. Thus, the probability distribution of the fluctuations of the velocity, rather than the root-mean-square of the velocity fluctuations, is needed to interpret the phenomenon. This explains why early theories on inertial impaction predicted that particles strike the wall at much smaller velocities than is observed.
(2) Particles start their free-flight to the wall from a number of locations.
(3) Particles can stop a free-flight before they hit the wall. This can lead to their being trapped in a region close to the wall where the fluid turbulence is small.
(4) Particles are defined as striking a smooth wall when their centers are at a distance of one radius from the wall, where the fluid turbulence is very small, but not zero. Although turbulent deposition is a possible fate of trapped particles (as suggested by several researchers), the most likely scenario is that trapped particles are caught in a large outward flow to regions away from the wall, where they have another chance to experience a free-flight to the wall.
(5) Kallio & Reeks (1989) and Brooke *et al.* (1992) have suggested that the buildup of particles in the viscous sub-layer ($y^+ < 5$) could be due to turbophoresis. However, computer studies show that turbophoresis aids the motion of particles toward the wall, but that accumulation is mainly due to free-flights.
(6) The optimum distance at which free-flight deposition starts is roughly $y^+ \approx 9$.

10.11 Stochastic representation of fluid turbulence

10.11.1 Prologue

Studies of particle motion in a direct numerical solution for the flow field are limited in that the complexity of the calculations requires a consideration of small Reynolds numbers, small diffusion times and a range of variables which is insufficient to capture the physics completely. This has led to the use of stochastic models to represent turbulence seen by the particles. By pursuing this approach, it has been possible to study a wide range of conditions, to identify several mechanisms for deposition and to explore methods to describe the concentration field. In the order of increasing inertial time constant, particles strike the wall by (1) Brownian motion, (2) turbulent diffusion, (3) free-flight, (4) free-flight from regions outside the viscous wall layer, (5) unidirectional motion from wall to wall.

10.11.2 The use of a modified Langevin equation in channel flow

The Langevin equation has been explored in studies by Perkins (1992), Sommerfeld *et al.* (1993), Pozorski & Minier (1988), Iliopoulos *et al.* (2003), Mito & Hanratty (2003).

In its simplest form, it consists of a damping force and a Wiener process which calculates velocities of passive tracers by assuming they are approximated by a Markov sequence. Lin & Reid (1963) and Obukhov (1959) applied the Langevin equation to homogeneous, isotropic turbulence. The change of the velocity of a tracer over a time interval dt is given by

$$du_i = -\frac{u_i}{\tau_i}dt + d\mu_i \tag{10.101}$$

where τ_i is a time constant. The forcing function, $d\mu_i$, has a zero mean, and fluctuations are given by a Gaussian function. This equation provides the same result as obtained by Taylor (1921) in his analysis of dispersion from a point source if the Lagrangian correlation is represented by $\exp(-t/\tau^{\mathrm{L}})$ and $\tau_i = \tau^{\mathrm{L}}$, the Lagrangian time-scale.

The Langevin equation has been adapted to describe dispersion of fluid particles in non-homogeneous fields by Durbin (1983, 1984), Hall (1975), Iliopoulos & Hanratty (1999), Legg & Raupach (1982), Reid (1979), Reynolds (1997), Thomson (1984, 1986, 1987), van Dop *et al.* (1985) and Wilson *et al.* (1981). The analysis by Mito & Hanratty (2004b) used the approach of Wilson *et al.* (1981) and of Thomson (1984),

$$d\left(\frac{u_i}{\sigma_i}\right) = -\frac{u_i}{\sigma_i\tau_i} + d\overline{\mu}_i + d\mu_i' \tag{10.102}$$

where the forcing function, the time-scale, τ_i, and the root-mean-square of the velocity fluctuations, σ_i, are functions of x_2. For a non-homogeneous flow, the forcing function, $d\mu_i$, consists of a mean component, $\overline{d\mu_i}$, and a fluctuating component, $d\mu_i'$. The fluctuating component is assumed to be Gaussian. A number of investigators have shown that $\overline{d\mu_i}$ must be non-zero in order to avoid spatial accumulations which are not physical. Equations for the mean drift, $\overline{d\mu_i}$, and the covariance, $\overline{d\mu_i' d\mu_j'}$, are derived from equation (10.102) by neglecting terms of higher order than dt (Iliopoulos & Hanratty, 1999; Mito & Hanratty, 2002).

$$\overline{d\mu_i} = \frac{\partial\left(\dfrac{\overline{u_2 u_i}}{\sigma_i}\right)}{\partial x_2} dt \tag{10.103}$$

$$\overline{d\mu_i' d\mu_j'} = \frac{\overline{u_i u_j}}{\sigma_i \sigma_j}\left(\frac{1}{\tau_i} + \frac{1}{\tau_j}\right) dt \tag{10.104}$$

A method for evaluating the unknowns in (10.102), (10.103) and (10.104) is outlined in Mito & Hanratty (2002, 2003, 2004a), who considered flow in a rectangular channel of infinite extent in the spanwise direction. The focus was vertical annular flow.

Woodmansee & Hanratty (1969) have shown that atomization of wall films occurs by rapid growth and removal of capillary waves which create drops that are entrained by the turbulence in a region outside the viscous wall layer. This process was represented by introducing drops at a short distance from the wall with a velocity characteristic of the velocity fluctuations outside the viscous wall layer. Injection velocities of $\left(V_1^{0+}, V_2^{0+}, V_3^{0+}\right) = (15, 1, 0)$ were used at the bottom wall and of $\left(V_1^{0+}, V_2^{0+}, V_3^{0+}\right) = (15, -1, 0)$ at the top wall. Equations (10.27), (10.52) were used to calculate the subsequent velocities and locations of these particles. The fluid velocity fluctuations encountered by the particles were represented by (10.102). The influence of gravity was ignored since vertical flow was considered. The influences of collisions and of feedback particle forces on the turbulence were not considered. For calculations in which Brownian motion is important, the approach adopted by Ounis *et al.* (1991) and by Chen & McLaughlin (1995) was used.

The drag coefficient, C_D, was represented by

$$C_D = \frac{24}{\mathrm{Re_P}}\left(1 + 0.15\,\mathrm{Re_P^{0.687}}\right) \tag{10.105}$$

where the particle Reynolds number, $\mathrm{Re_P}$, is defined with d_P and the magnitude of the relative velocity $|\vec{u} - \vec{v}_P|$. The dimensionless inertial time constant of a particle is defined as

$$\tau_P^+ = \frac{4 d_P^+ (\rho_P/\rho_f)}{3 C_D |\vec{u}^+ - \vec{v}_P^+|} \tag{10.106}$$

where $\vec{u}$ is the fluid velocity and $\vec{v}_P$ is the particle velocity. For a Stokes law resistance

$$\tau_{PS}^+ = \frac{d_P^{+2}(\rho_P/\rho_f)}{18} \tag{10.107}$$

In the non-Stokes region, τ_P^+ is a function of x_2. Thus, the volume-averaged inertial time constant, τ_P^+, is a more appropriate parameter to describe particle turbulence:

$$\tau^+_{PB} = \frac{1}{2HC_B}\int_0^{2H} \tau^+_P\,\overline{C}(x_2)dx_2 \qquad (10.108)$$

where C_B is defined as

$$C_B = \frac{1}{2H}\int_0^{2H} \overline{C}(x_2)\,dx_2 \qquad (10.109)$$

The rates of injection of particles at the two walls are calculated as $N_B/A\Delta t$, where A is the area of the wall over which particles are discharged and Δt is the time interval over which the N_B particles are admitted from a wall source. Particles are deposited (removed from the field) when their centers are located one particle radius from the wall. The run is terminated when a stationary state is reached. (There is no further change in the concentration field and the rate of injection, R_A, equals the rate of deposition, R_D.)

10.11.3 Identification of mechanisms for deposition

Stationary states were studied for two sets of calculations: In one of these, the dimensionless particle diameter was kept constant at $d^+_P = 0.368$ and the particle time constant was varied by changing ρ_P/ρ_f (see Table 10.2). Results were obtained for $\tau^+_{PS} = 1-25\,000$. This led to a consideration of unreasonably large ρ_P/ρ_f. In the second set of calculations,

Table 10.2 Stokesian and bulk-mean inertial time constants, bulk-mean particle Reynolds numbers for constant inertial time constant and for constant density ratio

$d^+_P = 0.368$				$\rho_P/\rho_f = 1000$			
τ^+_{PS}	τ^+_{PB}	Re_{PB}	ρ_P/ρ_f	τ^+_{PS}	τ^+_{PB}	Re_{PB}	d^+_P
1	0.977	0.0748	1.33×10^2	1	0.988	0.0274	0.134
3	2.91	0.118	4.00×10^2	3	2.92	0.0740	0.232
5	4.82	0.147	6.65×10^2	5	4.84	0.118	0.300
10	9.55	0.181	1.33×10^3	10	9.52	0.214	0.424
20	18.9	0.242	2.65×10^3	20	18.5	0.441	0.600
40	37.2	0.367	5.30×10^3	40	35.2	0.901	0.848
100	91.4	0.536	1.33×10^4	100	81.9	1.87	1.34
250	226	0.643	3.32×10^4	250	187	3.45	2.12
500	449	0.712	6.65×10^4	500	344	5.27	3.00
1000	888	0.810	1.33×10^5	1000	624	7.93	4.24
2500	2120	1.34	3.32×10^5	2500	1310	14.2	6.70
5000	4150	1.62	6.65×10^5	5000	2120	25.8	9.48
10,000	8260	1.71	1.33×10^6	10000	3180	52.1	13.4
25,000	20500	1.76	3.32×10^6	25000	6070	89.8	21.2
				50000	9960	131	30.0
				100000	16200	189	42.4

Source: Mito & Hanratty (2004b)

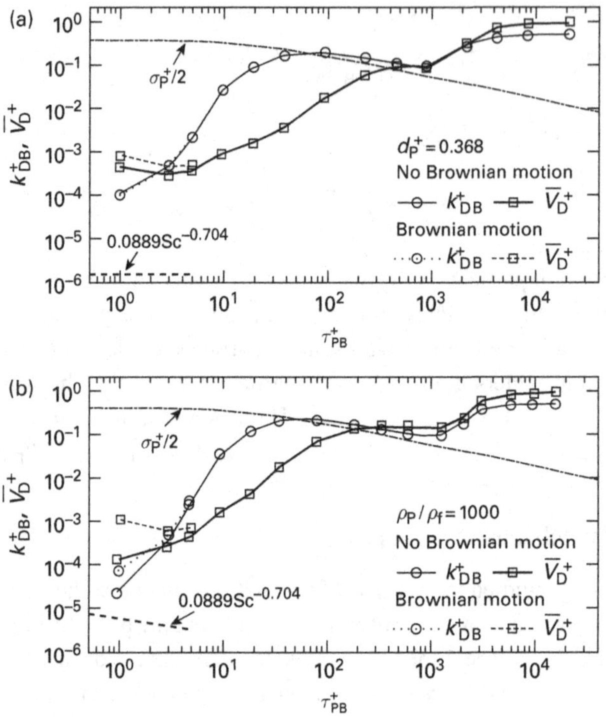

Figure 10.7 (a) Deposition constants for $d^+_P = 0.368$ (b) Deposition constants for $\rho_P/\rho_f = 1000$. The fluid turbulence is represented by a Langevin equation. Mito & Hanratty, 2004b.

ρ_P/ρ_f was kept constant and the dimensional constant was varied by changing d^+_P. This led to a consideration of unreasonably large particle sizes (compared with the mesh size used in the calculation). The influence of Brownian motion was considered only for $\tau^+_{PS} = 1, 3, 5$.

Figure 10.7a, for $d^+_P = 0.368$, presents calculations of mean values of the velocity with which particles are depositing, $\overline{V}_D$. For $\tau^+_{PS} = 1, 3, 5$, particles are striking the wall with very small velocities, $10^{-4} < \overline{V}^+_D < 10^{-3}$. These are of the order of the wall-normal fluid velocity fluctuations at $d^+_P/2$. For this range of variables, particles trapped close to the wall appear to be depositing by turbulent diffusion. For $\tau^+_{PS} >> 5$, the main contributors to deposition are particles which have velocities which are characteristic of the fluid outside the viscous sub-layer. Particles with these velocities are pictured to move in "free-flight" through the layer of trapped particles close to the wall.

In the diffusion regime, the mean deposition velocity of the particles is not terribly sensitive to changes in τ^+_{PB}. The slight decrease with increasing τ^+_{PB} reflects a decrease in the ability of the particles to follow fluid velocity fluctuations. The increase in V^+_D in the region $5 < \tau^+_{PB} < 250$ occurs because, on average, "free-flight" to the wall can be initiated at larger distances from the wall as τ^+_{PS} increases.

Values of $k^+_D = k_D/v^*$ are also plotted in Figure 10.7. A comparison of the calculated k^+_D in Figure 10.7 with the plot of experimental results in Figure 10.5 shows agreement within the spread of the experimental data.

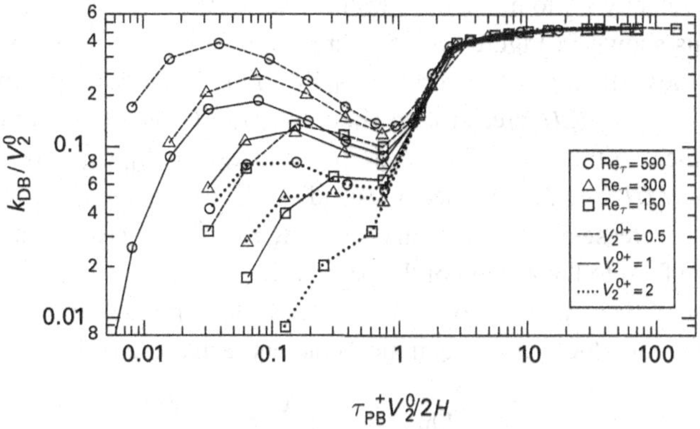

Figure 10.8 Effect of the inertial time constant $\tau_{PB}V_2^0/2H$ on the dimensionless deposition coefficient, for $d_P^+ = 0.368$. The fluid turbulence is represented by a Langevin equation. Mito & Hanratty, 2007.

Hay *et al.* (1996) and Hanratty *et al.* (2000) show that, for particles starting free-flights outside the viscous wall layer,

$$k_D = \frac{\sigma_P}{\sqrt{2\pi}} \tag{10.110}$$

where σ_P is the root-mean-square of the wall-normal velocity fluctuations of the particles. Equation (10.110) provides the average velocity of particles moving toward the wall at $y^+ = 40$. It decreases with increases in the inertial time constant since the particles experience increasing difficulties in following the fluid velocity fluctuations. Equation (10.110) is compared with the calculations in Figures 10.7 a, b. Good agreement is noted for a range of conditions characteristic of annular flow.

Studies of the behavior of single wall sources by Mito & Hanratty (2003) show that injected particles are transported to a distance slightly less than the stopping distance in a stationary fluid

$$x_{2\text{stop}}^+ \approx 0.65 V_2^{0+} \tau_{PS}^+ \tag{10.111}$$

where they mix with the turbulence.

Equation (10.111) suggests that the sudden change in the dependencies of V_D^+ and k_D^+ on the inertial time constant at $\tau_{PB}^+ \approx 1800$, shown in Figure 10.7, suggests a change in the behavior of the particles in that their motion is not influenced by fluid turbulence. Under these circumstances, the injection velocity is playing a strong role. At very large τ_{PB}^+, the deposition velocity equals the velocity at which the particles are injected into the field. Deposition then occurs by the trajectory mechanism discussed by Anderson & Russel (1970), Andreussi & Azzopardi (1983) and by Chang (1973). The system discussed above was used by Mito & Hanratty (2007) to provide quantitative criteria for determining when a trajectory mechanism will define the rate of deposition at very low concentrations.

They suggested that, in order to capture the trajectory regime, it is appropriate to use V_2^0 and $2H$ as the velocity and length scales, rather than v^* and υ/v^*. A plot of k_{DB}/V_2^0 versus $\tau_{PB}V_2^0/2H$ is presented in Figure 10.8. Note that the time it would take for a particle with a velocity V_2^0 to move from wall to wall is $2H/V_2^0$, if fluid drag is ignored. The calculations shown in Figure 10.8 for three particle Reynolds numbers and three injection velocities fall on a single curve if $\tau_{PB}V_2^0/2H$ is greater than a number slightly larger than 1. At $\tau_{PB}V_2/2H$ greater than about 10, k_{DB}/V_2^0 is a constant equal to 0.5. This follows since, in this limit, $R_D = 0.5V_2^0 C_B$ and $k_{DB} = R_D/C_B = 0.5V_2^0$. Thus, in the region $\tau_{PB}V_2^0/2H > 1.2$, the trajectories and the deposition are not affected by fluid turbulence. The decrease of k_{DB}/V_2^0 with decreasing $\tau_{PB}V_2^0/2H$ is associated with mean resistance of the fluid to the motion of the particles.

The influence of Brownian motion is treated the same as molecular motion in that the flux in the i-direction, due to a concentration gradient, is represented by Fick's law

$$\text{Flux}_x = -D\frac{\partial C}{\partial x} \tag{10.112}$$

As indicated in Table 10.2, the series of calculations carried out by Mito & Hanratty (2004a) for constant ρ_P/ρ_f offer the opportunity to study the effect of Brownian motion. Chen & McLaughlin (1995) show that the Schmidt number is given as

$$\text{Sc} = \frac{v}{D} = \frac{3\pi v^3 \rho_f d_P^+}{C_c k T v^*} \tag{10.113}$$

where C_c is the Cunningham slip factor, k is the Boltzmann constant, T is the absolute temperature, D is the diffusion coefficient for Brownian motion and the gas is assumed to be air at atmospheric conditions. This indicates increases in D with decreasing particle diameter. Because of this, Brownian motion could show an important effect at $\tau_{PS}^+ = 1, 3, 5$ in Figure 10.7b. At small enough τ_{PB}^+ Brownian motion could be controlling deposition. The rate should then be similar to that found for turbulent mass transfer of molecular species at large Schmidt numbers. A comparison of (10.98) with calculations of the type summarized in Table 10.2 is awkward because D need not be a constant. Thus, from (10.113), it is seen that Sc varies with d_P^+ and v^*. Mito & Hanratty (2004b) followed the lead of Chen & McLaughlin(1995) by considering a constant friction velocity of 0.6 m/s. The Schmidt number is considered to be a constant in the calculations for $d_P^+ = 0.368$ given in Figure 10.7a.

Chen & McLaughlin (1995) studied particle deposition in a DNS of turbulent flow in a channel at a very low Reynolds number, $\text{Re}_\tau = 125$. The analysis included the effects of Brownian motion. The friction velocity was assumed to be 0.6 m/s. Calculations of k_D^+ are comparable with calculations of Mito & Hanratty (2004b) for $\rho_P/\rho_f = 1000$, $v^* = 0.6$ m/s. (This corresponds to a Schmidt number variation of 1.6×10^6 to 5×10^6 for $\tau_{PS} = 1, 3, 5$.) The dashed curve on the bottom of Figure 10.7a represents (10.98). The calculations of Chen & McLaughlin extrapolate to (10.98) at $\tau_P^+ = 1$. The Mito & Hanratty calculations suggest that a smaller τ_P^+ is needed for Brownian motion to control.

This could reflect the very small Reynolds number used by Chen & McLaughlin. Indirect evidence for this is that their predicted k_D^+ at large τ_{PB}^+ are smaller than the calculations by Mito & Hanratty and available laboratory measurements.

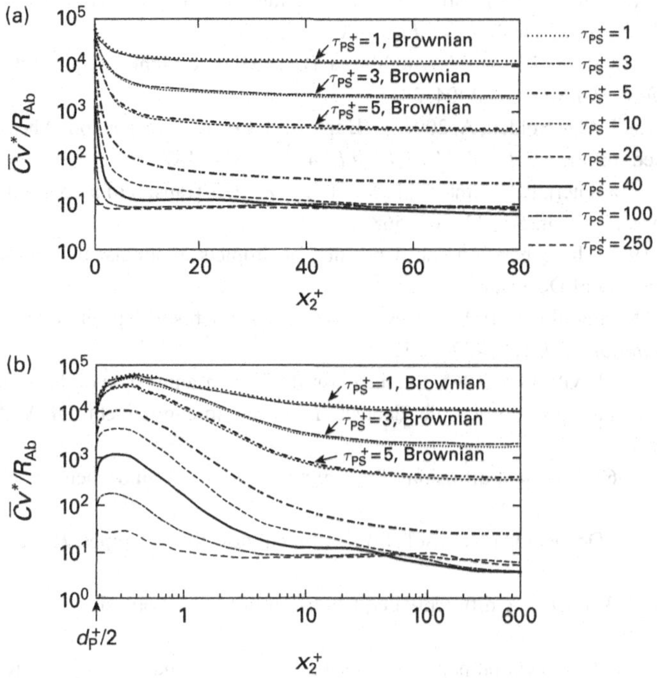

Figure 10.9 Concentration profiles for $\rho_P/\rho_f = 1000$. The fluid turbulence is represented by a modified Langevin equation. Mito & Hanratty, 2004b.

10.11.4 Concentration profiles

Calculated concentration profiles are presented in Figure 10.9a,b for $\rho_P/\rho_f = 1000$. A logarithmic abscissa is used in Figure 10.9b. A range of $\tau_{PS}^+ = 1-250$ is covered. The injection process causes small bumps at $x^+ = 13, 26, 65$ for $\tau_{PS}^+ = 20, 40, 100$. No evidence of this influence is seen for $\tau_{PS}^+ \leq 10$. A striking feature of these profiles is the accumulation of particles at x_2^+ less than 1.

References

Anderson, R. J. & Russel, T. W. F. 1970 Circumferential variation of interchange in horizontal annular two-phase flow. *Ind. Eng. Chem. Fund.* 9, 340–344.

Andreussi, P. & Azzopardi, B. J. 1983 Droplet deposition and interchange in annular two-phase flow. *Int. J. Multiphase Flow* 9, 681–695.

Basset, A. B. 1888 *Treatise on Hydrodynamics*, Vol. 2. New York: Dover.

Beyerlein, S. W., Cossman, R. K. & Richter, H. J. 1985 Prediction of bubble concentration profiles in vertical turbulent two-phase flow. *Int. J. Multiphase Flow* 11, 629–641.

Boothroyd, R. G. 1967 Turbulence characteristics of the gaseous phase in duct flow of a suspension of fine particles. *Trans. IChemE* 45, T297.

Brenner, H. 1961 Slow motion of a sphere through viscous fluid toward a plane surface. *Chem. Eng. Sci.* 16, 242.

Brooke, J.W., Kontomaris, K., Hanratty, T.J. & McLaughlin, J.B. 1992 Turbulent deposition and trapping of aerosols at a wall. *Phys. Fluids* A 4, 825–834.

Brooke, J.W., Hanratty, T.J. & McLaughlin, J.B. 1994 Free-flight mixing and deposition of aerosols. *Phys. Fluids* 6 (10), 3404–3415.

Cerbelli, S., Giusti, A. & Soldati, A. 2001 ADE approach to predict dispersion of heavy particles in wall-bounded turbulence. *Int. J. Multiphase Flow* 27, 1861–1879.

Caporaloni, M., Tampieri, R., Trombetti, F. & Vittori, O. 1975 Transfer of particles in nonisotropic air turbulence. *J. Atmos. Sci.* 32, 565–568.

Chang, D.R. 1973 The generation and deposition of droplets in annular two-phase flow. Ph.D. thesis, University of Delaware.

Chen, M. & McLaughlin, J. 1995 A new correlation for aerosol deposition in vertical ducts. *J. Colloid Interface Sci.* 169, 437–455.

Cousins, L.B. & Hewitt, G.F. 1968 Liquid mass transfer in annular two-phase flow to obtain interchange and entrainment: droplet deposition and liquid entrainment. UKAEA Report No. AERE-R5657.

Csanady, G.T. 1963 Turbulent diffusion of heavy particles in the atmosphere. *J. Atmos. Sci.* 20, 201–208.

Davies, C.N. 1966 Deposition of aerosols from turbulent flow through pipes. *Proc. R. Soc. Lon.* A 239, 235.

Durban, P.A. 1983 Stochastic differential equations on turbulent dispersion. NASWA Reference Publication 1103.

Durban, P.A. 1984 Comment on papers by Wilson, *et al.* 1981 and by Legg and Raupach 1982. *Boundary-Layer Meteor.* 29, 409–411.

Einstein, A. 1905 On the movement of small particles suspended in stationary liquids required by the molecular-kinetic theory of heat. *Ann. Physik* (Ser. 5) 17, 549–560.

Farmer, R.A. 1969 Liquid droplet trajectories in two-phase flow. Ph.D. thesis, Massachusetts Institute of Technology.

Forney, L.F. & Spielman, L.A. 1974 Deposition of coarse aerosols from turbulent flow. *J. Aerosol Sci.* 5, 257–271.

Friedlander, S.K. & Johnstone, H.F. 1957 Deposition of suspended particles from turbulent gas streams. *Ind. Eng. Chem.* 49, 1151–1156.

Goldman, A.I., Cox, N.G. & Brenner, H.B. 1967 Slow viscous motion of a sphere parallel to a plane wall: II Couette flow. *Chem. Eng. Sci.* 22, 653–660.

Groenhof, H.C. 1970 Eddy diffusion in the central region of turbulent pipe flows in pipes and between parallel plates. *Chem. Eng. Sci.* 25, 1005.

Hall, C.D. 1975 The simulation of particle motion in the atmosphere by a numerical random walk model. *Q. J. R. Met. Soc.* 101, 235–244.

Hanratty, T.J., Woods, B.D., Iliopoulos, I. & Pan, L. 2000 The roles of interfacial stability and particle dynamics in multiphase flows: a personal viewpoint. *Int. J. Multiphase Flow* 26, 169–190.

Hay, K.J., Liu, Z.C. & Hanratty, T.J. 1996 Relation of deposition to drop size when the rate law is non-linear. *Int. J. Multiphase Flow* 22, 829–848.

Hinze, J.O. 1959 *Turbulence*. New York: McGraw-Hill.

Hinze, J.O. 1975 *Turbulence*, 2nd edn. NewYork: McGraw-Hill.

Iliopoulos, I. & Hanratty, T.J. 1999 Turbulent dispersion in a non-homogeneous field. *J. Fluid Mech.* 392, 45–71.

Iliopoulos, I., Mito, Y. & Hanratty, T.J. 2003 A stochastic model for solid particle dispersion in a non-homogeneous turbulent field. *Int. J. Multiphase Flow* 29, 375–394.

Jagota, A.K., Rhodes, E. & Scott, D.S. 1973 Tracer measurements in two-phase annular flow to obtain interchange and entrainment. *Can. J. Chem. Engng* 51, 139–148.

Kallio, G.A. & Reeks, M.W. 1989 A numerical simulation of particle deposition in boundary layers. *Int. J. Multiphase Flow* 15, 433–446.

Kneen, T. & Straus, W. 1969 Deposition of dust from turbulent gas streams. *Atmos. Env.* 3, 55–67.

Laufer, J. 1954 The structure of turbulent fully developed flow. NACA Report No. 1174.

Lee, M.M., Hanratty, T.J. & Adrian, R.J. 1989 An axial viewing photographic technique to study turbulence characteristics of particles. *Int. J. Multiphase Flow* 15, 787–802.

Legg, B.J. & Raupach, M.R. 1982 Markov-chain simulation of particle dispersion in homogeneous flows: The mean drift velocity induced by a gradient in Eulerian velocity variance. *Boundary-Layer Met.* 24, 3–13.

Lin, C.C. & Reid, W.H. 1963. Turbulent flow, theoretical aspects. In *Encyclopedia of Physics*, Vol. 8/2. Berlin: Springer-Verlag, pp. 438–542.

Liu, B.Y. & Agarwal, J.K. 1974 Experimental observations of aerosol deposition in turbulent flow. *J. Aerosol Sci.* 5, 145–155.

Llori, T.A. 1971 Turbulent deposition of aerosol particles inside pipes. Ph.D. thesis, University of Minnesota.

Maude, A.D. 1961 End effects in a falling film viscometer. *Brit. J. Appl. Phys.* 12, 293–295.

Maxey, M.R. & Riley, J.J. 1983 Equation of motion for a small rigid sphere in a non-uniform flow. *Phys. Fluids* 26, 883–889.

McCoy, D.D. & Hanratty, T.J. 1977 Rate of deposition of droplets in annular two-phase flow. *Int. J. Multiphase Flow* 3, 319–331.

McLaughlin, J.B. 1989 Aerosol particle deposition in numerically simulated channel flow. *Phys. Fluids* A1, 1211.

McLaughlin, J.B. 1991 Inertial migration of a small sphere in linear shear flows. *J. Fluid Mech.* 22, 261–274.

Meek, C.C. & Jones, B.G. 1973 Studies of the behavior of heavy particles in a turbulent fluid flow. *J. Atmos. Sci.* 30, 239–244.

Mei, R., Adrian, R.J. & Hanratty, T.J. 1991 Particle dispersion in isotropic turbulence under Stokes drag and Basset force with gravitational settling. *J. Fluid Mech.* 225, 481–495.

Mito, Y. & Hanratty, T.J. 2002 Use of a modified Langevin equation to describe turbulent dispersion of fluid particles in a channel flow. *Flow, Turbulence, Combustion* 68, 1–26.

Mito, Y. & Hanratty, T.J. 2003 A stochastic description of wall sources in a turbulent field: Part 1. Verification. *Int. J. Multiphase Flow* 29, 1373–1394.

Mito, Y. & Hanratty, T.J. 2004a A stochastic description of wall sources in a turbulent field: Part 2. Calculation for a simplified model of horizontal annular flow. *Int. J. Multiphase Flow* 30, 803–825.

Mito, Y. & Hanratty, T.J. 2004b Concentration profiles in a turbulent suspension when gravity is not affecting deposition. *Int. J. Multiphase Flow* 30, 1311–1336.

Mito, Y. & Hanratty, T.J. 2007 Trajectory mechanism for particle deposition in turbulent flows. *Int. J. Multiphase Flow* 33, 101–107.

Na, Y., Papavassiliou, D.V. & Hanratty, T.J. 1999 Use of direct numerical simulation to study the effect of Prandtl number on temperature fields. *Int. J. Heat Fluid Flow* 20, 187–195.

Obukhov, A.M. 1959 Description of turbulence in terms of Lagrangian variables. *Adv. Geophys.* 6, 113–116.

Ounis, H., Ahmadi, G. & McLaughlin, J.B. 1991 Dispersion and deposition of Brownian particles from point sources in a simulated turbulent channel flow. *J. Colloid Interface Sci.* 147, 233–250.

Perkins, R.J. 1992 The entrainment of heavy particles into a plane turbulent jet. *Proceedings of the 6th Workshop on Two-phase Flow Predictions*, ed. M. Sommerfeld. Erlangen: Forschungszentrum Julich, pp. 18–33.

Pozorski, J. & Minier, J.P. 1998 On the Lagrangian turbulent dispersion models based on the Langevin equation. *Int. J. Multiphase Flow* 24, 913–945.

Reeks, M.W. 1977 On the dispersion of small particles suspended in an isotropic turbulent fluid. *J. Fluid Mech.*, 83, 529–546.

Reeks, M.W. 1983 The transport of discrete particles in inhomogeneous Turbulence. *J. Aerosol Science* 14, 729–739.

Reid, J.D. 1979 Markov chain simulations of vertical dispersion in a neutral surface layer for surface and elevated releases. *Boundary-layer Meteorol.* 16, 3–22.

Reynolds, A.M. 1997 On the application of Thomson's random flight model to the prediction of particle dispersion within a ventilated air pace. *J. Wind. Eng. Ind. Aerodyn.* 67–68, 627–638.

Saffman, P.G. 1965 The lift of a small sphere in a slow shear. *J. Fluid Mech.* 2, 340–385.

Saffman, P.G. 1968 Corrigendum to the lift on a small sphere in a slow shear flow. *J. Fluid Mech.* 31, 624.

Schlichting, H. 1979 *Boundary-layer Theory.* New York: McGraw-Hill.

Schwendiman, L.C. & Postma, A.K. 1961 USAEC Report. HW-65309.

Sehmel, G.A. 1973 Particle eddy diffusivities and deposition velocities for isothermal flow and smooth surfaces. *J. Aerosol Sci.* 4, 145–155.

Shaw, D.A, & Hanratty, T.J. 1977 Turbulent mass transfer rates to a wall for large Schmidt numbers. *AIChE Jl* 23, 28–37.

Sommerfeld, M., Kohnen, G. & Ruger, M. 1993 Some open questions And inconsistencies of Lagrangian particle dispersion models. *Proceedings of the Ninth Symposium on Turbulent Shear Flows*, Kyoto, Japan, Paper No. 15–1.

Sun, Y.F. & Lin, S.P. 1986 Aerosol concentration in a turbulent boundary-layer flow. *J. Colloid Interface Sci.* 113, 315–320.

Taylor, A.R. & Middelman, S. 1974 Turbulent dispersion in drag reducing fluids. *AIChE Jl* 20, 454–461.

Taylor, G.I. 1921 Diffusion by continuous movements. *Proc. Lond. Math. Soc.* 151, 196–211.

Thomson, D.J. 1984 Random walk modeling in inhomogeneous turbulence. *Q. J. R. Met. Soc.*110, 1107–1129.

Thomson, D.J. 1986 A random walk model of dispersion in turbulent flows and its application to dispersion in a valley. *Q. J. R. Met. Soc.* 112, 511–530.

Thomson, D.J. 1987 Criteria for the selection of stochastic models of particle trajectories in turbulent flows. *J. Fluid Mech.* 180, 529–556.

Towle, W.L. & Sherwood, T.K. 1939 Mass transfer in the central portion of a turbulent air stream. *Ind. Eng. Chem.* 31, 457.

van Dop, H., Nieustadt, F.T.M. & Hunt, J.C. R. 1985 Random walk models for particle displacements in inhomogeneous turbulent flows. *Phys. Fluids* 28, 1639–1653.

Vasiliadou, E. & Hanratty, T.J. 1988 Turbulent transfer to a wall at large Schmidt numbers. In *Transport Phenomena in Turbulent Flows* Washington, DC: Hemisphere.

Wells, A.C. & Chamberlain, A.C. 1967 Transport of small particles to vertical surfaces. *Brit. J. Appl. Phys.* 18, 1793–1799.

Wilson, J.D., Thurtell, G.W. & Kidd, G.E. 1981 Numerical simulation of particle trajectories in inhomogeneous turbulence II: Systems with variable turbulent velocity scale. *Boundary-layer Meteorol.* 21, 423–441.

Woodmansee, D.E. & Hanratty, T.J. 1969 Mechanism for the removal of a droplet from a liquid surface by a parallel air flow. *Chem. Eng. Sci.* 24, 299–307.

Young, J.B. & Hanratty, T.J. 1991 Transport of solid particles at a wall in a turbulent flow. *AIChE Jl* 37, 1529–1536.

Young, J. & Leeming, A. 1997 A theory of particle deposition in turbulent pipe flow. *J. Fluid Mech.*, 340, 129–159.

Yudine, M.I. 1959 Physical considerations on heavy-particle diffusion. *Adv. Geophys.* 6, 185–191.

11 Vertical annular flow

11.1 Prologue

In the annular pattern, part of the liquid flows as a film along the wall and part flows as drops entrained in the gas. The interfacial stress varies with the flow rate of the film. Thus, the pressure gradient depends on the fraction of the liquid flow, E, entrained as drops in the gas. A predictive approach is to view E as resulting from a balance between the rate of atomization of the liquid film, R_A, and the rate of deposition of drops, R_D. Thus, measurements of R_A and R_D are a priority. A knowledge of drop size is of importance since it is needed to predict drop turbulence and the influence of gravity on the motion of drops.

Discussions of the initiation of annular flow, the initiation of atomization, the properties of waves at the interface, the prediction of interfacial stress and the prediction of film height have been presented in previous chapters. It is useful to give a brief review.

Annular flow is observed for air and water flowing upward in a vertical pipe for

$$V^*_{GS} = u_{GS}\rho_G^{1/2}[gd_t(\rho_L - \rho_G)]^{-1/2} > \sim 1 \tag{11.1}$$

Thus, for air–water flow at atmospheric conditions, transition occurs at u_{GS} of the order of 10 m/s. As discussed in Sections 2.7.4 and 3.2.3, this criterion can be explained by suggesting that the transition from vertical annular flow to churn flow is associated with a flooding mechanism whereby stresses associated with gas flow over the film cannot balance the force of gravity on the wall film.

For very small liquid flows, the thin liquid film on the wall is covered with capillary waves, as described in Section 7.7. These waves received their energy from wave-induced variations of the shear stress over the interface, which cause the waves to tumble. Thus, this instability does not lead to the removal of liquid from the film. Since this laminar film is disturbed by waves, the average film height is given by (3.46), rather than the classical relation for an undisturbed rectilinear shear flow. The waves create a roughened interface; the interfacial friction factor is given by (3.44) or (3.47).

Highly disturbed ringlike patches appear when the film Reynolds number reaches a value of approximately 280 (see Figure 3.3). These disturbances look like collections of capillary waves. Their thicknesses are much larger than the base film over which they propagate. Because of this, the waves within a disturbance induce pressure variations along the wave surface (in addition to shear stress variations), which can cause them to grow to form drops by a Kelvin–Helmholtz mechanism.

11.2 Distribution function for drop size

Thus, atomization of the wall film is initiated with the appearance of disturbance waves. The mechanism for the creation of drops was studied in a transparent horizontal rectangular channel by obtaining high-speed photographs (3500–6000 frames per second) with a camera located beneath the bottom wall (Woodmansee & Hanratty, 1969). Woodmansee gives the following description of the atomization process, shown in Figure 11.1.

One of these ripples suddenly accelerates to the front of the disturbance or roll wave. The central section of the ripple is lifted by the air stream having one or both ends connected to the flowing liquid film. As the detached region of the tube of liquid is lifted, it is blown into an arc which narrows as it stretches downstream until it ruptures into a number of pieces. Though difficult to

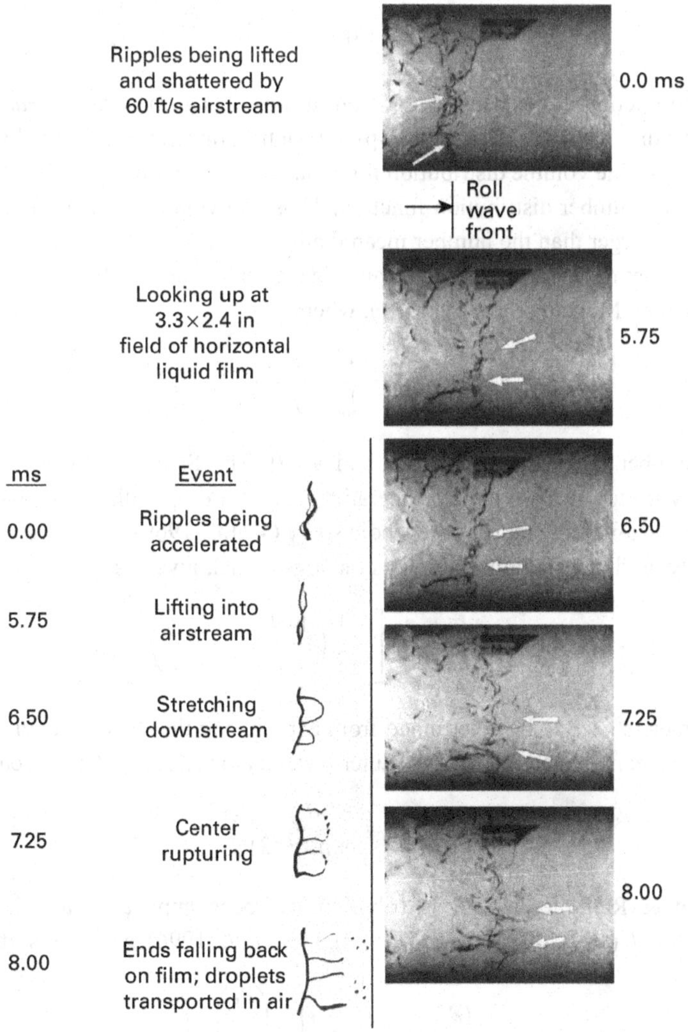

Figure 11.1 Photographs from Woodmansee & Hanratty (1969) showing atomization of a wall film.

count, as many as twenty drops can be sent streaming off into the flowing air after the filament rupture. Finally the end attached to the water film lies down ahead of the roll wave and is passed under the passing disturbance.

The entrained liquid in annular flows is carried along as droplets with a large range of sizes. A number distribution function, $f_n(d_P)$, is defined so that $\int_0^{d_P} f_n dd_p$ is the fraction of the drops with diameters less than d_P. The number mean diameter is given as

$$d_{10} = \int_0^\infty d_P f_n(d_P) d(d_P) \tag{11.2}$$

A normalized volume distribution is also used. It is defined by

$$\frac{dv}{dd_P} = f_v(d_P) \tag{11.3}$$

$$\int_0^\infty f_v(d_P) dd_P = 1 \tag{11.4}$$

It gives the fraction of the volume made up of drops between d_P and $d_P + dd_P$. Since the volume of a drop varies as d_P^3, larger drops are contributing more to the total volume than smaller drops. The volume distribution function is skewed more in the direction of large d_P than is the number distribution function. Thus, the volume median diameter, $d_{v\mu}$, is considerably larger than the number mean diameter, d_{10}.

Another way to characterize the droplet size would be one of the average diameters, d_{qm}, defined by Mugele & Evans (1951), where

$$d_{qm} = \frac{\int_0^\infty d_P^q f_n dd_P}{\int_0^\infty d_P^m f_n dd_P} \tag{11.5}$$

For the number mean diameter, $q = 1$ and $m = 0$. The Sauter mean diameter, d_{32}, is particularly useful in mass transfer operations, since a drop with this diameter has the same surface-to-volume ratio as the whole spray (Wallis, 1969).

A widely used distribution function is the log-normal, given as

$$f_v(d_P) = \frac{1}{\sigma\sqrt{2\pi}} \exp\left[-\frac{1}{2}\left(\frac{\ln d_P - \ln d_{v\mu}}{\sigma}\right)^2\right] \frac{dd_P}{d_P} \tag{11.6}$$

Two parameters need to be determined from experiment: the standard deviation, σ, and the volume mean diameter, $d_{v\mu}$. The Sauter mean diameter can then be calculated as

$$d_{32} = \frac{d_{v\mu}}{\exp(\sigma^2/2)} \tag{11.7}$$

The Rosin & Rammler (1933) distribution has been applied to annular flows by Azzopardi *et al.* (1991), Jepson *et al.* (1989), Hay *et al.* (1996), and others. It is given as

$$f_v(d_P) = 1 - \exp\left[-\left(\frac{d_P}{\overline{X}}\right)^N\right] \tag{11.8}$$

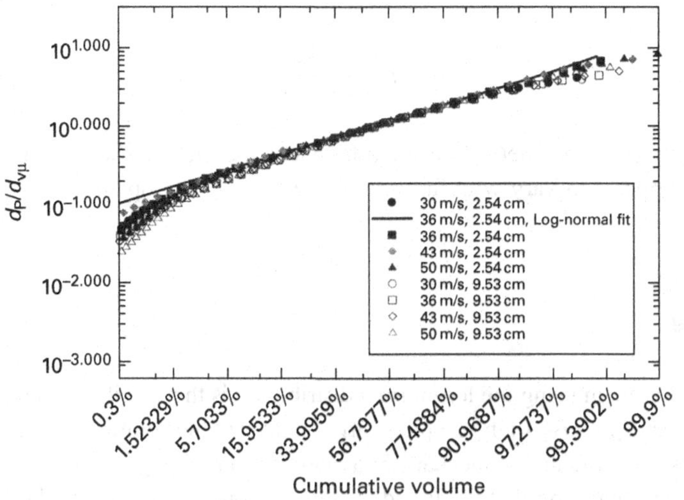

Figure 11.2 Log-probability plot of drop sizes in horizontal 2.54 cm and 9.53 cm pipes, $u_{LS} = 0.041$ m/s. Al-Sarkhi & Hanratty, 2002.

where N and $\overline{X}$ are obtained from experiment. The log-normal distribution is the main focus of this chapter.

The parameter σ in equation (11.6) characterizes the spread of drop sizes around $d_{v\mu}$. It does not vary greatly with flow conditions. This is not the case for $d_{v\mu}$, which represents the size of larger drops that carry most of the mass. Figure 11.2 shows measurements of $d_P/d_{v\mu}$ for air and water flowing in 2.54 cm and 9.53 cm horizontal pipes by Simmons & Hanratty (2001) and by Al-Sarkhi & Hanratty (2002). The plot uses log-normal paper (a log-probability plot) for which equation (11.6) produces straight lines that give a value of $d_P/d_{v\mu} = 1$ at a cumulative volume fraction of 50%. The parameter σ determines the slope. Note that a value of $\sigma = 0.82$ produces a good fit for a wide range of conditions. The measured $d_{v\mu}$ varied from 81.8 μm to 181 μm and the liquid flow varied from 30 g/s to 170 g/s. This includes changes of the superficial liquid velocity of 0.041–0.125 m/s (not indicated in Figure 11.2). Equation (11.6) fits the data, except for small and large drop diameters (about 4.5% of the drop volume for the 2.54 cm pipe). The departure from the log-normal distribution at small and large drop sizes is more severe in the larger pipe. A comparison of the log-normal and Rosin–Rammler equations with the data of Simmons & Hanratty (2001) indicates that the Rosin–Rammler equation overpredicts the volume concentration of small drops and underpredicts the volume contribution of large drops. The log-normal provides a better fit of the measurements that they examined.

Wicks & Dukler (1966), and others, have used the upper limit log-normal distribution (proposed by Mugele & Evans, 1951) for annular flows

$$\frac{dv}{dd_P} = \frac{\delta d_m}{\sqrt{\pi} d_P (d_m - d_P)} \exp\left[-\delta^2 \left(\ln \frac{d_P}{d_m - d_P} - \ln a\right)^2\right] \tag{11.9}$$

where

$$a = \frac{d_{v\mu}}{d_m - d_{v\mu}} \tag{11.10}$$

A third empirical parameter, the maximum drop diameter, d_m, is introduced. This parameter appears to vary with the size of the sample (Azzopardi, 1997), so it is not easily defined.

11.3 Drop size

The critical issue in using the log-normal distribution is the specification of a characteristic diameter, $d_{v\mu}$ or d_{32}, of the droplets carrying most of the volume. Taylor (1940) made two proposals regarding the mechanism of atomization.

One is the suggestion that the liquid is sheared from a viscous boundary layer at the interface by the shear stress imposed by a high-velocity gas stream. The drop size is related to the thickness of this boundary layer. Experimental studies of annular flow are inconsistent with this interpretation in that measurements of drop size and the rate of atomization are not strongly affected by liquid viscosity.

The second proposal is that atomization occurs by the removal of waves and that drop size scales with the wavelength of the unstable waves. Taylor used the inviscid theory outlined in Chapter 4 (see Lane & Green, 1956) to calculate the wave which is growing the fastest:

$$\lambda_m = \frac{2\pi\sigma}{\rho_G \overline{u_G}^2} f(\theta) \tag{11.11}$$

where $\overline{u_G}$ is the relative velocity of the gas to the liquid and

$$\theta = \frac{\rho_L}{\rho_G} \frac{\sigma^2}{\mu_L^2 \overline{u_G}^2} \tag{11.12}$$

Here μ_L is the liquid viscosity, σ is the surface tension and $f(\theta)$ is a function which corrects for effects of liquid viscosity. Values of $f(\theta)$ are given by Lane & Green (1956). For liquids with small viscosity, $f(\theta) = 1.5$. Thus, the assumption that the drop diameter scales with the wavelength yields the following relation

$$\frac{\rho_G \overline{u_G}^2 d_P}{\sigma} = C \tag{11.13}$$

where C is a constant. This equation is attractive since it also represents the gas velocity at which a drop with a diameter d_P will break up. See Lane & Green (1956). However, (11.13) is not consistent with measurements, which indicate that $d_{v\mu} \sim u_G^{-1.1}$. This disagreement led Tatterson *et al.* (1977) to consider another scaling.

The wavelets observed to be atomizing in annular flows are too small, relative to the thickness of the wall layer, to be described by a deep liquid analysis. Consequently, the

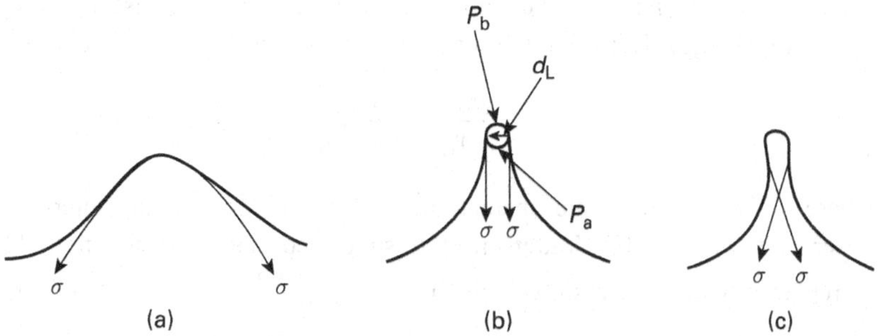

Figure 11.3 Kelvin–Helmholtz mechanism for the formation of drops. Tatterson *et al.*, 1977.

above approximation for the scaling of λ_m is not appropriate. Tatterson *et al.* suggest that the unstable wavelengths scale with the height of the wall layer.

The unstable wave shown in Figure 11.3a is pictured to grow until the crest adopts the configuration shown in Figure 11.3b. Here, the crest resembles a two-dimensional tube with a diameter d_L, which is removed from the wave because the suction pressure of the gas, $\Delta P = P_b - P_a$, is just balanced by surface tension (see Section 4.3).

$$\Delta P \approx \sigma \Big/ \frac{d_L}{2} \tag{11.14}$$

By using Kelvin–Helmholtz theory for the flow of an inviscid fluid over a small-amplitude wave Tatterson *et al.* (1977) suggested that the pressure gradients in a gas flowing over a wavy surface scale as $\rho_G \overline{u_G}^2 k$ and that ΔP in (11.14) scales as $\rho_G \overline{u_G}^2 k d_L$, where $k = 2\pi/\lambda$ is the wave number of the wave producing the ligament. If the drop diameter, d_P, is assumed to be proportional to d_L (rather than the wavelength) and if the above scaling for ΔP is substituted into (11.14), one obtains

$$d_P \left(\frac{\rho_G \overline{u_G}^2 k}{\sigma} \right)^{1/2} = \text{constant} \tag{11.15}$$

It is expected that k would be the wave number of the fastest-growing wave, k_m. For wavelets on thick layers $k_m \sim \rho_G \overline{u_G}^2/\sigma$. If this is substituted into (11.15), equation (11.13) is obtained. As mentioned above, this is not consistent with experiments. Tatterson *et al.* (1977) argue that the deep liquid assumption is not appropriate since the wall film is too thin. They assumed that the wavelength of the unstable wave scales with the film height, m. If this is used in (11.15) the following relation for d_P is obtained:

$$\frac{d_P}{m} \left(\frac{\rho_G m \overline{u_G}^2}{\sigma} \right)^{1/2} = C_1 \tag{11.16}$$

where C_1 is a constant. The use of (11.16) requires the specification of an appropriate value for $\overline{u}_G$ since the analysis was carried out on the assumption that $\overline{u}_G$ is uniform. The distance over which a surface disturbance penetrates a flowing gas stream is proportional to the wavelength. The velocity profile over a wavy surface can be represented as

$$\frac{\overline{u_G}(y)}{v_G^*} = \frac{1}{\kappa}\ln\frac{y}{k_S} + B \tag{11.17}$$

where k_S is a length scale characterizing the roughness of the interface. From this observation and (11.17), Tatterson *et al.* suggested that $\overline{u}_G$ in equation (11.16) should be replaced with the friction velocity, $v_G^* = (\tau_i/\rho_G)^{1/2}$. Thus, (11.16) changes to

$$\frac{d_P}{m}\left(\frac{\rho_G v_G^{*2} m}{\sigma}\right)^{1/2} = C_2 \tag{11.18}$$

where

$$\tau_i = f_i \frac{1}{2}\rho_G \overline{u}_G^2 \tag{11.19}$$

and

$$v_G^* = \left(\frac{f_i}{2}\right)^{1/2}\overline{u}_G = \left(\frac{f_S}{2}\right)^{1/2}\overline{u}_G\left(\frac{f_i}{f_S}\right)^{1/2} \tag{11.20}$$

The term f_S is the friction factor for a smooth surface,

$$f_S = \frac{0.046}{\mathrm{Re}_G^{0.2}} \tag{11.21}$$

The work of Henstock & Hanratty (1976), described in Chapter 3, gives an approximation for m:

$$\frac{m}{d_t} = \frac{6.59}{(1 + 1400F)^{1/2}} \tag{11.22}$$

where

$$F = \frac{\gamma(\mathrm{Re}_{LF})}{\mathrm{Re}_G^{0.9}}\frac{\upsilon_L}{\upsilon_G}\sqrt{\frac{\rho_L}{\rho_G}} \tag{11.23}$$

Henstock & Hanratty also give

$$\frac{f_i}{f_S} = 1 + 1400F \tag{11.24}$$

Tatterson *et al.* used these equations to eliminate m and f_i from (11.18). They rewrote (11.18) as

$$\frac{d_P}{d_t}\left[\frac{\rho_G \overline{u}_G^2 (f_S/2) d_t}{\sigma}\right]^{1/2} = C_2\left(\frac{m}{d_t}\right)\left(\frac{f_i}{f_S}\right)^{-1/2} \tag{11.25}$$

where (m/d_t) is given by (11.22) and (f_i/f_S) is given by (11.24). The correlations of Henstock & Hanratty suggest that the right side of (11.25) is a function of F. (Somewhat different expressions are obtained if equations (3.47) and (3.49) in Chapter 3 are used to calculate m/d_t.)

Pogson *et al.* (1970) used still photography to measure drop size in an upward flow of steam and water in a 0.32 cm pipe. Namie & Ueda (1972) measured drop size for air–water flow in a horizontal 1 × 6 cm rectangular channel by collecting samples of the spray in a silicone oil. The sample sizes, in both studies, were not large enough to measure the volume median drop size, so Tatterson *et al.* (1977) calculated the average drop size, d_{10}, from their data on drop size distribution. These are plotted in Figure 11.4, as suggested by (11.25). The ordinate is the left side of (11.25). The abscissa is the flow parameter. Terms m/d_t and f_i/f_S are calculated from (11.22) and (11.24). The curve, calculated from (11.25) with $C_2 = 0.58$, provides a good fit. Note that the data represent studies with surface tensions of 40 dynes/cm and 70 dynes/cm.

Thus, the paper of Tatterson *et al.* tentatively suggests that d_{10} can be estimated by

$$\left[\frac{\rho_G \bar{u}_G^2 d_t (f_S/2)}{\sigma}\right]^{1/2} \frac{d_{10}}{d_t} \approx 2.4 \times 10^{-3} \tag{11.26}$$

The above analysis is speculative. Clearly, more measurements need to be considered.

Simmons & Hanratty (2001) give the following account of drop size measurements:

Several techniques have been employed: Tatterson *et al.* (1977) used charge removal from an insulated probe. Wicks & Dukler (1966) used a needle bridging method, where two needles, placed a small distance apart, were connected to a resistance battery. Drops larger than the gap

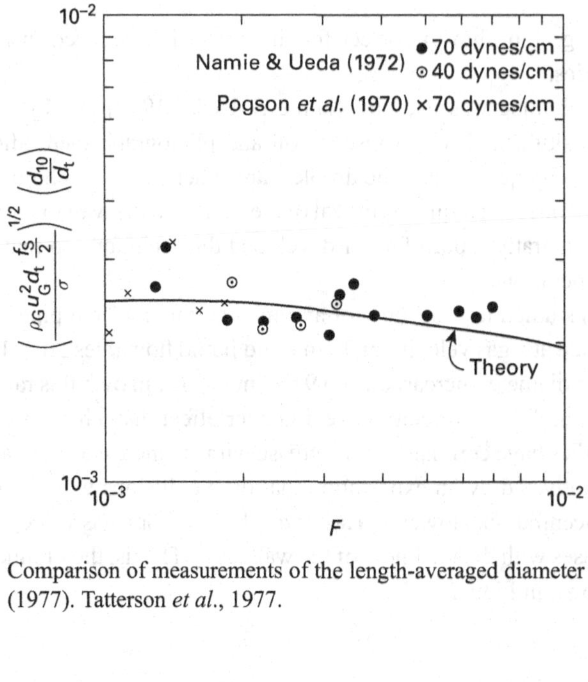

Figure 11.4 Comparison of measurements of the length-averaged diameter with the theory of Tatterson *et al.* (1977). Tatterson *et al.*, 1977.

completed the circuit and caused an electric pulse. By varying the needle gap, the cumulative size distribution was obtained. Semiat & Dukler (1981) and Lopez & Dukler (1985) used a laser gradient technique, which produced a local measurement of drop size and velocity. All these techniques were unable to detect very small drops, below 100–200 μm. Subsequent work has shown that these techniques lead to significant error in measuring the volume median size since a significant portion of the volume is carried by drops smaller than 100 μm. Further refinements were made to the laser-grating technique by Fore & Dukler (1995). These allowed measurements of droplets down to a size of 10 μm.

Several workers have used photographic techniques (Cousins & Hewitt, 1968; Pogson *et al.*, 1970; Andreussi *et al.*, 1978). A problem with these techniques is that they, too, can favour large drops, since it is more probable that part of a large droplet may be within the field of the camera lens. Furthermore, smaller droplets may not be detected since they could be out of focus, particularly if the resolution of the photographs is poor. Hay *et al.* (1996) addressed this problem by using a sheet of laser light to illuminate the flow field. The laser sheet was thinner than the focal depth of the camera lens, so only droplets illuminated by the laser sheet appeared in focus. Droplets out of focus were eliminated from analysis by examining changes in the light gradient at the edges of the drops.

A laser diffraction technique, invented by Switherbank *et al.* (1976) has been used by several workers. This work has been reviewed by Azzopardi (1997). Instruments based on this approach are marketed by Malvern instruments. Early versions required the assumption of a distribution function for drop size, either the equation of Rosin and Rammler (1933) or the upper limit log-normal distribution of Mugele & Evans (1951). Current versions employ a fifteen parameter "model independent" algorithm which does not impose any unimodal function. The detection range of the instrument is governed by the focal length of the lens used in the detector. Combellack and Matthews (1981) and Azzopardi (1985) indicated the importance of using the optimal focal length.

The use of the diffraction technique has been limited to situations in which the concentration of droplets is low. The algorithms used to analyze the measurements assume that the light is scattered only by one drop. At higher concentrations, multiple scattering of laser light by many drops can exist. This will cause the scattered light to enter the detector at a larger angle; the distributions obtained will overpredict the number of smaller drops.

A multiple scattering algorithm to correct for this effect is employed in the Malvern Spraytec R5000 series.

An immersion technique was used by Namie & Ueda (1972) and by Okada *et al.* (1995) to collect in situ samples in a viscous oil and photograph them after removal. Advantages of this technique are that the droplets are spherical when photographed and that the measurement does not require removal of the film from the wall to provide optical access. Hurlburt & Hanratty (2002) further developed this technique so that it might be employed in field operations.

Hay *et al.* (1996) studied upward flow of air and water in a 4.2 cm pipe. The experiments were performed at a gas velocity of 36 m/s and liquid flow rates 30 to 170 g/s. The volume median drop diameter increased from 91.2 μm to 192 μm over this range of liquid flow rates. Changes in the gas velocity have a greater effect than changes in the liquid flow rate. Many studies have correlated this increase with the increase in entrained liquid. This seems to be motivated by an expectation that drop collisions would increase with increasing drop concentration. However, Hay *et al.* showed, for a gas velocity of 36 m/s, that drop size increases with the thickness of the wall layer. That is, they found that $d_{\nu\mu}/m$ is a constant, as shown in Figure 11.5.

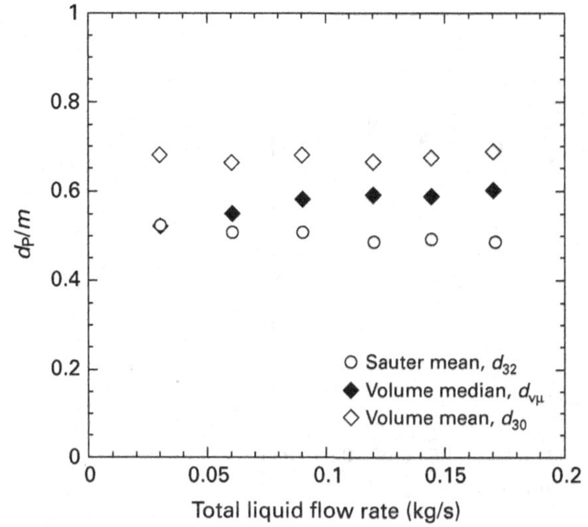

Figure 11.5 Scaling of drop size with the average thickness of the wall layer. Measurements of Hay *et al.* in a 4.2 cm pipe. Hay *et al.*, 1996.

It is useful, initially, to ignore the influence of liquid flow rate. Values of

$$\left[\frac{\rho_G \overline{u}_G^2 d_t (f_S/2)}{\sigma}\right]^{1/2} \left(\frac{d_{v\mu}}{d_t}\right) = 2.7\text{–}6.1 \times 10^{-3} \tag{11.27}$$

are calculated for $d_{v\mu}$ varying from 91.2 to 192 μm.

Equation (11.27) seems consistent with measurements reported in the literature on the influence of $\overline{u}_G$, ρ_G and σ on drop size. The prediction that $d_{v\mu}$ increases with pipe diameter is more contentious, since many calculations (Azzopardi, 1985; Pan & Hanratty, 2002) presume that drop size is independent of pipe diameter. This is an important issue because it is desirable to use measurements of drop size in small-diameter pipes to predict the behavior of large commercial pipes, such as used to transport natural gas/condensate. This prompted two studies (Simmons & Hanratty, 2001; Al-Sarkhi & Hanratty, 2002) in horizontal 2.54 cm and 9.53 cm pipes, which use the Malvern Spraytec to measure drop size. This has the advantage that comparison is made in the same flow system with the same measurement technique.

One of the difficulties is that drop size decreases with height in a horizontal pipe because of the gravitational force on the drops and because the height of the liquid film varies around the circumference. The drop size was found to be constant along any horizontal chord in the pipe. Figure 11.6 shows measurements made in a 9.53 cm pipe for $u_{GS} = 43$ m/s at the pipe center, 1.9 cm above the center and 1.9 cm below the center. Note that the influence of gravity on drop size measurements is not great below the pipe center. However, differences of about 10% were found at gas velocities of 30 m/s and 36 m/s. Figure 11.7 compares plots of the data obtained at the center of the 2.54 cm pipe and at 1.9 cm below the center in the 9.53 cm pipe. These show that the drop size roughly increases with $d_t^{1/2}$ and decreases with $u_{GS}^{1.1}$.

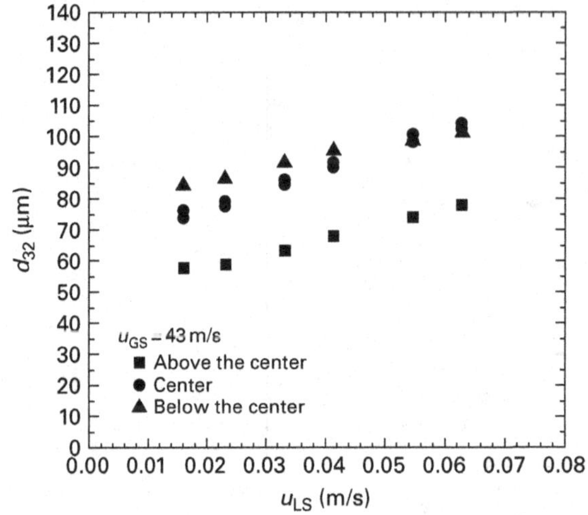

Figure 11.6 Comparison of measurements of the Sauter mean diameter at different positions in a 9.53 cm pipe at u_{GS}. Simmons & Hanratty, 2001.

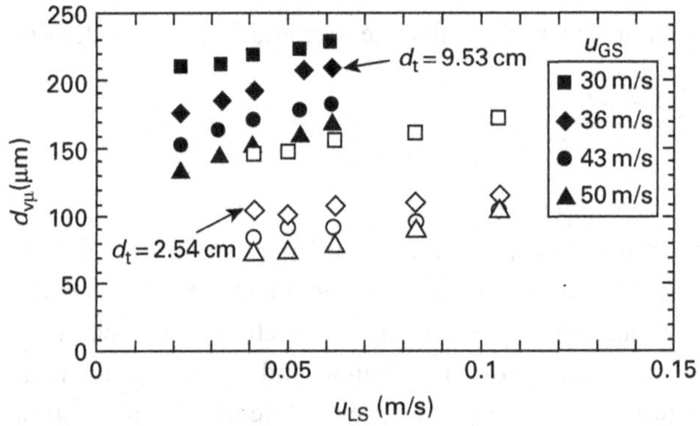

Figure 11.7 Comparison of volume median drop diameters obtained at the center of a 2.54 cm pipe (open symbols) and at 1.9 cm below the center of a 9.53 cm pipe. Al-Sarkhi & Hanratty, 2002.

The following equation is given by Al-Sarkhi & Hanratty if the influence of liquid flow is ignored:

$$\left(\frac{d_t \rho_G \overline{u}_G^2}{\sigma}\right)^{1/2} \left(\frac{d_{v\mu}}{d_t}\right) = 0.1072 \tag{11.28}$$

It is of interest to compare this result with that obtained by Hay *et al.* (1996) for upward flow of air and water at 36 m/s in a 4.2 cm pipe, for which $(f_S/2)^{1/2} = 4.74 \times 10^{-2}$. Hay *et al.* obtained

$$\left(\frac{\rho_G \overline{u}_G^2 d_t}{\sigma}\right)^{1/2} \left(\frac{d_{v\mu}}{d_t}\right) = 0.057 - 0.129 \tag{11.29}$$

The average of the right side is 0.093. Thus, (11.28) agrees, approximately, with the measurements of Hay *et al.*

Equation (11.29) offers a first approximation of data. It does not exactly match measurements since the predicted dependency of drop diameter on $\overline{u}_G{}^{-1}$ does not agree with the dependency of $\overline{u}_G{}^{-1.1}$ found in measurements. Furthermore, the role of liquid flow is not defined.

Two interpretations of the influence of liquid flow have been proposed: the increase of coalescence with increasing flow of entrained drops, G_{LE}, and changes of the drops generated by atomization of the wall layer. (A mechanism which has not been considered is the decrease of gas turbulence with increasing particle concentration.)

11.4 Rate of deposition

For small drop concentrations, the rate of deposition of drops in vertical annular flow can be represented by the linear rate equation

$$R_D = k_D C_B = k_D \frac{W_{LE}}{SQ_G} \tag{11.30}$$

where C_B is the average drop concentration over the pipe cross-section, R_D is the rate of deposition per unit area, Q_G is the volumetric gas flow and S is the ratio of the drop velocity and the axial velocity of the gas flow. The drops, in annular flow, are large enough that $\tau_P^+ > 20$. Thus, the dominant mechanism involves drops starting a free-flight to the wall from a region outside the viscous wall layer (see Chapter 10).

The rate of deposition can be described as the average velocity with which droplets strike the wall, V_W, so that

$$R_D = V_W C_W = V_W C_B \frac{C_W}{C_B} \tag{11.31}$$

where C_W is the concentration at the wall. Measurements of C show that it is constant over the pipe cross-section for a vertical annular flow (Hay *et al.*, 1996). Thus $C_W \approx C_B$ and $k_D = V_W$. If the turbulent velocities are Gaussian, the average velocity is given as $V_W = \sigma_P/\sqrt{2\pi}$, so

$$k_D = \sigma_P/\sqrt{2\pi} \tag{11.32}$$

where $\sigma_P = \left(\overline{v_{Pr}^2}\right)^{1/2}$ is the root-mean-square of the radial component of the particle velocity fluctuations outside the viscous wall layer. Equation (11.32) has been verified for dilute concentrations (Dykhno & Hanratty, 1996) and by Lee *et al.* (1989).

Measurements of $(\overline{v_{Pr}^2})$ are discussed in Section 10.8. The particle turbulence is related to the fluid turbulence by equation (10.59). Thus, $\overline{v_{Pr}^2} = \overline{u_r^2}[\beta/(\alpha+\beta)]$ with $\alpha = 0.7/\tau_i^{path}$ where v_{Pr} and u_r are the r-components of the particle velocity fluctuations and the fluid velocity fluctuations, respectively (see Figure 10.3). The term $\beta = 1/\tau_P$, where τ_P is the inertial time constant of the drops.

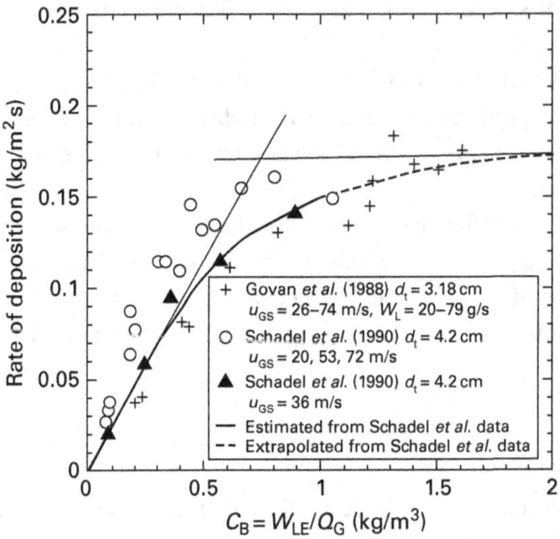

Figure 11.8 Measurements of rates of deposition. Hay & Hanratty, 1996.

Measurements of R_D by Govan *et al.* (1988) and by Schadel *et al.* (1990) are shown in Figure 11.8 as a function of C_B. Equation (11.30) describes the limiting slope at small W_{LE}/Q_G. An interesting feature of these measurements is the breakdown of the limiting law; that is, k_D depends on concentration. At large C_B (volume fractions roughly equal to 0.6×10^{-3}) a limiting behavior is suggested such that $k_D \propto C_B^{-1}$, so that R_D is independent of C_B. This behavior has also been observed for downward flow in a vertical pipe by Andreussi *et al.* (1983). See the review article by Dykhno & Hanratty (1996). The data shown in Figure 11.8 are roughly approximated by fitting them with two straight lines. At small drop concentrations, the rate of deposition is given by (11.30). For $W_{LE}/Q > 0.75$ (which corresponds to a volume fraction of drops of 0.75×10^{-3}), $k_D \propto C_B^{-1}$ and the rate of deposition is constant at 0.17 kg/m^2 s.

If k_D is represented by (11.32) the measurements of R_D shown in Figure 11.8 suggest the dependency on $\left(\overline{v_{Pr}^2}\right)^{1/2}$ displayed in Figure 11.9. Several explanations for the decrease of particle turbulence with increasing drop concentration have been proposed: (1) Drop size increases with liquid flow. This leads to a more sluggish behavior of the drops and, therefore, a decrease in $\overline{v_{Pr}^2}$. However, this effect is not large enough to explain Figure 11.9 (see Hay *et al.*, 1996). (2) The number of encounters per unit time increases with the number of drops per unit volume. Therefore, it is plausible to explore the notion that particle–particle encounters result in a decrease in particle turbulence through inelastic interactions. Calculations (Hay *et al.*, 1996) show that this could lead to a significant decrease in the particle turbulence and to the rate of deposition. (3) Namie & Ueda (1972) suggest that the decrease in particle turbulence with increasing concentration is due to a decrease in the fluid turbulence. The next section describes numerical experiments in a turbulent velocity field, which support this suggestion.

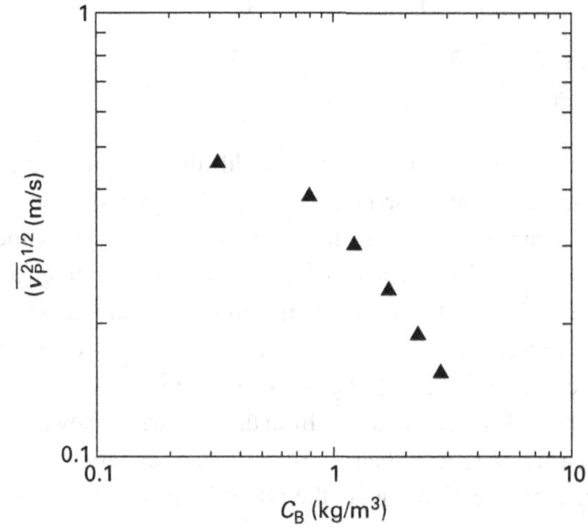

Figure 11.9 Root-mean-square particle velocity needed to explain deposition data. Hay *et al.*, 1996.

11.5 Experiments in turbulence generated by a DNS

Direct numerical solutions (DNS) of the Navier–Stokes equations have provided results on the behavior of a turbulent flow. Particles embedded in the turbulence experience a drag force which is related to the difference in velocity between the particle and the fluid

$$f_{Pi} = -\frac{3\rho_f C_D}{4 d_P \rho_P} |\vec{v}_P - \vec{u}_G| (v_{Pi} - u_{Gi}) \tag{11.33}$$

The force of the particle on the fluid is $-f_{Pi}$. This feedback effect of particles on fluid flow has been modeled with a point force method using the assumption that the particles are small enough for the eddy shedding process not to occur. Squire & Eaton (1990) and Elgobashi & Truesdell (1993) used this approach in a DNS of an isotropic field. Pan & Banerjee (1996), Li *et al.* (2001), Mito & Hanratty (2006), and Hanratty & Mito (2009) used it in a DNS of turbulent flow in a channel. The approach involves the introduction of forces of the particles on the fluid into the Navier–Stokes equations, which are solved numerically.

The computational domain is divided into a number imaginary cells which contain one grid point. The force of particles on the fluid motion at a grid point is the sum of the reaction forces exerted by particles whose centers exist in the computational cell surrounding the grid point (Li *et al.*, 2001).

$$F_i = -\frac{\rho_P}{\rho_f}\frac{V_P}{V_{cell}}\sum_{k=1}^{N_{cell}} f_{Pik} = -\frac{C}{\rho_f}\langle f_{Pi}\rangle_{cell} \tag{11.34}$$

where V_P is the volume of the particle, V_{cell} is the volume of the computational cell, N_{cell} is the number of particles in the cell, f_{Pik} is the f_{Pi} for the kth particle, $C = \rho_P V_P N_{cell}/V_{cell}$ is the concentration of the particles and $\langle f_{Pi}\rangle_{cell}$ is the average point force of the N_{cell} particles.

The equation for fluid motion is then represented as

$$\rho_f\left(\frac{\partial u_{Gi}}{\partial t}+u_{Gj}\frac{\partial u_{Gi}}{\partial x_j}\right)=-\frac{\partial P}{\partial x_i}+\mu\frac{\partial^2 u_{Gi}}{\partial x_j\partial x_j}+\rho_f F_i \tag{11.35}$$

The method for introducing particles into the flow field differs from author to author. Mito & Hanratty considered the annular configuration for upward flow in a vertical channel. The ratio of the particle density to the fluid density was the same as for an air–water flow, 1000 : 1. The atomization process is mimicked by injecting particles from $x_2 = d_P/2$ with a velocity of $(15v_0^*, v_0^*, 0)$ and a rate of atomization per unit area of R_{ib} from the bottom wall. Particles were also injected from $x_2 = 2H - d_P/2$ with a velocity of $(15v_0^*, -v_0^*, 0)$ and a rate per unit area of R_{it}, where v_0^* is the friction velocity in the absence of particles. The particles were removed from the field when they hit a wall. If the rate of introduction of particles is less than a critical value, a stationary state can be reached for which the rate of injection equals the rate of deposition. If it is above this critical value, the concentration continues to increase with time since the rate of deposition is not sufficient to balance the rate of injection.

The calculations of Mito & Hanratty (2006) are of interest because they mimic the atomization process. However, they have limitations in that the walls are smooth. (The direct influence of wall roughness was not considered.) The stress on the wall, τ_W, is obtained from the calculated velocity gradient at the wall, $\tau_W = \mu(d\bar{u}/dy)_W$. The friction velocity is defined as $v^* = \sqrt{\tau_W/\rho_f}$. The calculations were done for a fixed volumetric flow of fluid. The friction is found to decrease with an increasing concentration of particles. The Reynolds number, defined with the bulk mean velocity and the half-height of the channel, H, was 2260. The Reynolds number, Re_τ, defined with the friction velocity in the absence of particles, v_0^*, and the half-height of the channel, H, was 150.

Figure 11.10 presents their calculations of the rate of deposition for cases in which stationary states were reached. The left ordinate is the calculated rate of deposition made dimensionless with v_0^*. The abscissa is the volume fraction of spheres, α_P. The right

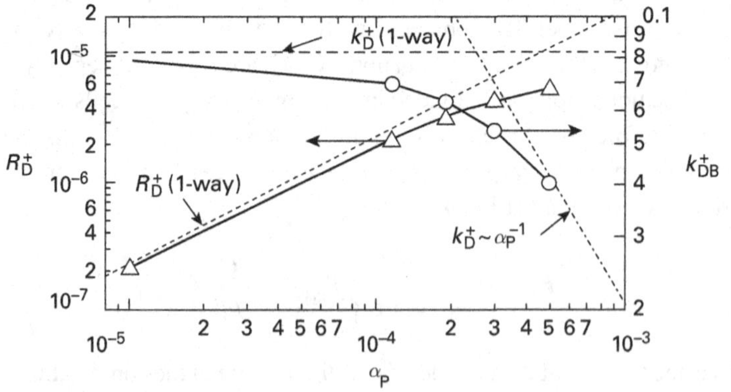

Figure 11.10 Effect of feedback, as demonstrated by the effect of α_P on the rate of deposition, R_D^+, and on the deposition constant, k_D^+. Mito & Hanratty, 2006.

ordinate is the deposition coefficient made dimensionless with the friction velocity that existed when no particles were present, $k_D^+ = k_D/v_0^*$. The values of R_D^+ and k_{DB}^+ are also shown for the one-way coupling situation, for which the effect of α_P on the fluid turbulence is ignored. The results for two-way coupling capture the trend of measurements in annular flows.

Large changes in the normal velocity fluctuations at very small volume fractions of the particles were observed. Instantaneous velocity fluctuations calculated at the center plane are shown in Figure 11.11a for single-phase flow, in Figure 11.11b for a stationary case with $\alpha_P = 4.9 \times 10^{-4}$ and in Figure 11.11c for a non-stationary case with $\alpha_P = 3.0 \times 10^{-3}$, where α_P is the volume fraction of particles embedded in the fluid. Dots show the locations of particles. (They do not represent the actual size of the particles.) Significant attenuation of the fluid turbulence is noted in Figure 11.11b. The particles are seen to be almost uniformly distributed and to be affecting all turbulence structures. The small-scale turbulence is seen almost to vanish for the non-stationary case shown in Figure 11.11c. Instantaneous velocity fluctuations in a cross-section perpendicular to the

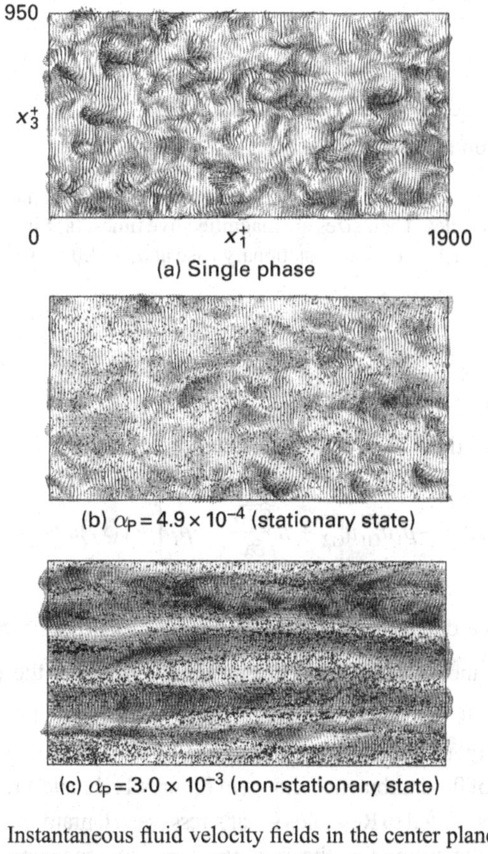

Figure 11.11 Instantaneous fluid velocity fields in the center plane. Dots represent locations of particles. (a) Single-phase flow. (b) A statistical stationary case at $\alpha_P = 4.9 \times 10^{-4}$. (c) A non-stationary case at $\alpha_P = 3.0 \times 10^{-3}$. Mito & Hanratty, 2006.

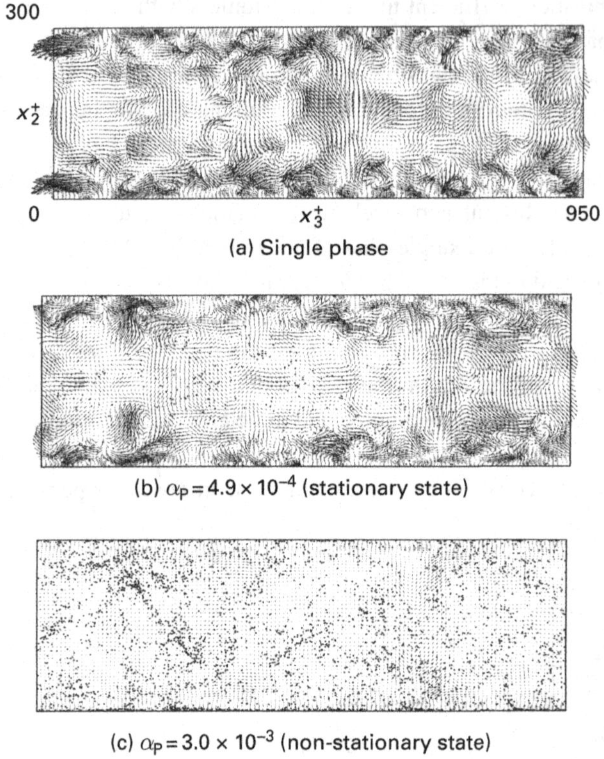

Figure 11.12 Instantaneous fluid velocity fields in a cross-section perpendicular to the direction of mean flow. Dots represent locations of the particles. Their sizes are magnified five times. (a) Single-phase flow. (b) A stationary case at $\alpha_P = 4.9 \times 10^{-4}$. (c) A non-stationary case at $\alpha_P = 3.0 \times 10^{-3}$. Mito & Hanratty, 2006.

direction of mean flow are shown in Figure 11.12. The vertical structures observed close to the wall for single-phase flows are seen to disappear.

Under fully developed conditions the average of equation (11.35) gives

$$-\frac{\partial \overline{P}}{\partial x_1}(H - x_2) = -\rho_f \overline{u_{G1}^t u_{G2}^t} + \mu \frac{\partial \overline{u_{G1}}}{\partial x_2} - \rho_f \int_{x_2}^{H} \overline{F_1}(x_2) dx_2 \tag{11.36}$$

For single-phase flow, the force due to the pressure gradient is balanced by the Reynolds shear stress $\left(-\rho_f \overline{u_1^t u_2^t}\right)$ and the viscous shear stress $(\mu \partial \overline{u_{G1}} / \partial x_2)$. In the presence of particles the pressure gradient is balanced by particle forces, as well as the viscous and turbulent stresses. This can explain the calculated decrease of Reynolds stress shown in Figure 11.13 and the damping of fluid turbulence shown in Figures 11.11 and 11.12, since the production of fluid turbulence is related to Reynolds shear stress (see Hanratty & Mito, 2009).

These calculations support the notion that the reduction of k_D at large concentrations is associated with the damping of fluid turbulence.

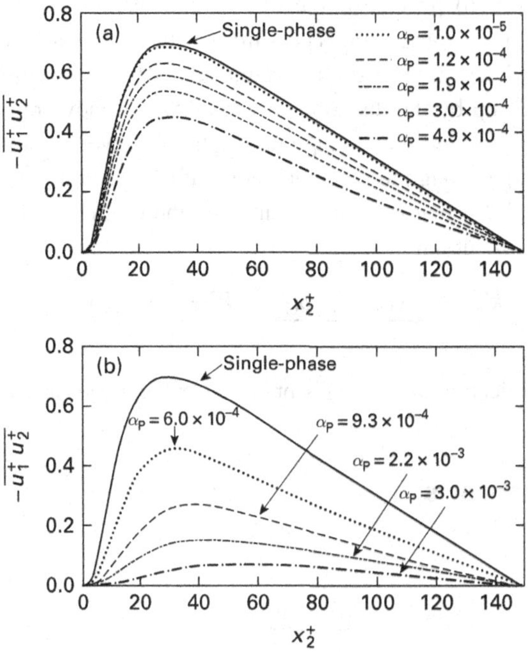

Figure 11.13 Effect of feedback on the fluid Reynolds shear stress. (a) Statistically stationary cases. (b) Non-stationary cases when 2.7 particles are injected per time step. Mito & Hanratty, 2006.

11.6 Rate of atomization

An important phenomenon in considering the rate of atomization and the entrainment is the existence of a critical film flow rate, W_{LFC}, below which atomization does not occur (Hewitt & Hall-Taylor, 1970; Dallman *et al.*, 1979; Asali *et al.*, 1985a, b; Andreussi *et al.*, 1983; Schadel *et al.*, 1990). The air–water data in small- and large-diameter pipes give average values of the critical film Reynolds number in the range 241–280.

The rate of atomization has been represented by the empirical equation

$$R_A = \frac{k'_A \bar{u}_G^2 (\rho_L \rho_G)^{0.5}}{\sigma} \frac{(W_{LF} - W_{LFC})}{P} \tag{11.37}$$

where k'_A is a dimensionless constant. Equation (11.37) shows that the rate of atomization varies linearly with the excess film flow rate at small W_{LF} (Lopez de Bertodano *et al.*, 1997; Assad *et al.*, 1998; Pan & Hanratty, 2002). For $W_{LF} < W_{LFC}$, $R_A = 0$. The linear relation in (11.37) breaks down at large W_{LF} in that dimensionless k'_A decreases with increasing film flow, as has been found for k_D (see Figure 11.10). At large W_{LF}, this is particularly evident in the measurements of Andreussi & Zanelli (1976, 1979) for downward annular flow. In contrast to upward flow, gravity is aiding, so that thicker wall flows can be realized. A maximum R_A at large W_{LF} is suggested.

Dykhno & Hanratty (1996) provide a speculative explanation for this non-linear behavior: At low liquid flows, atomization occurs from disturbance waves which appear intermittently in the wall layer. Deposition occurs, mainly, on the base film. This deposited liquid is picked up by the disturbance waves, which move rapidly over the base film. Thus, the base film does not change its height. However, at large deposition rates this pickup is not large enough to keep the height of the base film constant.

A form of equation (11.37) that recognizes, directly, the existence of a critical gas velocity for the initiation of entrainment is

$$R_{\mathrm{A}} = \frac{k'_{\mathrm{A}}(\bar{u}_{\mathrm{G}} - \bar{u}_{\mathrm{GC}})^2(\rho_{\mathrm{L}}\rho_{\mathrm{G}})^{0.5}}{\sigma}\frac{(W_{\mathrm{LF}} - W_{\mathrm{LFC}})}{P} \tag{11.38}$$

A discussion of this modification of (11.37) is provided in the next section.

11.7 Entrainment in vertical flows

Entrainment is defined as

$$E = \frac{W_{\mathrm{L}} - W_{\mathrm{LF}}}{W_{\mathrm{L}}} \tag{11.39}$$

This section develops expressions for entrainment, except for cases for which the inertial time constant is so large that the drops have unidirectional paths.

Plots by Willetts (1987) of W_{LF} versus $\bar{u}_{\mathrm{G}}$ for a fixed liquid flow, W_{L}, provide easily identified critical gas velocities, u_{GC}, above which $W_{\mathrm{L}} > W_{\mathrm{LF}}$. His results are fitted approximately with the relation

$$\frac{d_{\mathrm{t}}^{0.5} u_{\mathrm{GC}}(\rho_{\mathrm{L}}\rho_{\mathrm{G}})^{0.25}}{\sigma^{0.5}} \approx 40 \tag{11.40}$$

(see Pan & Hanratty, 2002).

By equating (11.38) to (11.30), the following equation is obtained

$$\frac{E/E_{\mathrm{M}}}{1 - E/E_{\mathrm{M}}} = \frac{k'_{\mathrm{A}} d_{\mathrm{t}} \bar{u}_{\mathrm{G}}(\bar{u}_{\mathrm{G}} - u_{\mathrm{GC}})^2 S(\rho_{\mathrm{L}}\rho_{\mathrm{G}})^{0.5}}{4k_{\mathrm{D}}\sigma} \tag{11.41}$$

where k_{D} has the units of velocity and

$$E_{\mathrm{M}} = 1 - \frac{W_{\mathrm{LFC}}}{W_{\mathrm{L}}} \tag{11.42}$$

Equation (11.41) requires that W_{LF} and $\bar{u}_{\mathrm{G}}$ are above the critical values of W_{LFC} and u_{GC} that are needed for drops to appear, so $E = 0$ if $u_{\mathrm{G}} < u_{\mathrm{GC}}$ or if $W_{\mathrm{LF}} < W_{\mathrm{LFC}}$. Note that the liquid flow does not appear in (11.41).

Strictly, (11.41) is valid only for small entrainments where linear rate equations are observed. Non-linearities could be captured by allowing k'_{A} and k_{D} to decrease with increasing W_{L}. The inclusion of these effects makes (11.41) clumsy. Pan & Hanratty

(2002) used these equations outside the range of liquid flows where they should be valid by making the assumption that k'_A/k_D is approximately independent of W_L. They substituted $\bar{u}_G$ for k_D/S and obtained the following rough fit to data of Schadel *et al.* (1990), Binder (1991), Lopez de Bertodano & Jan (1998), Lopez de Bertodano *et al.* (1997,1998), Willetts (1987), Andreussi & Zanelli (1976,1979):

$$\frac{E/E_M}{1-E/E_M} = \frac{A_1(\bar{u}_G - u_{GC})^2(\rho_G\rho_L)^{1/2}d_t}{\sigma} \tag{11.43}$$

with $A_1 = 6\times10^{-5}$.

Figure 11.14 shows measurements of E/E_M for air and water flowing up a vertical 4.2 cm pipe. These indicate that E/E_M is more sensitive to changes in the gas velocity than to changes in the liquid flow. These data are plotted as E versus W_L in Figure 11.15. For a constant gas velocity, an influence of liquid velocity, in (11.41), is included in the term $(E/E_M)/(1 - E/E_M)$. The solid curves in Figure 11.15 were calculated using (11.43).

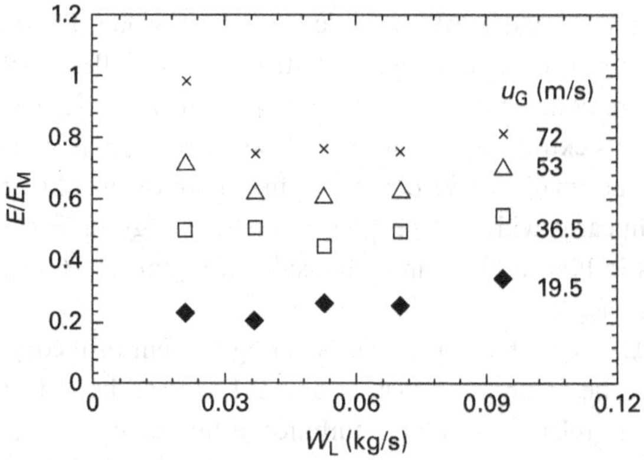

Figure 11.14 Plots of E/E_M for air and water flowing up a vertical 4.2 cm pipe. Pan & Hanratty, 2002.

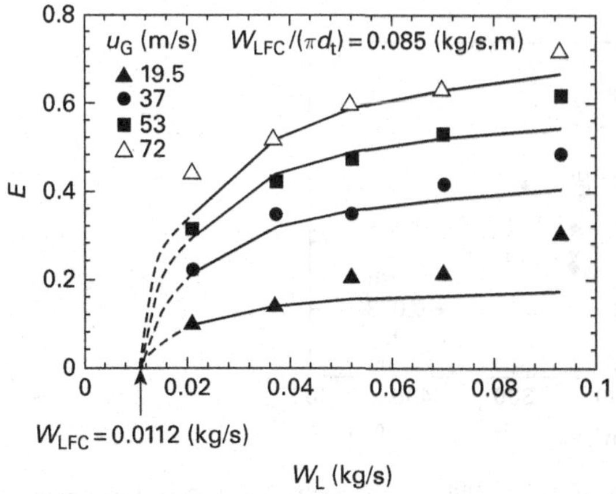

Figure 11.15 Plots of entrainment, E, for air and water flowing up a vertical 4.2 cm pipe. Pan & Hanratty, 2002.

At $W_{LF} = W_{LFC} = 0.0112$ kg/s, the atomization rate is zero, so $E = 0$. Thus, there is a sharp drop in E close to $W_L = W_{LFC}$, as indicated by the dashed lines. From (11.43), it would be expected that entrainment reaches an asymptotic value at large W_L. This type of behavior is only approximately realized. The entrainment actually tends to increase at large W_L. This is not surprising since (11.43) was derived by using linear relations for R_A and R_D. It is expected that k'_A/k_D should change with W_L at large values of this quantity. The increase of E at large W_L suggests that k_D is decreasing more with increasing W_L than is k'_A/k_D.

It is of interest to see how well (11.43) describes observed effects of surface tension and gas density:

Figure 11.16 shows results obtained by Willetts (1987) for air–water ($\sigma = 0.073$ N/m) and for air–genkelene ($\sigma = 0.01$ N/m) flowing up a 1.026 cm pipe. The critical gas velocity, calculated with (11.40), is included in the figure. Entrainments are seen to be larger for the liquid with lower surface tension. The slight differences in the plot reflect the observation that the constant in (11.40) is slightly different for the two systems (see Pan & Hanratty, 2002, Table 2).

All measurements show that entrainment increases with increasing gas density. This is illustrated in Figure 11.17 where data of Lopez de Bertodano & Jan (Private communication, 1998) for air–water flow for three gas densities and by Willetts (1987) for air–water & helium–water are compared by plotting E/E_M versus $\bar{u}_G\rho_G^{0.25}$, as suggested by (11.43). The measurements extrapolate to the critical gas velocity, u_{GC}, and E/E_M varies linearly with $\bar{u}_G - u_{GC}$ at small gas velocities. A first impulse would be to assume entrainment increases linearly with the gas-phase kinetic energy or that E/E_M varies with $\rho_G^{0.5}\bar{u}_G$. The results in Figure 11.17 and the results in Figure 11.16 for a 1 cm pipe show that this is not the case.

Pan & Hanratty (2002) show that somewhat better agreement of theory and experiments can be realized if the assumption that k_D/S in (11.41) can be replaced by $\bar{u}_G$ is abandoned. Instead, k_D is related to particle turbulence by equation (11.32). Particle turbulence is related to fluid turbulence by the equations outlined in Section 11.4. A value of $k'_A = 1.4 \times 10^{-6}$ was used to correlate data for air and water flowing up pipes with

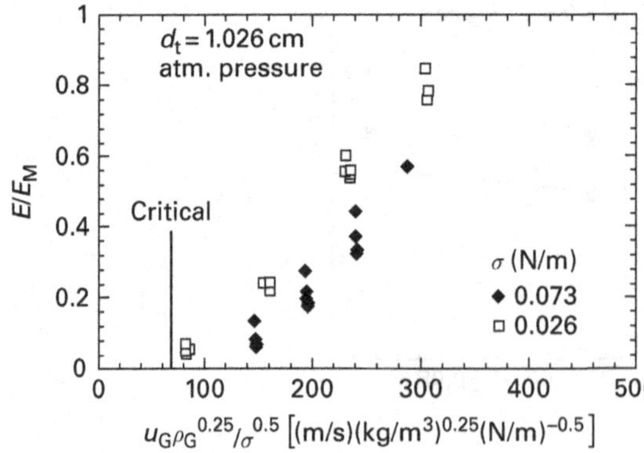

Figure 11.16 The effect of surface tension on E/E_M. The line intersecting the abscissa represents u_{GC}. Pan & Hanratty, 2002.

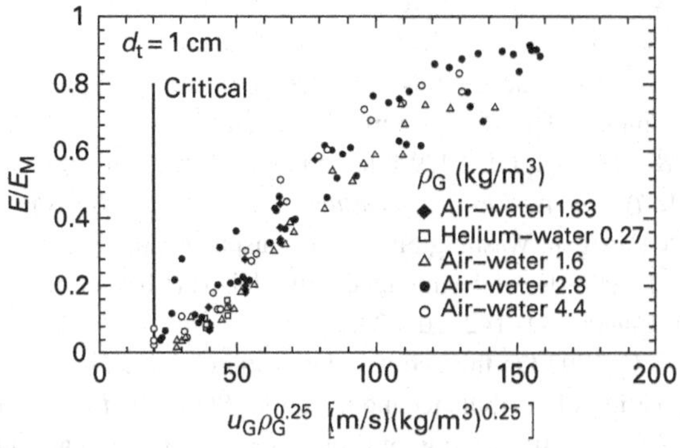

Figure 11.17 An examination of the influence of gas density on E/E_M. The line intersecting the abscissa represents u_{GC}. Pan & Hanratty, 2002.

diameters of 0.0254 m, 0.042 m, 0.00953 m (Lopez de Bertodano & Jan, 1998; Lopez de Bertodano *et al.*, 1997, 1998; Schadel *et al.*, 1990) and for air and water flowing up and down in a 0.024 m pipe (Andreussi & Zanelli, 1976, 1979). This calculation is somewhat more complicated than (11.43) because drop diameter needs to be known.

References

Al-Sarkhi, A. & Hanratty, T.J. 2002 Effect of pipe diameter on drop size in a horizontal annular gas–liquid flow. *Int. J. Multiphase Flow* 28, 1617–1629.

Andreussi, P. & Zanelli, S. 1976 Liquid phase mass transfer in annular two-phase flow. *Ing. Chim.* 12, 132–136.

Andreussi, P. & Zanelli, S. 1979 Downward annular-mist flow of air–water mixtures. In *Two-phase Flow, Momentum, Heat and Mass Transfer*, Vol. 2, ed. F. Durst, G.V. Tsiklauri & N.H. Afgan. Washington, DC: Hemisphere.

Andreussi, P., Romano, P. & Zanelli, S. 1978 Drop size distribution in annular-mist flow. In *Proceedings of the First Conference on Liquid Atomization in Spray Systems*, Tokyo, August 27–31.

Andreussi, P., Asali, J.C. & Hanratty, T.J. 1983 Initiation of small waves in gas–liquid flows. *AIChE Jl* 31, 119–126.

Asali, J.C., Hanratty, T.J. & Andreussi, P. 1985a Interfacial drag and film height for vertical annular flow. *AIChE Jl* 31, 895–902.

Asali, J.C., Leman, G.W. & Hanratty, T.J. 1985b Entrainment measurements and their use in design equations. *Phys-Chem. Hydrodyn.* 6, 207–221.

Assad, A., Jan, C.S., Lopez de Bertodano, M. & Beuss, S. 1998 *Nucl. Eng. Des.* 184, 437–447.

Azzopardi, B.J. 1985 Drop sizes in annular two-phase flow. *Exp. Fluids* 3, 53–59.

Azzopardi, B.J. 1997 Drops in annular two-phase flow. *Int. J. Multiphase Flow* 23, 1–53.

Azzopardi, B.J., Piercey, A. & Jepson, D.M. 1991 Drop size measurements for annular two-phase flow in a vertical tube. *Exp. Fluids* 11, 191–192.

Binder, J.L. 1991 Use of Lagrangian methods to describe particle deposition and distribution in dispersed flows. Ph.D. thesis, University of Illinois.

Combellack, J.H. & Matthews, G.A. 1981 Droplet spectra measurements of fan and cone atomizers using a laser-diffraction technique. *J. Aerosol Sci.* 12, 529–540.

Cousins, L.B. & Hewitt, G.F. 1968 Liquid mass transfer in annular two-phase flow; droplet deposition and liquid entrainment. UKAEA Report AERE-R5657.

Dallman, J.C., Jones, B.J. & Hanratty, T.J. 1979 Interpretation of entrainment measurements in annular gas–liquid flow. In *Two-phase Flow, Momentum, Heat and Mass Transfer*, Vol. 2, ed. F. Durst, G.V. Tsiklauri & N.H Afgan. Washington, DC: Hemisphere, pp. 681–693.

Dykhno, L.A. & Hanratty, T.J. 1996 Use of the interchange model to predict entrainment in vertical annular flow. *Chem. Eng. Comm.* 141–142, 207–235.

Elgobashi, S. & Truesdell, G.C. 1993 On the two-way interaction between homogeneous turbulence and dispersed solid particles I: Turbulence modification. *Phys. Fluids* A5, 1101–1203.

Fore, L.B. & Dukler, A.E. 1995 The distribution of drop size and velocity in gas–liquid annular flow. *Int. J. Multiphase Flow*, 21, 137–149.

Govan, A.H., Hewitt, G.F., Owen, D.G. & Bott, T.R. 1988 An improved CHD modeling code. Paper presented at the Second UK National Conference, Strathclyde University, Glasgow, pp. 33–52.

Hanratty, T.J. & Mito, Y. 2009 A unifying explanation for the damping of turbulence by additives and external forces. *Flow, Turbulence Combustion* 83, 293–303.

Hay, K.J., Liu, Z.C. & Hanratty, T.J. 1996 Relation of deposition rate to drop size when the rate law is non-linear. *Int. J. Multiphase Flow* 22, 829–848.

Henstock, W.H. & Hanratty, T.J. 1976 Interfacial drag and film height in annular flows. *AIChE Jl* 22, 990–1000.

Hewitt, G.F. & Hall-Taylor, N.S. 1970 *Annular Two-Phase Flow.* Oxford: Pergamon Press.

Hurlburt, E.T. & Hanratty, T.J. 2002 Measurement of drop size in horizontal annular flow with the immersion technique. *Exp. Fluids* 32, 692–699.

Jepson, D.M., Azzopardi, B.J. & Whalley, P.B. 1989 The effects of gas properties on drops in annular flow. *Int. J. Multiphase Flow* 15, 327–339.

Lane, W.R. & Green, H.L. 1956 The mechanics of drops and bubbles. In *Surveys in Mechanics*, ed. G.K. Batchelor, Cambridge: Cambridge University Press, pp. 162–215.

Lee, M.M., Hanratty, T.J. & Adrian, R.J. 1989 The interpretation of droplet measurements with a diffusion model. *Int. J. Multiphase Flow* 15, 459–469.

Li, Y., McLaughlin, J.B., Kontomaris, K. & Portelo, L. 2001 Numerical simulation of particle-laden turbulent laden flow. *Phys. Fluids* 13, 2957–2967.

Lopez, J.C.B. & Dukler, A.E. 1985 Droplet sizes, dynamics and deposition in vertical annular flow. U.S. Nuclear Regulatory Commission, Washington DC, Report NUREG/CR-4424.

Lopez de Bertodano, M.A., Jan, C.S. & Beus, S.G. 1997 Annular flow entrainment rate experiment in a small, vertical pipe. *Nucl. Eng. Des.* 178, 61–70.

Lopez de Bertodano, M.A., Jan, C.S. Assad, A. & Beus, S. 1998 Entrainment rate of droplets in ripple-annular regime for small diameter vertical ducts. Paper presented at the Third International Conference on Multiphase Flow, Lyon, France.

Mito, Y. & Hanratty, T.J. 2006 Effect of feedback and inter-particle collisions in an idealized gas–liquid annular flow. *Int. J. Multiphase Flow*, 32, 692–717.

Mugele, R.A. & Evans, H.D. 1951 Droplet size distribution in sprays. *Ind. Eng. Chem.* 43, 1915–1931.

Namie, S. & Ueda, T. 1972 Droplet transfer in two-phase annular-mist flow. *Bull. JSME* 15, 1568–1580.

Okada, O., Fujimatsu, T., Fujita, H. & Nakajima, Y. 1995 Measurement of droplet size distribution in an annular-mist flow in a vertical pipe by immersion liquid method. In *Proceedings of the 2nd International Conference on Multiphase Flow*, Kyoto, April, 3–7, IP2–11.

Pan, L. & Hanratty, T.J. 2002 Correlation of entrainment for annular flow in vertical pipes. *Int. J. Multiphase Flow* 28, 363–384.

Pan, Y. & Banerjee, S. 1996 Numerical simulation of particle interactions with wall turbulence. *Phys. Fluids* 8, 2733–2755.

Pogson, J.T., Roberts, J.H. & Waibler, P.J. 1970 An investigation of the liquid distribution in annular-mist flow. *J. Heat Mass Transfer* 92, 651–658.

Rosin, P. & Rammler, E. 1933 Laws governing the fineness of powdered coal. *J. Inst. Fuel* 7, 29–36.

Schadel, S.A., Leman, G.W., Binder, J.L. & Hanratty, T.J. 1990 Rates of atomization and deposition in vertical annular flow. *Int. J. Multiphase Flow* 16, 363–374.

Semiat, R. & Dukler, A.E. 1981 Simultaneous measurements of size and velocity of bubbles and drops, a new optical technique. *AIChE Jl* 27, 148–159.

Simmons, M.J.H. & Hanratty, T.J. 2001 Droplet size measurement in horizontal gas–liquid flow. *Int. J. Multiphase Flow* 27, 861–883.

Squire, K.D. & Eaton, J.K. 1990 Particle response and turbulent modification in isotropic turbulence. *Phys. Fluids* A2, 1191–1203.

Switherbank, J., Beer, J.M., Taylor, D.S., Abbot, D. & McCreath, G.C. 1976. A laser diagnostic for the measurement of droplet and particle size distributions. *Prog. Astronaut. Aeronaut.* 1, 421–427.

Tatterson, D.F., Dallman, J.C. & Hanratty, T.J. 1977 Drop sizes in gas–liquid flows. *AIChE Jl* 23, 68–76.

Taylor, G.I. 1940 Generation of ripples by wind flowing over a viscous fluid. Reprinted in *The Scientific Papers of Sir Geoffrey Ingram Taylor*, Vol. 3, ed. G.K. Batchelor. Cambridge: Cambridge University Press, 1963, p. 244.

Wallis, G.B. 1969. *One-dimensional Two-Phase Flow*. New York: McGraw-Hill.

Wicks, M. & Dukler, A.E. 1966 In-situ measurements of drop size distribution in two-phase flow: a new method for electrically conducting liquids. Paper presented at the Third International Heat Transfer Conference, Chicago.

Willetts, I. 1987 *Non-aqueous annular two-phase flow*. Ph.D. thesis, University of Oxford.

Woodmansee, D.E. & Hanratty, T.J. 1969 Mechanism for the removal of droplets from a liquid surface by a parallel air flow. *Chem. Eng. Sci.* 24, 299–307.

12 Horizontal annular flow

12.1 Prologue

Horizontal annular flows differ from vertical annular flows in that gravity causes asymmetric distributions of the liquid in the wall layer and of droplets in the gas flow. The understanding of this behavior is a central problem in describing this system. Because of these asymmetries, entrainment can increase much more strongly with increasing gas velocity than is found for vertical flows.

Theoretical analyses of the influence of gravity on the distribution of liquid in the wall film and on the distribution of droplets in the gas phase are reviewed. As with vertical annular flows, entrainment is considered to be a balance between the rate of atomization of the wall film and rate of deposition of droplets. Because of the asymmetric film distribution, the local rate of atomization varies around the pipe circumference. This is treated theoretically by assuming that the local rate is the same as would be observed for vertical annular flow. Gravitational settling contributes directly to deposition so that the rate of deposition is enhanced. Thus, at low gas velocities, entrainment can be much smaller for horizontal annular flows than for vertical annular flows.

Equating the rates of atomization and deposition produces a prediction of the entrainment for fully developed flow, equation (12.82). This equation is used in a few calculations but, because of its complexity, simplified versions are used. Thus, in correlating measurements of entrainment, Pan & Hanratty (2002b) use it to guide the development of an empirical method for correlating data (equation (12.84)). A more attractive approach is to match relations for low gas velocities, where gravity is having a dominant role, and for high gas velocities, where equations developed for vertical annular flow can be used.

The prediction of droplet distribution is an important consideration in predicting film thickness at the top of the pipe. This can be used to predict the transition to annular flow. It is also of importance in understanding the operation of large-diameter pipelines, where it might be desirable to keep the top of the pipe wetted.

An important application of the theory, developed in this chapter, is the understanding of the behavior of petroleum gas–condensate pipelines. These offer a challenge in applying experimental results, obtained mainly for air–water flows, to conditions which are quite different: pipes with diameters of 2–3 feet, high pressures, low surface tensions.

12.2 Drop distribution and local gas velocity

12.2.1 Measurements

Impact tubes have been used by a number of researchers to measure the local drop flux, the local gas velocity, and the slip ratio in vertical flows (Hewitt & Hall-Taylor, 1970; Asali *et al.*, 1985; Williams, 1990).

In the experiments of Asali *et al.*, the tube was constructed with 1/8 inch stainless steel tubing, which entered through the wall and faced upstream. The end of the tube was beveled; its inside diameter was 1.457 mm. Local mass fluxes, G_{LE}, were determined by withdrawing liquid through the tube. The flux was calculated as the mass flow of the liquid divided by the inside area of the tube. Williams (1990) and Asali *et al.* (1985) tested the accuracy of the method by comparing fluxes obtained with tubes that have different diameters.

The total entrainment, E, was calculated from measured droplet fluxes at different locations over traverses from the pipe center to the time-averaged location of the wall layer. Measurements could not be made too close to the layer, so an extrapolation of G_{LE} to the average location of the interface had to be made. This was the chief source of error in determining E (about 6%).

Measurements of the difference in pressure at the opening of the impact tube and a pressure tap on the wall of the pipe were used to calculate the local gas velocity. The following equation, derived by Anderson & Mantzouranis (1960), was used:

$$\Delta p = \frac{1}{2}\alpha\rho_G\bar{u}_G^2 + e^c S\bar{u}_G G_{LE} \tag{12.1}$$

where $\bar{u}_G$ is the local time-mean gas velocity, α is the void fraction and e^c is the capture efficiency. For experiments in annular flow, all drops moving toward the impact tube opening were captured, so $e^c = 1$. The slip ratio, S, in equation (12.1) is the ratio of the drop velocity to the gas velocity. It was assumed to be constant over the entire cross section of the pipe; it was adjusted until the integrated gas-phase velocity profile agreed with the measured gas flow rate into the system. Values of S determined in this way for air–water flow in a 9.53 cm horizontal pipe (Williams, 1990) are presented in Figure 12.1. (Figure 9.8 provides slip ratios of gas entrained in slugs, that are obtained from measurements of void fractions in slugs.)

Concentrations were calculated by dividing the local drop flux by the local drop velocity

$$C = \frac{G_{LE}}{S\bar{u}_G} \tag{12.2}$$

12.2.2 Maps of gas velocity and droplet flux

Impact tubes have been used to map the spatial variation of the gas velocity and droplet flux in horizontal pipes. The impact tube was moved to obtain a traverse along a diameter.

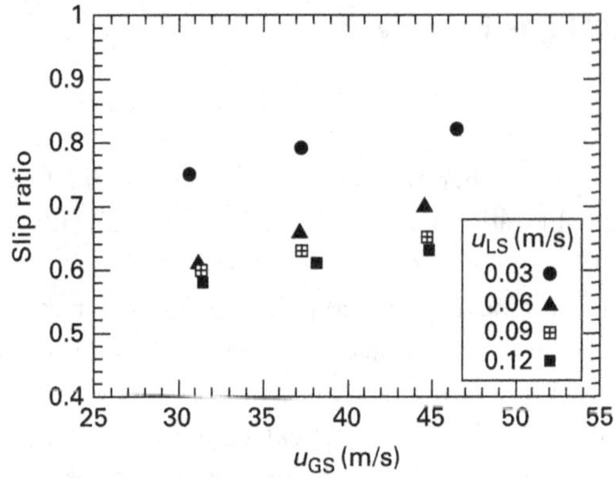

Figure 12.1 Slip ratios for air and water flowing in a vertical annular flow, estimated from (12.1) and the measured gas velocity. Williams, 1990.

Since the distributions of gas velocities and droplet fluxes are asymmetric in a horizontal annular flow, a device was needed to rotate the test section. This provided measurements along diameters which were oriented at different angles to the vertical.

Maps of mean gas velocity for stratified air–water flows with and without atomization were made by Dykhno *et al.* (1994). Measured velocity profiles along the vertical diameter, for conditions at which little or no entrainment was observed, are presented in Figure 12.2. The symbol y represents the distance from the crests of waves at the bottom of the traverse. The lines in Figure 12.2a represent the average location of the interface, h_W. Figure 12.2b normalizes y with $(d_t - h_W)$, rather than with d_t. The maximum velocities for the three runs were located in the lower half of the gas space. If the maximum is the locale of zero shear stress, a force balance indicates that this maximum would be located above the center of the gas space for a simple flow, because of the roughness of the interface. However, the gas flow is not simple.

The layer on the bottom of the pipe is rougher than the layer on the top. This gives rise to a variation of the drag of the gas on the liquid around the circumference of the pipe. Hinze (1967) has demonstrated that this can cause a secondary flow in the gas, which has associated with it an upward-oriented circumferential component of the interfacial stress and a downward flow in the center of the pipe. Darling & McManus (1969) developed an expression for the variation of the circumferential component of the shear stress and argued for its importance in redistributing the wall layer. The results presented in Figure 12.2 have been used to support the proposal by Darling & McManus.

However, when a significant amount of entrainment is present, a different picture emerges. Figure 12.3 shows velocity profiles measured under these circumstances. The maximum of the mean velocity is located above the center of the gas space. Dykhno *et al.* suggest, for these cases, that a secondary flow exists in the gas whereby the flow is downward at the wall and upward at the center. They propose that this occurs because the concentration of drops is higher close to the wall. The force of gravity on

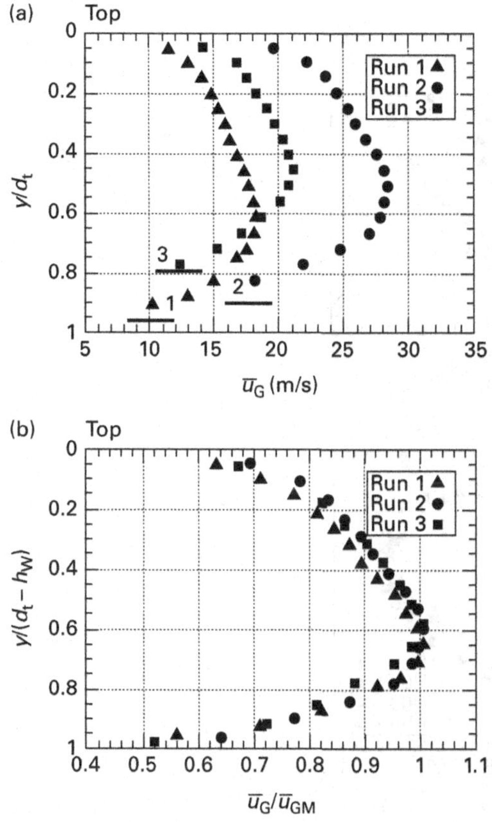

Figure 12.2 Regular and normalized velocity profiles for conditions with no and small amounts of atomization. Run 1, $u_{GS} = 14.4$ m/s, $u_{LS} = 1$ cm/s. Run 2, $u_{GS} = 23.2$ m/s, $u_{LS} = 4$ cm/s. Run 3, $u_{GS} = 17$ m/s, $u_{LS} = 8$ cm/s. Dykhno *et al.*, 1994.

these drops is associated with a downward pull on the gas, which is larger close to the wall.

Figure 12.4 shows contours of the streamwise component of the velocity of the gas (isotachs), determined by Dykhno *et al.* (1994), where the velocities are normalized with the maximum. Run 1 (u_{GS} = 14.4 m/s, u_{LS} = 0.01 m/s) is a stratified flow for which atomization is not occurring. Run 2 (u_{GS} = 23.2 m/s, u_{LS} = 0.04 m/s) and Run 3 (u_{GS} = 17 m/s, u_{LS} = 0.08 m/s) are for stratified flows for which only a small amount of atomization is occurring. These represent situations where the concentration of drops is zero or close to zero.

The use of the method of Prandtl (1927) indicates a secondary pattern in the top of the pipe with downward flow at the center plane and upward flow at the wall. The upward bulges in the bottom part of the gas space are not understood. Dykhno *et al.* (1994) discuss possible interpretations.

Figure 12.5 presents measurements of isotachs for situations in which a significant amount of liquid is entrained in the gas. The dotted lines in Figures 12.4 and 12.5 represent the locations of the tops of the waves.

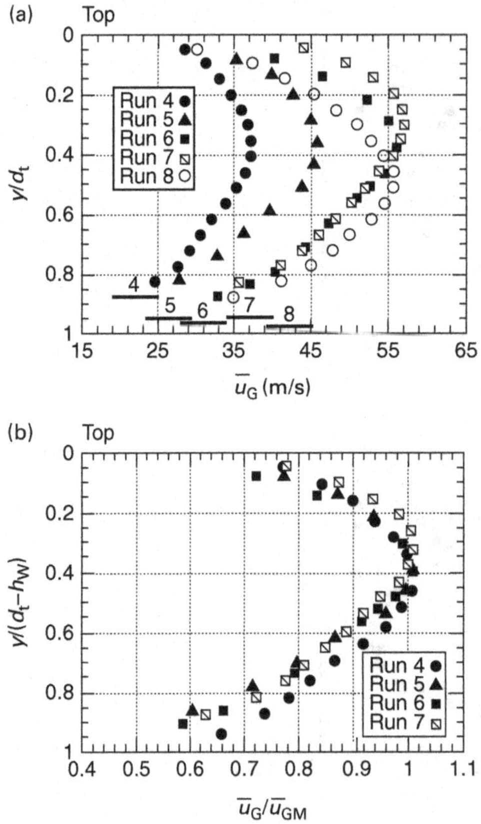

Figure 12.3 Regular and normalized velocity profiles for conditions with a large amount of atomization. Run 4, $u_{GS} = 25.3$ m/s, $u_{LS} = 9$ cm/s. Run 5, $u_{GS} = 37.2$ m/s, $u_{LS} = 6$ cm/s. Run 6, $u_{GS} = 44.5$ m/s, $u_{LS} = 6$ cm/s. Run 7, $u_{GS} = 44.5$ m/s, $u_{LS} = 9$ cm/s. Run 8, $u_{GS} = 43.2$ m/s, $u_{LS} = 3$ cm/s. Dykhno *et al.*, 1994.

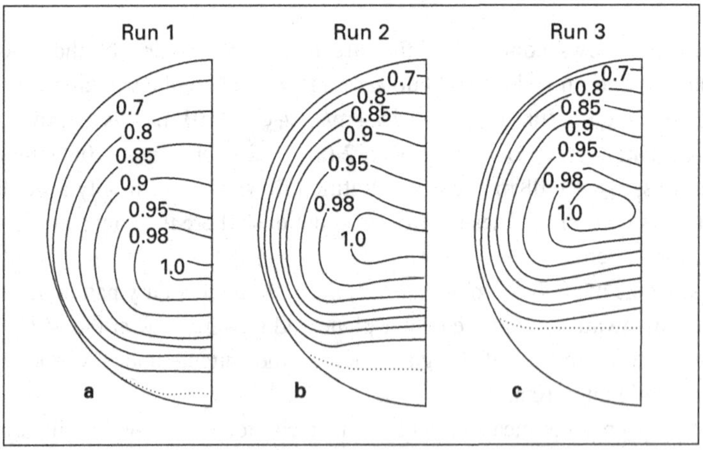

Figure 12.4 Velocity contours (isotachs) for conditions with no or small amounts of atomization, normalized with the maximum velocity. Dykhno *et al.*, 1994.

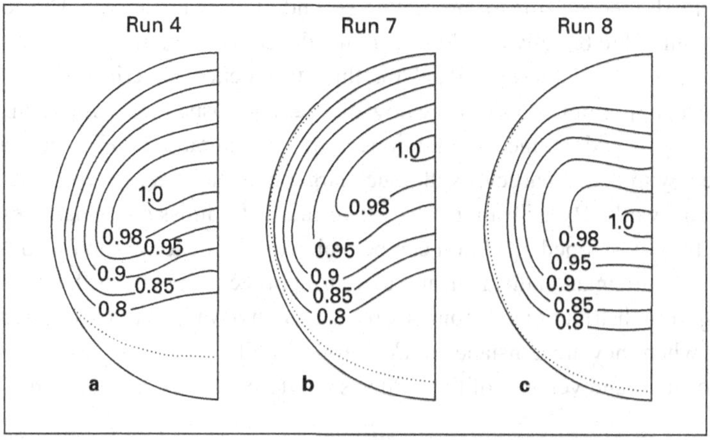

Figure 12.5 Velocity contours (isotachs) for conditions with a large amount of atomization, normalized with the maximum velocity. Dykhno *et al.*, 1994.

Lin & Hanratty (1987) defined annular flow as a situation for which the top of the pipe is wetted with a turbulent film (which need not contain disturbance waves). Run 4 (u_{GS} = 25.3 m/s, u_{LS} = 0.09 m/s) represents a situation close to the transition from stratified to annular flow. Run 7 (u_{GS} = 44.5 m/s, u_{LS} = 0.09 m/s) represents an annular flow for which the wall is wetted with disturbance waves over the bottom two-thirds of the pipe wall. The bulges of the isotachs in the bottom of the pipe for runs 4 and 7 are consistent with the existence of an upward secondary flow in the center plane of the pipe.

Run 8 was performed at the same gas velocity as run 7 but at a much lower liquid flow (u_{GS} = 43.2 m/s, u_{LS} = 0.03 m/s). It represents conditions close to the transition to annular flow. It is a situation for which the maximum is located approximately at the center of the pipe. This seems to be a condition for which entrainment is small, but significant. Secondary flows due to density gradients could be just balancing a secondary flow due to the circumferential variation of the wall roughness.

Measurements of droplet fluxes by Williams (1990) in a 9.53 cm horizontal pipe and by Paras & Karabelas (1991) in a 5.08 cm horizontal pipe show much larger variations in the vertical direction than in the horizontal direction. This prompted the development of a diffusion model, by Paras & Karabelas, for which the concentration is constant in planes perpendicular to the direction of gravity. The determination of $C(r, \theta)$ then requires only a calculation of $C(y)$, where y is the vertical axis.

12.2.3 Stochastic calculations for a 2-D rectangular channel

A simple system to begin a discussion of the diffusion model is the flow of a dilute suspension of spheres with uniform size in an infinitely wide rectangular channel that is horizontal. The particle density is considered to be much larger than the fluid density. The coordinates perpendicular to the wall and in the direction of mean flow are x_2 and x_1.

For this field, mechanisms for promoting secondary flow are absent. The concentration field is assumed to be fully developed, so that the concentration varies only with x_2. The development of a stochastic model for fluid turbulence, described in Section 10.11, provided an opportunity to study the flow of a gas and suspended particles in a horizontal channel over a wide range of conditions (Mito & Hanratty, 2005). Annular flow was simulated by injecting particles with much larger densities than the continuum at the top and bottom walls. By adjusting the relative rates of admission of particles at the two walls, a fully developed condition can be realized. The concentration field is calculated by using a Lagrangian, rather than an Eulerian, approach whereby the field is described as resulting from distributions of point sources at the two walls. Particles are removed from the field when they are a distance of $d_P/2$ from a wall.

The location and velocity of the particles are described by the equations

$$\frac{dx_i}{dt} = v_{Pi} \tag{12.3}$$

$$\frac{dv_{Pi}}{dt} = -\frac{3\rho_f C_D}{4d_P\rho_P}|\vec{u} - \vec{v}_P|(v_{Pi} - u_i) + g_i \tag{12.4}$$

where $\vec{u}$ is the instantaneous fluid velocity and $\vec{v}_P$ is the instantaneous particle velocity. Coordinate x_2, perpendicular to the walls, has a value of zero at the bottom wall. Unlike the approach outlined in Chapter 10, the force of gravity is included in the equation of motion of a particle.

Calculated dimensionless concentration fields are presented for $\tau_P^+ = 3$ in Figure 12.6, and for $\tau_P^+ = 20$ in Figure 12.7, where $\tau_P^+ = \tau_P v^*/\upsilon_f$ is the dimensionless time constant of a particle

$$\tau_P^+ = \frac{4d_P^+(\rho_P/\rho_f)}{3C_D|v_P^+ - u^+|} \tag{12.5}$$

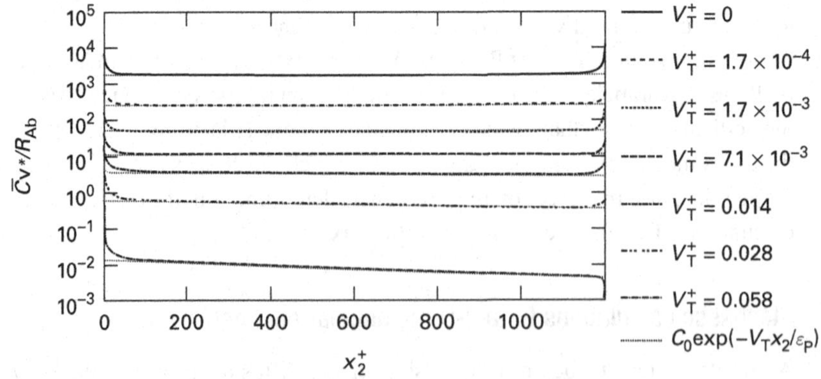

Figure 12.6 Concentration profiles in a channel for $\tau_P^+ = 3$, calculated by Mito & Hanratty (2005), where x_2 is the distance from the bottom wall. Mito & Hanratty, 2005.

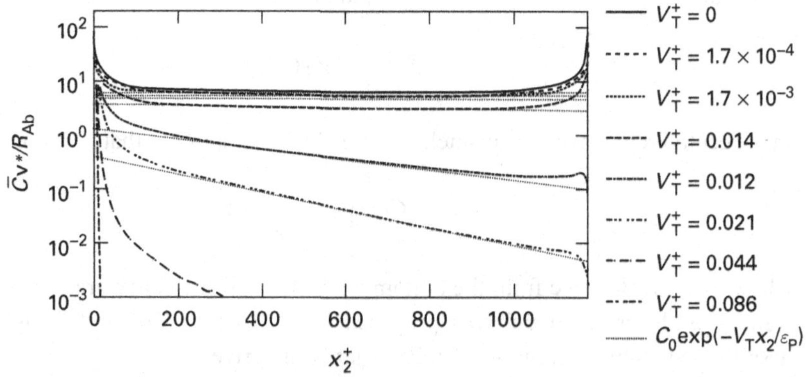

Figure 12.7 Concentration profiles in a channel for $\tau_P^+ = 20$, calculated by Mito & Hanratty (2005), where x_2 is the distance from the bottom wall. Mito & Hanratty, 2005.

For a Stokes law resistance

$$\tau_{PS}^+ = \frac{d_P^{+2}(\rho_P/\rho_f)}{18} \tag{12.6}$$

The dimensionless acceleration of gravity and free-fall velocity are defined as $g^+ = g\upsilon/v^*$ and $V_T^+ = V_T/v^*$, where $V_T^+ = \tau_P^+ g^+$.

Since the concentration field is fully developed, the rates of atomization and deposition at the top and bottom walls satisfy the condition that

$$R_{Ab} = R_{Db} \quad \text{and} \quad R_{At} = R_{Dt} \tag{12.7}$$

Since the net fluxes are zero at all x_2 , the representation of the concentration field in terms of fundamental mechanisms responsible for particle transport and mixing is simplified. Thus, the flux of particles is not dependent on x_2 , so

$$\frac{d}{dx_2}(\overline{Cv_{P2}}) = 0 \tag{12.8}$$

The concentration and the velocity are represented as the sum of an average and a turbulent fluctuation. Thus

$$\overline{Cv_{P2}} = \bar{C}\bar{v}_{P2} + \overline{C^t v^t{}_{P2}} \tag{12.9}$$

Turbophoresis (see Section 10.9) is ignored. The mean velocity in the x_2-direction is then given as

$$\overline{v_{P2}} = -g\tau_P = -V_T \tag{12.10}$$

If a diffusion model is used

$$\overline{C^t v_{P2}^t} = -\varepsilon_P^E \frac{d\bar{C}}{dx_2} \tag{12.11}$$

Substitute (12.10) and (12.11) into (12.9) to get

$$-\varepsilon_{\mathrm{P}}^{\mathrm{E}}\frac{d\overline{C}}{dx_2} - V_{\mathrm{T}}\overline{C} = 0 \tag{12.12}$$

In the central regions of the channel, a reasonable assumption is that $\varepsilon_{\mathrm{P}}^{\mathrm{E}}$ is constant so that

$$\overline{C} = C_0 \exp\left(-\frac{V_{\mathrm{T}}}{\varepsilon_{\mathrm{P}}^{\mathrm{E}}}x_2\right) \tag{12.13}$$

where x_2 is the distance from the bottom wall and C_0 is the concentration at the bottom wall. This diffusion model was proposed by O'Brien (1930) and Rouse (1937) to describe sediment distribution in open channels and rivers.

From equation (8.4), the free-fall velocity is given as

$$V_{\mathrm{T}}^2 = U_{\mathrm{S}}^2 = \frac{4d_{\mathrm{P}}g(\rho_{\mathrm{L}} - \rho_{\mathrm{G}})}{3C_{\mathrm{D}}\rho_{\mathrm{G}}} \tag{12.14}$$

where C_{D} is the drag coefficient. For Stokes law,

$$C_{\mathrm{D}} = \frac{24}{\mathrm{Re_P}} \tag{12.15}$$

For $1.92 < \mathrm{Re_P} < 500$, the drag coefficient can be approximated as

$$C_{\mathrm{D}} = \frac{18.5}{\mathrm{Re_P^{0.6}}} \tag{12.16}$$

where $\mathrm{Re_P} = d_{\mathrm{P}}V_{\mathrm{T}}\rho_{\mathrm{f}}/\mu_{\mathrm{f}}$.

The dispersion of fluid and solid particles downstream of a point source in a pipe at large times can be represented with a turbulent diffusivity. Section 10.7 shows that the diffusivities of fluid and solid particles in such a situation are approximately the same. Experiments give equation (10.22) for pipe flows:

$$\varepsilon_{\mathrm{P}} = 0.074 r_{\mathrm{t}} v^* \tag{12.17}$$

For channel flows, H, the half-height of the channel, would be substituted for r_{t}

$$\varepsilon_{\mathrm{P}} = \xi H v^* \tag{12.18}$$

with $\xi = 0.074$. Since the use of an eddy diffusivity model in an Eulerian formulation is not fundamentally correct (see Hanratty, 1956), ξ must be looked upon as an empirical constant. It need not be the same for different systems.

If (12.18) is substituted into (12.13),

$$\frac{\overline{C}}{C_0} = \exp\left(\frac{-V_{\mathrm{T}}x_2}{\xi H v^*}\right) \tag{12.19}$$

The dotted lines below the curves in Figures 12.6 and 12.7 represent (12.19) with $\xi = 0.094$. (This is close to the value of 0.074, estimated from studies of point source diffusion.)

Note that particles injected at the bottom wall do not reach the top wall for $V_T^+ \geq 0.11, 0.16, 0.31, 0.5$ at $\tau_P^+ = 5, 10, 20, 40$. These results indicate that a limit for the existence of annular flow can be roughly defined as $g^+ \leq 0.012$ for this range of τ_P^+.

In the saltation regime, $g^+ > \sim 0.04$, the trajectories of the particles are not affected by fluid turbulence and the behavior depends strongly on the velocity with which the particles enter their field (see Mito & Hanratty, 2005).

12.2.4 Formulation of the diffusion equation for particles in a pipe

The formulation of the proposal of Paras & Karabelas for a circular pipeline produces a more complicated result than presented in the previous subsection. Concentrations are assumed to be constant in horizontal planes. The location of these planes is given by the distance above the bottom of the pipe. A mass balance over a differential volume considers that the net flux due to free-fall and turbulent mixing equals the net contribution of sources and sinks of drops that result from atomization and deposition at the boundary. Thus

$$d\left[V_T\overline{C}w + w\varepsilon_P\frac{d\overline{C}}{dy}\right] = (R_A - R_D)r_t d\theta \tag{12.20}$$

where r_t is the pipe radius, w is the width of the control volume and θ is measured from the bottom of the pipe. The average of $(R_A - R_D)$ around the pipe circumference is zero since the field is fully developed. However, atomization is larger at the bottom of the pipe because the wall film is thicker at that location. Thus, the wall is a source of drops at the bottom and a sink at the top. The term on the right side of (12.20) causes a net transport of liquid from the bottom of the pipe to the top.

The concentration $\overline{C}$ can be considered as an average over w for a given y. In order to develop a solution of (12.20) either y or θ needs to be eliminated. The width w is given as

$$w = 2r_t \sin\theta \tag{12.21}$$

Furthermore,

$$r_t - y = r_t \cos\theta \tag{12.22}$$

The following relation is obtained from (12.22):

$$d\theta = \frac{\delta y}{r_t \sin\theta} \tag{12.23}$$

so

$$\frac{d\overline{C}}{dy} = \frac{d\overline{C}}{d\theta}\frac{1}{r_t \sin\theta} \tag{12.24}$$

Thus, y or θ can be eliminated from (12.20) by using (12.24) and (12.21).

The implementation of (12.20) introduces complications not present in (12.12). One needs to develop relations for local rates of atomization and deposition around the pipe circumference. The width of the control volume, w, varies with y; this could have an impact on the spatial variation of C.

The integration of (12.20) gives

$$V_{\mathrm{T}}\overline{C} + \varepsilon_{\mathrm{P}}\frac{d\overline{C}}{dy} = \int_0^y \frac{(R_{\mathrm{A}} - R_{\mathrm{D}})}{w}\frac{dy}{\sin\theta} + B \tag{12.25}$$

where B is the integration constant and R_{D}, R_{A}, w are functions of y. The right side can be considered as a flux of droplets due to atomization and deposition. Paras & Karabelas (1991a, b) represented the right side by parameter a, so that

$$V_{\mathrm{T}}\overline{C} + \varepsilon_{\mathrm{P}}\frac{d\overline{C}}{dy} = a \tag{12.26}$$

They assumed, for a given set of conditions, that a may be considered independent of y. They chose $\varepsilon_{\mathrm{P}} = 0.1 r_{\mathrm{t}} v^*$ and developed an empirical correlation for a.

12.2.5 Concentration distributions measured in a pipe

Baik & Hanratty (2003) presented measurements of Paras & Karabelas (1991a, b) for air–water flow in a 5.08 cm pipe and of Williams *et al.* (1996) in a 9.53 cm pipe. They used equation (12.19) to correlate the data by ignoring the effects described by the right side of (12.25) so that $C(y)$ is given by the same equation as for a rectangular channel,

$$\frac{\overline{C}}{C_0} = \exp\left(-\frac{V_{\mathrm{T}}y}{\xi r_{\mathrm{t}} v^*}\right) \tag{12.27}$$

The terminal velocity is given by (12.14). Thus, the application of (12.27) to annular flow requires information on the drop size, d_{P}, in order to evaluate V_{T}.

For dilute concentrations droplet–droplet interactions can be neglected so (12.27) can be combined with a relation for the probability distribution function for the drop size to predict $\overline{C}(y)$. Baik & Hanratty (2003) considered the suspension as having a single drop size, the volume median diameter. The equation chosen to predict $d_{\mathrm{v\mu}}$ was (11.28), that is, $\left(\frac{d_{\mathrm{t}}\rho_{\mathrm{G}}\overline{u}_{\mathrm{G}}^2}{\sigma}\right)^{1/2}\left(\frac{d_{\mathrm{v\mu}}}{d_{\mathrm{t}}}\right) = 0.1072$. An important aspect of this choice is that it predicts that $d_{\mathrm{v\mu}}$ depends on the pipe size.

Concentrations were calculated from measured droplet fluxes by using (12.2), $C = G_{\mathrm{LE}}/S\overline{u}_{\mathrm{G}}$, where $\overline{u}_{\mathrm{G}}$ is the local gas velocity. "To simplify the analysis," Baik & Hanratty substituted u_{GS} for $\overline{u}_{\mathrm{G}}$. From Figure 12.1, a value of $S = 0.7$ was used. A bulk mean concentration is calculated from the entrainment, E, as

$$\langle\overline{C}\rangle = \frac{EW_{\mathrm{L}}}{Q_{\mathrm{G}}S} \tag{12.28}$$

where Q_G is the volume flow of gas and W_L is the mass flow of liquid. The concentration, C_0, that appears in (12.27) is related to $\langle C \rangle$ by integrating (12.27) over the cross-section and assuming a plug flow

$$\langle \overline{C} \rangle = \frac{1}{A} \int C_0 w \exp\left(-\frac{V_T y}{\xi r_t v^*} \right) dy \tag{12.29}$$

The measurements by Williams *et al.* and by Paras & Karabelas are presented in Figures 12.8 and 12.9. The numbers in parentheses are values of V_T/v^*. The lines were calculated using (12.27), where C_0 is given by (12.29) and $\xi = 0.165$. This result is to be compared with $\xi = 0.094$, obtained from a stochastic analysis of flow in a two-dimensional channel presented in Section 12.2.3, and 0.074, obtained from measurements of dispersion from point sources located in the center of a pipe. (See equation (10.22), recognizing that it uses $2r_t$ as a length scale.)

The behavior defined by (12.27) and (12.29) indicates a strong influence of drop size because of its influence on V_T (see equation (12.14)). Previous papers by Pan & Hanratty (2002a, b) used an equation for drop size which indicates no effect of pipe diameter. They needed to use different values of ξ (0.04 and 0.08) to fit the data of Williams in a 9.53 cm pipe and of Paras & Karabelas in a 5.08 cm pipe. The results presented in Figures 12.8 and 12.9 provide support for the finding that drop size depends on pipe diameter.

The rough fits displayed in Figures 12.8 and 12.9 suggest that there is room for improvement both in theory and in measurements. A correlation that includes, directly, the influences of R_A and R_D could introduce new dimensionless groups. A possible improvement can also be obtained by abandoning the assumption of plug flow, so that $\overline{C}$ is defined by using the local velocity, rather than u_{GS}. Some justification for using the simplified approach can be obtained from the study in a 9.53 cm pipe. Williams *et al.*

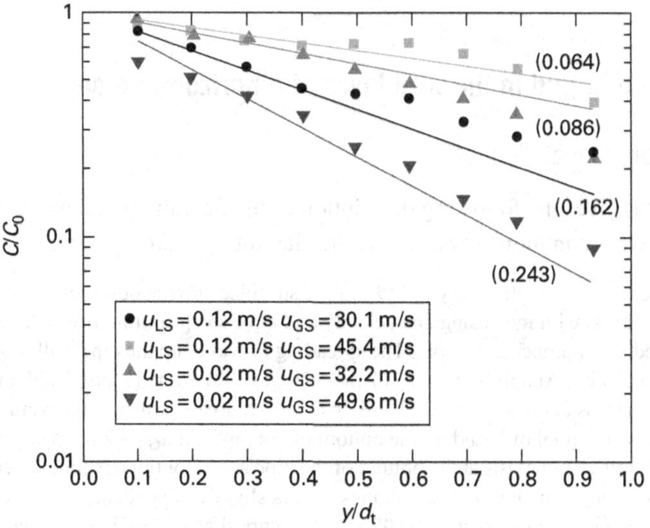

Figure 12.8 Concentration measurements for air–water flow by Williams *et al.* (1996). The solid lines are (12.27), with $\xi = 0.165$. The values of V_T/v^* are shown in parentheses. Williams *et al.*, 1996.

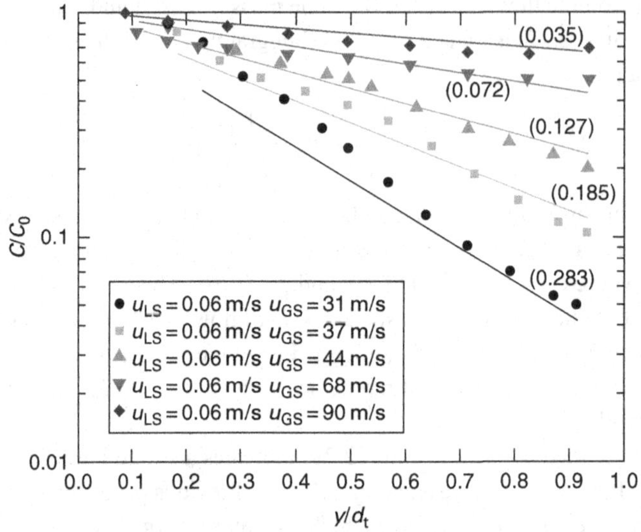

Figure 12.9 Comparison of concentration measurements for air–water flow in a 5.08 cm pipe, by Paras & Karabelas (1991a,b), where y is the distance from the bottom wall. The solid lines are equation (12.27). Paras & Karabelas, 1991b.

(1996) measured the velocity profiles for some of the runs in which droplet fluxes were measured, so that concentration profiles could be calculated with local gas velocities. Baik & Hanratty (2003) reveal that results obtained in this way "were not appreciably different from those obtained by using the plug flow assumption."

The finding that drop size in a horizontal pipe depends on location in large-diameter pipes (see Chapter 11) encourages an adaptation of the analysis so as to consider drop size distribution.

12.3 Distribution of liquid in the wall layer of a horizontal pipe

12.3.1 Description of the wall film

Dykhno (1996) gives the following description of the flow of air and water in a 9.53 cm pipe in an internal communication within the Hanratty research group.

At low superficial gas velocities (u_{GS} = 5–9 m/s), a stratified flow is observed. At 12 m/s, waves increase in amplitude with increasing gas velocity and have steep spilling fronts. They are the roll waves described by a number of authors. The spreading of waves up the pipe wall is observed with increasing gas velocity. At approximately 17 m/s, the gas flow starts to atomize the waves on the liquid interface. Drops deposit on the wall and, in this way, greatly enhance the wetting of the wall at locations above the pool of liquid on the bottom of the pipe. At u_{GS} = 24 m/s, u_{LS} = 0.09 m/s, a thick layer with roll waves exists at the bottom of the pipe and a continuous film covered the upper part of the pipe. The formation of a continuous film is slightly dependent on the liquid velocity. Thus, at a superficial liquid velocity of 0.02 m/s it occurred at a greater superficial gas velocity, 30 m/s. As the height of the wall layer increased, disturbance waves of the type seen in vertical annular flows exist over part of the film. These are haphazard packets of capillary waves which could be considered packets of turbulence. Atomization occurs over these disturbance waves. This enhanced the amount of liquid entrained in the gas. Eventually, at high enough gas velocities, disturbance waves covered the whole circumference.

Dye injected into the film was observed to move downward (due to gravitational pull) when no disturbance waves were present. However, when the dye encountered a disturbance wave, it was observed to disperse explosively, in the circumferential direction, as would be expected in a turbulent patch.

Pan & Hanratty (2002) and Williams *et al.* (1996) developed a framework from which to describe entrainment in horizontal pipes. Let $\theta = 0$ be the bottom of the pipe. The local rate of atomization, R_A, is assumed to be the same as for a symmetric flow (equation (11.40))

$$R_A = \frac{k' A \bar{u}_G^2 (\rho_G \rho_L)^{1/2}}{\sigma} (\Gamma - \Gamma_c) \tag{12.30}$$

where Γ is the local mass flow rate per unit perimeter length and Γ_c is the critical film flow below which atomization does not occur.

The asymmetry of the wall layer is usually characterized by the ratio of the kinetic energy of the gas, represented by $\rho_G u_{GS}^2$, to the work against gravity which is needed to lift liquid from the bottom of the pipe to the top, $\rho_L g d_t$, that is, the dimensionless group $\rho_G^{1/2} u_{GS} / (\rho_L g d_t)^{1/2}$. Thus, the asymmetry can be characterized by a modified Froude number, $(\rho_G/\rho_L)^{1/2}$ Fr, where $\text{Fr} = u_{GS}/(g d_t)^{1/2}$.

Conductance techniques have been used to measure the variation of the height of the wall layer around the pipe circumference. (See the thesis by Williams (1990) for a detailed description of the technique.) Hurlburt & Newell (2000) suggest that the asymmetry of the wall layer can be characterized by the ratio of the height of the layer at the bottom of the pipe to the average height around the circumference, $h_{LB}/\langle h_L \rangle$.

Plots of $h_{LB}/\langle h_L \rangle$ for studies of air–water flow in horizontal pipes with diameters of (a) $d_t = 9.53$ cm, (b) $d_t = 5.08$ cm, (c) $d_t = 2.31$ cm have been constructed by Pan & Hanratty (2002b). These are presented in Figure 12.10. Note that, for large Froude numbers, the measurements are close to what is expected for a symmetric flow, $h_{LB}/\langle h_L \rangle \approx 1$. For small Fr, $h_{LB}/\langle h_L \rangle$ approaches a value close to that suggested by the idealized model of a stratified flow used in Chapter 5. The transition between these two extremes is seen to be sharp for a 9.53 cm pipe. For a ρ_G/ρ_L characteristic of an air–water flow, it occurs at $\text{Fr} \approx 47$.

12.3.2 Asymmetric wall layer model

Henstock & Hanratty (1976, Section 3.3) used the following relation for the height of the wall layer for vertical annular flow (equation (3.46)), where m signifies the average of the height of the wall layer around the circumference:

$$\frac{m}{d_t} = \frac{6.59}{(1 + 1400F)^{1/2}} \tag{12.31}$$

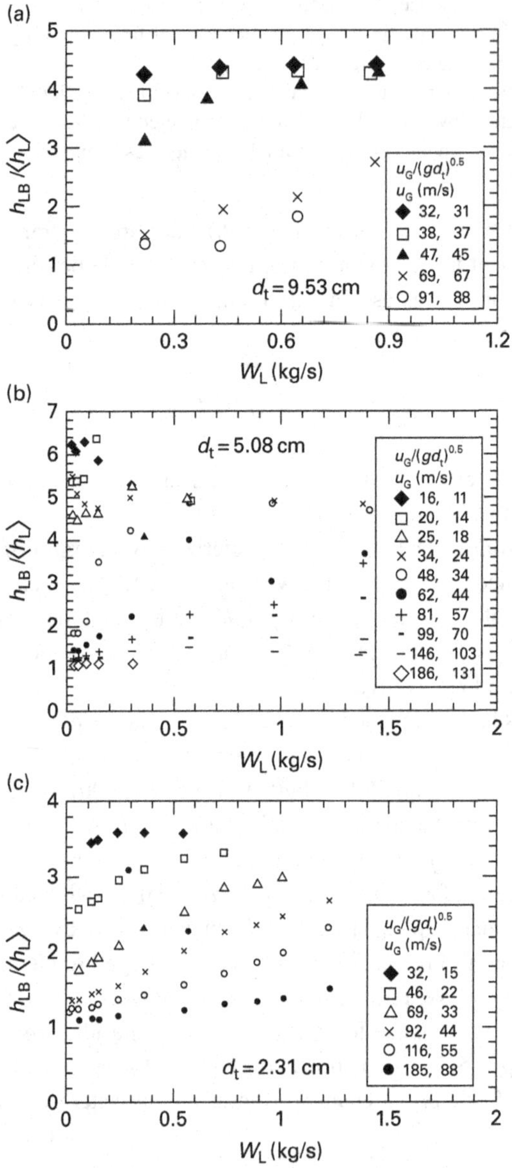

Figure 12.10 Measurements of the ratio of the liquid height at the bottom of the pipe to the average around the circumference: (a) Air–water in a horizontal pipe with $d_t = 9.53$ cm. (b) Air–water with $d_t = 5.08$ cm. (c) Air–water with $d_t = 2.31$ cm. Pan & Hanratty, 2002b.

The flow factor, F, is defined as

$$F = \frac{\gamma(\mathrm{Re}_{\mathrm{LF}})}{\mathrm{Re}_{\mathrm{G}}^{0.9}} \frac{\upsilon_{\mathrm{L}}}{\upsilon_{\mathrm{G}}} \left(\frac{\rho_{\mathrm{L}}}{\rho_{\mathrm{G}}} \right)^{1/2} \tag{12.32}$$

and

$$m^+ = \gamma(\mathrm{Re_{LF}}) = \left[\left(0.707\mathrm{Re}_{\mathrm{LF}}^{0.5}\right)^{2.5} + \left(0.0379\mathrm{Re}_{\mathrm{LF}}^{0.9}\right)^{2.5}\right]^{0.4} \tag{12.33}$$

(equation (3.17)). The friction factor, defined in the equation below

$$\tau_\mathrm{i} = \frac{1}{2}\rho_\mathrm{G}\bar{u}_\mathrm{G}^2 f_\mathrm{i} \tag{12.34}$$

is given as

$$\frac{f_\mathrm{i}}{f_\mathrm{S}} = 1 + 1400F \tag{12.35}$$

(equations (3.41) and (3.44)), where f_S is the friction for a smooth wall

$$f_\mathrm{S} = 0.046\mathrm{Re}_\mathrm{G}^{-0.2} \tag{12.36}$$

These expressions were modified by Dallman (1978) so as to describe horizontal annular flow for large Fr (see Laurinat *et al.*, 1984).

For horizontal flow, it is desired to calculate the average height around the circumference, $m = \langle h_\mathrm{L} \rangle$. The argument is made that, locally, the relation of the film height to the volumetric flow per unit pipe perimeter, $\Gamma(\theta)$, is the same as for vertical annular flow.

Equation (12.33), for vertical annular flow, is an interpolation formula where the first and second terms on the right side represent laminar and turbulent behavior, respectively. For laminar flow (equation (3.7))

$$\frac{\Gamma}{\upsilon_\mathrm{L}} = \frac{h_\mathrm{L}^{+2}}{2} \tag{12.37}$$

where

$$h_\mathrm{L}^+ = \frac{h_\mathrm{L}(\tau_\mathrm{i}/\rho_\mathrm{L})^{1/2}}{\upsilon_\mathrm{L}} \tag{12.38}$$

Integrate (12.37) around the circumference to obtain

$$\frac{\langle \Gamma \rangle}{\upsilon_\mathrm{L}} = \frac{1}{2}\langle h_\mathrm{L}^{+2} \rangle \tag{12.39}$$

where

$$\frac{\langle \Gamma_L \rangle}{\upsilon_\mathrm{L}} = \frac{1}{4}\mathrm{Re_{LF}} \tag{12.40}$$

The dimensionless spatially averaged film thickness is defined as

$$m^+ = \frac{\langle h_\mathrm{L} \rangle \langle \tau_\mathrm{i}/\rho_\mathrm{L} \rangle^{1/2}}{\upsilon_\mathrm{L}} \tag{12.41}$$

Thus, from (12.39) and (12.40),

$$\mathrm{Re}_{\mathrm{LF}} = 2m^{+2}\frac{\langle h_{\mathrm{L}}^{+2}\rangle}{m^{+2}} \tag{12.42}$$

for laminar flow since, for an asymmetric wall layer, the average of the square, $\langle h_{\mathrm{L}}^{2}\rangle$, is not equal to the square of the average, m^2.

For high $\mathrm{Re}_{\mathrm{LF}}$, the local film height is defined by the second term on the right side of (12.33)

$$h_{\mathrm{L}}^{+} = 0.0379\left(\frac{4\Gamma_{\mathrm{L}}}{\upsilon_{\mathrm{L}}}\right)^{0.9} \tag{12.43}$$

or

$$\left(h_{\mathrm{L}}^{+}\right)^{1.1} = (0.0379)^{1.1}\left(\frac{4\Gamma_{\mathrm{L}}}{\upsilon_{\mathrm{L}}}\right) \tag{12.44}$$

The integration of (12.44) around the pipe circumference gives

$$m^{+} = \frac{0.0379\mathrm{Re}_{\mathrm{LF}}^{0.9}}{\langle h_{\mathrm{L}}^{+1.1}\rangle^{0.9}/m^{+}} \tag{12.45}$$

For a horizontal pipe, the asymptotic behaviors for small and large $\mathrm{Re}_{\mathrm{LF}}$ are (12.42) and (12.45). Measurements show that an interpolation formula of the same form as (12.33) can be used to describe m^+. Dallman (1978) approximated $\langle h_{\mathrm{L}}^{+2}\rangle/m^{+2}$ and $\langle h_{\mathrm{L}}^{+1.1}\rangle^{0.9}/m^{+}$, respectively, as $\langle h_{\mathrm{L}}^{2}\rangle^{1/2}/m$ and $\langle h^{1.1}\rangle^{0.9}/m$. Experiments in a 2.54 cm pipe indicate that $\langle h_{\mathrm{L}}^{2}\rangle^{1/2}/m$ approaches 1.4 at small $\mathrm{Re}_{\mathrm{LF}}$. Thus, for $\mathrm{Re}_{\mathrm{LF}} \to 0$,

$$m^{+} = 0.50\mathrm{Re}_{\mathrm{LF}}^{0.5} \tag{12.46}$$

in a horizontal pipe. Dallman (1978) also presented measurements which suggest that a value of $\langle h_{\mathrm{L}}^{+1.1}\rangle^{0.9}/m^{+}$ of 1.38 can be used for the high Reynolds number correction so that for $\mathrm{Re}_{\mathrm{LF}} \to \infty$

$$m^{+} = 0.028\mathrm{Re}_{\mathrm{LF}}^{0.9} \tag{12.47}$$

Laurinat *et al.* (1984) suggested the following interpolation formula to represent measurements of m^+ (shown in Figure 12.11), which provides asymptotic behaviors close to (12.46) and (12.47)

$$m^{+} = \gamma_{\mathrm{H}}(\mathrm{Re}) = \left[\left(0.566\,\mathrm{Re}_{\mathrm{LF}}^{0.5}\right)^{2.5} + \left(0.0303\,\mathrm{Re}_{\mathrm{LF}}^{0.9}\right)^{2.5}\right]^{0.4} \tag{12.48}$$

Note that 0.566 replaces 0.50 in (12.46) and that 0.0303 replaces 0.028 in (12.47). Good agreement between (12.48) and measurements is noted. For comparison, equation (12.33), representing data in vertical pipes, is also presented.

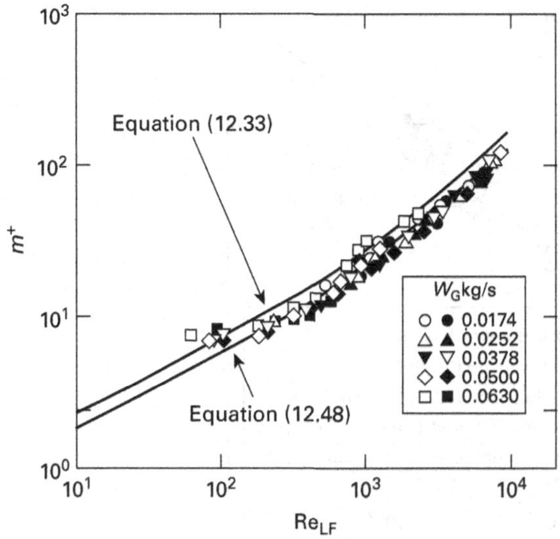

Figure 12.11 Correlation of film height measurements taken in a 2.54 cm pipe. Open points are for $\rho_G = 1.34\ \text{kg/m}^3$. Filled points are for $\rho_G = 2.73\ \text{kg/m}^3$. Laurinat *et al.*, 1984.

A flow factor analogous to F can be defined for horizontal flows

$$F_H = \frac{\gamma_H(\text{Re}_{LF})}{\text{Re}_G^{0.9}}\left(\frac{\upsilon_L}{\upsilon_G}\right)\sqrt{\frac{\rho_L}{\rho_G}} \tag{12.49}$$

A plot of m/d_t against F_H is presented in Figure 12.12. The data are represented quite well by

$$\frac{m}{d_t} = \frac{6.59F_H}{\left[(2.3)^{0.5} + (90F_H)^{0.9}\right]^{0.2}} \tag{12.50}$$

(See Laurinat *et al.* (1984) for the development of equation (12.50).)

A correlation for the frictional pressure loss

$$\left|\frac{dp}{dx}\right|_f = \frac{2f_i}{d_t}\rho_G u_{GS}^2 \tag{12.51}$$

is also given by Laurinat. Measurements of f_i/f_S are plotted against m/d_t in Figure 12.13, where f_S is the friction factor for a smooth wall, for air–water flow in a 2.54 cm horizontal pipe. For a given m/d_t, gas velocity increases with increasing W_L. Such a plot has been successful for vertical annular flows. At high gas velocities (greater than 30 m/s), where the wall layer is described as an asymmetric film, the friction factor approaches an asymptotic behavior. However, it fails to do this at small gas velocities, where the wall layer is highly asymmetric, resembling a stratified flow with an agitated surface. The friction factor relation at low gas velocities is a strong function of liquid Reynolds number. This region is best treated as a stratified-annular flow (see Section 12.5).

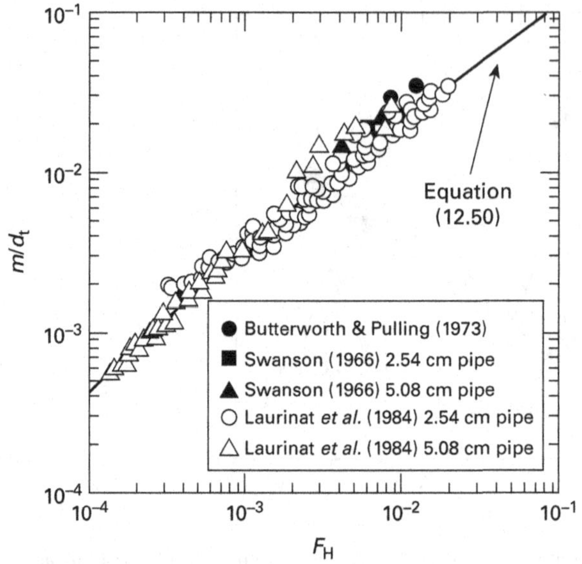

Figure 12.12 Film holdup correlation. Laurinat *et al.*, 1984.

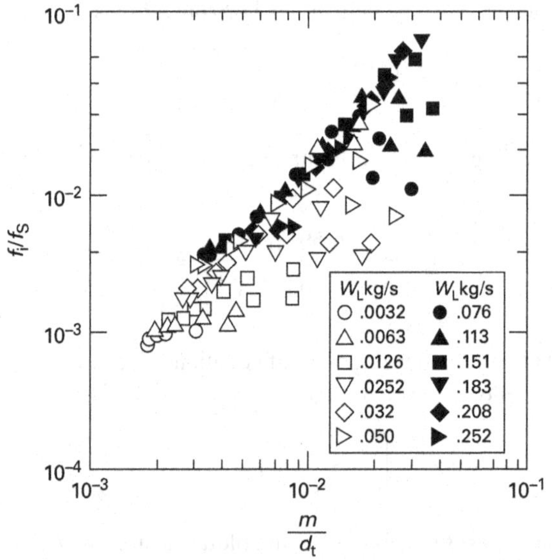

Figure 12.13 Comparisons of the friction factor ratio taken in a 2.54 cm pipe at 500 diameters from the entry. Laurinat *et al.*, 1984.

12.3.3 Distribution of the liquid in an asymmetric wall layer

A number of investigators have considered the modification of the behavior of the wall layer by gravitational draining (McManus, 1961; Dallman, 1978; Laurinat, 1982; Pletcher & McManus, 1965; Lin *et al.*, 1985; Fukano & Ousaka, 1989;

Butterworth, 1972; Butterworth & Pulling, 1973; Anderson & Russell, 1970; Hutchinson *et al.*, 1974; Fisher & Pearce, 1979; James & Burns, 1979; Wilkes *et al.*, 1980; Swanson, 1966). Several mechanisms for redistributing the wall layer have been proposed. These are incorporated into a single equation by Laurinat *et al.* (1985).

In formulating the problem, it is assumed that the wall layer is thin enough compared with the pipe radius for the momentum and mass balance equations to be written in Cartesian, rather than polar, coordinates. The x-coordinate is in the circumferential direction; the y-coordinate is the distance from the wall; the z-coordinate is in the flow direction. The goal is to predict $\overline{h}_{\mathrm{L}}(\theta)$ from mass and momentum balances in the x-direction, where $x = r_{\mathrm{t}}\theta$ and r_{t} is the pipe radius.

The instantaneous volumetric flow in the x-direction, per unit circumferential length, is given as $\Gamma_x = \langle u_{\mathrm{L}x} \rangle h_{\mathrm{L}}$.

The analytic framework used by Laurinat *et al.* (1985) is analogous to the formulation for droplet concentration. Since $u_{\mathrm{L}x} = \overline{u}_{\mathrm{L}x} + u^{\mathrm{t}}_{\mathrm{L}x}$ and $h_{\mathrm{L}} = \overline{h}_{\mathrm{L}} + h^{\mathrm{t}}_{\mathrm{L}}$

$$\overline{\Gamma}_x = \langle \overline{u}_{\mathrm{L}x} \rangle \overline{h}_{\mathrm{L}} + \langle \overline{u^{\mathrm{t}}_{\mathrm{L}x} h^{\mathrm{t}}_{\mathrm{L}}} \rangle \tag{12.52}$$

The inclusion of the turbulence term is motivated by visual observation of the dispersion of dye by the disturbance waves (Section 12.3.1). This term can be represented by a diffusivity model so that

$$\overline{u^{\mathrm{t}}_{\mathrm{L}x} h^{\mathrm{t}}_{\mathrm{L}}} = -\varepsilon_{\mathrm{h}} \frac{d\overline{h}_{\mathrm{L}}}{dx} \tag{12.53}$$

Deposition increases the volumetric flow in the film. Atomization decreases the volumetric flow, so that atomization and deposition contribute to a local change of the flow in the film.

$$\frac{d\overline{\Gamma}_x}{dx} = \frac{R_{\mathrm{D}}}{\rho_{\mathrm{L}}} - \frac{R_{\mathrm{A}}}{\rho_{\mathrm{L}}} \tag{12.54}$$

$$\overline{\Gamma}_x = \int_0^x \left(\frac{R_{\mathrm{D}} - R_{\mathrm{A}}}{\rho_{\mathrm{L}}} \right) dx \tag{12.55}$$

At the top of the pipe, $x = \pi r_{\mathrm{t}}$. Since a fully developed flow is assumed

$$\int_0^{\pi r_{\mathrm{t}}} \left(\frac{R_{\mathrm{D}} - R_{\mathrm{A}}}{\rho_{\mathrm{L}}} \right) dx = 0 \tag{12.56}$$

Thus, (12.55) describes a redistribution of the liquid flowing in the wall layer. Conservation of mass gives

$$\int_0^x \left(\frac{R_{\mathrm{A}} - R_{\mathrm{D}}}{\rho_{\mathrm{L}}} \right) dx = \langle \overline{u}_{\mathrm{L}x} \rangle \overline{h}_{\mathrm{L}} - \varepsilon_{\mathrm{h}} \frac{d\overline{h}_{\mathrm{L}}}{dx} \tag{12.57}$$

where $\langle \overline{u}_{\mathrm{L}x} \rangle$ is the spatially averaged value of $\overline{u}_{\mathrm{L}x}$ at a given x.

A momentum balance in the x-direction is used to define $\langle \overline{u}_{\mathrm{L}x} \rangle$. Surface tension effects are ignored. From equation 12 in Laurinat *et al.* (1985)

$$\frac{\partial \tau_{xy}}{\partial y} + \frac{\partial \tau_{xx}}{\partial x} - \rho_{\mathrm{L}} g \sin\theta - \rho_{\mathrm{L}} g \cos\theta \frac{dh_{\mathrm{L}}}{dx} = 0 \tag{12.58}$$

where τ_{xy} and τ_{xx} are local shear and normal stresses in the liquid, and

$$\overline{\tau}_{xx} = -\rho_{\mathrm{L}} \overline{u^{\mathrm{t}2}_{\mathrm{L}x}} \tag{12.59}$$

Integrate (12.58) between y and $\overline{h}_{\mathrm{L}}$. The variation of $\overline{u^{\mathrm{t}2}_{\mathrm{L}x}}$ with y is ignored. (A constant value is used.) Then

$$\left(\overline{h}_{\mathrm{L}} - y\right) \frac{d\tau_{xx}}{dx} + \overline{\tau}_{\mathrm{i}x} - \overline{\tau}_{xy}(y) - \left(\overline{h}_{\mathrm{L}} - y\right)\rho_{\mathrm{L}} g \left(\sin\theta + \cos\theta \frac{d\overline{h}_{\mathrm{L}}}{dx} \right) = 0 \tag{12.60}$$

where $\overline{\tau}_{\mathrm{i}x}$ is the stress imposed at the interface in the x-direction. Take

$$\frac{\overline{\tau}_{xy}}{\rho_{\mathrm{L}}} = \left(\frac{\mu_{\mathrm{L}}}{\rho_{\mathrm{L}}} + \upsilon^{\mathrm{t}} \right) \frac{d\overline{u}_{\mathrm{L}x}}{dy} \tag{12.61}$$

where υ^{t} is the turbulent eddy viscosity. Substitute (12.61) into (12.60). Integrate once to obtain $\overline{u}_{\mathrm{L}x}(y)$. Integrate again to obtain $\langle \overline{u}_{\mathrm{L}x} \rangle$.

$$\langle \overline{u}_{\mathrm{L}x} \rangle \overline{h}_{\mathrm{L}} = I_1 \left(\frac{\overline{\tau}_{\mathrm{i}x}}{\rho_{\mathrm{L}}} \right) + \frac{I_2}{\rho_{\mathrm{L}}} \left(\frac{d\overline{\tau}_{xx}}{dx} \right) + I_2 g \left(\sin\theta + \cos\theta \frac{d\overline{h}_{\mathrm{L}}}{dx} \right) \tag{12.62}$$

$$I_1^{(N)} = \int_0^{h_1} \int_0^{y_2} \frac{N}{\left[(\mu_{\mathrm{L}}/\rho_{\mathrm{L}}) + \upsilon^{\mathrm{t}}\right]} dy_1 dy_2 \tag{12.63}$$

$$I_2^{(N)} = \int_0^{h_1} \int_0^{y_2} \frac{N(h_{\mathrm{L}} - y)}{\left[(\mu_{\mathrm{L}}/\rho_{\mathrm{L}}) + \upsilon^{\mathrm{t}}\right]} dy_1 dy_2 \tag{12.64}$$

Substitute (12.62) into (12.57) to obtain

$$\underset{(1)}{\varepsilon \frac{d\overline{h}_{\mathrm{L}}}{dx}} + \underset{(2)}{\int_0^x \frac{(R_{\mathrm{A}} - R_{\mathrm{D}})}{\rho_{\mathrm{L}}}} = \underset{(3)}{I_1 \left(\frac{\overline{\tau}_{\mathrm{i}x}}{\rho_{\mathrm{L}}} \right)} + \underset{(4)}{\frac{I_2}{\rho_{\mathrm{L}}} \left(\frac{d\overline{\tau}_{xx}}{dx} \right)} + \underset{(5)}{I_2 g} \left(\sin\theta + \cos\theta \frac{d\overline{h}_{\mathrm{L}}}{dx} \right) \tag{12.65}$$

This equation defines $\overline{h}_{\mathrm{L}}(\theta)$. The left side represents mixing due to turbulence and to atomization/deposition. The right side represents contributions to $\langle \overline{u}_{\mathrm{L}x} \rangle$ due to forces acting on the film. Term (3) considers forces due to gas-phase drag on the interface in the x-direction. These could result from secondary flow in the gas and/or an orientation of the interfacial waves to the direction of mean flow. Term (4) represents the influence of gradients of turbulent normal stresses in the circumferential direction. Term (5) represents the contribution of the force of gravity on the layer. The cosine term in (5) is important only in the neighborhood of the bottom of the pipe where $\sin\theta = 0$. Equation (12.65) has not been fully exploited.

A number of investigators have used only (2) and (5) (Anderson & Russell, 1970; Hutchinson *et al.*, 1974; Fisher & Pearce, 1979; James & Burns, 1979; Wilkes *et al.*, 1980). These and other papers show that a deposition-atomization mechanism, alone, cannot account for the variation of film height. Darling & McManus (1969) used (3) and (5) to calculate $\overline{h}_L(\theta)$. Laurinat *et al.* (1985) used (2), (3), (4) and (5). Their motivation was to use term (4) to describe the influence of turbulent velocity fluctuations. They concluded that this approach was not adequate.

This led Pan & Hanratty to introduce term (1) to represent turbulent mixing in the wall layer. For cases where disturbance waves cover the whole circumference, they used only terms (1) and (5) in equation (12.65):

$$\varepsilon_h \frac{dh_L}{dx} = I_2 g\left(\sin\theta + \cos\theta \frac{dh_L}{dx} \right) \tag{12.66}$$

The turbulent mixing coefficient, ε_h, can be pictured as the product of a velocity and a length. It is reasonable to assume that the velocity scales with the friction velocity. The mixing in the disturbance waves can be pictured as a large-scale phenomenon. Pan & Hanratty, therefore, modeled the turbulent diffusivity as

$$\varepsilon_h = \xi_h r_t v^* \tag{12.67}$$

where $v$is the friction velocity, r_t is the pipe radius and ξ_h is a dimensionless constant. Equation (12.66) can be made dimensionless using

$$r_t^+ = \frac{r_t v^*}{\upsilon} I_2^+ = I_2 \frac{v^{*3}}{\upsilon_2} h_L^+ = \frac{h_L v^*}{\upsilon} g^+ = g\frac{\upsilon}{v^{*3}} \tag{12.68}$$

The following relation is obtained from (12.66)

$$\xi_h \frac{dh_L^+}{d\theta} = g^+ I_2^+ \left(\sin\theta + \cos\theta \frac{dh_L^+}{dx^+} \right) \tag{12.69}$$

Pan & Dykhno (2001, private communication) compared (12.69) with measurements (in unpublished calculations). The results are presented in Figure 12.14 and in Figure 12.15. From (12.64), it is seen that I_2 varies with $h_L(\theta)$ for given flow conditions. One expects that $h_L \propto r_t$ or that $h_L^+ \propto r_t^+$. Thus, $I_2^+ \propto r_t^+$. Plots of I_2^+ calculated from measurements by Dallman in a 2.54 cm pipe (Run 67, $u_{GS} = 78.18$ m/s, $u_{LS} = 0.0572$ m/s, $r_t^+ = 1626$, $\rho_G = 1.538$ kg/m^3) and by Laurinat in a 5.08 cm pipe (Run 46, $u_{GS} = 56.8$ m/s, $u_{LS} = 0.0752$ m/s, $r_t^+ = 2775$, $\rho_G = 2.039$ kg/m^3 are shown in Figure 12.14. Here, θ is measured from the top of the pipe. It is seen that $I_2^+(\theta = 180°)$ minus $I_2^+(\theta = 0°)$ are, respectively, 2600 and 1200 for the 2.54 cm and 5.08 cm pipes. The change in I_2^+ scales linearly with r_t^+.

Thus, (12.69) can be written as

$$\frac{dh_L^+}{d\theta} = \frac{Cg^+ I_2^+ (\sin\theta)}{r_t^+} \tag{12.70}$$

with C = constant, if the term containing cos θ is ignored. This solution should be matched to a solution valid at the top part of the channel, $\theta = 0^0$. For Run 67 of Dallman

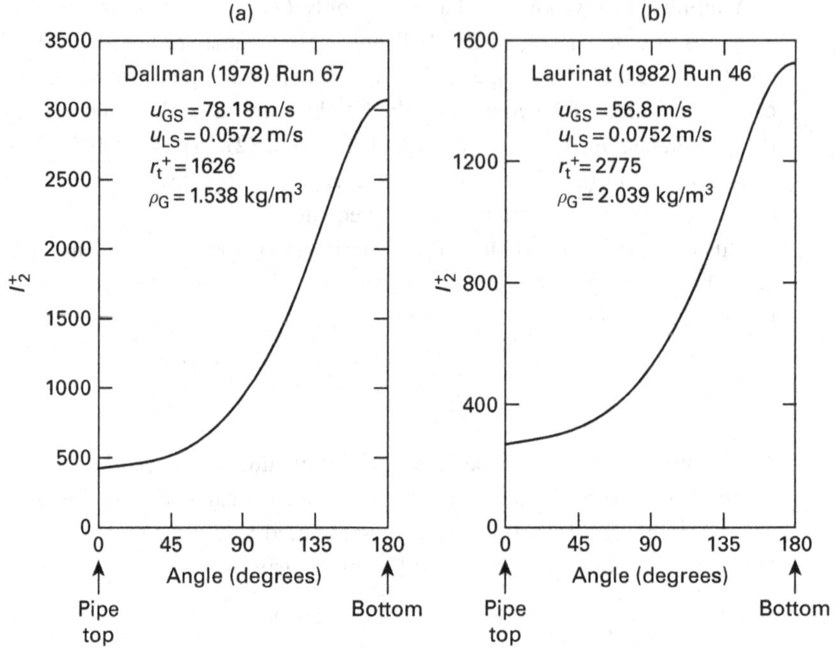

Figure 12.14 Angle θ is measured from the top of the pipe. (a) Calculated I_2^+ (equation (12.64)) versus θ for Dallman Run 67 in a 2.54 cm pipe. (b) Calculated I_2^+ versus θ for Laurinat Run 46 in a 5.08 cm pipe. L. Pan & L.A. Dykhno, Private communication.

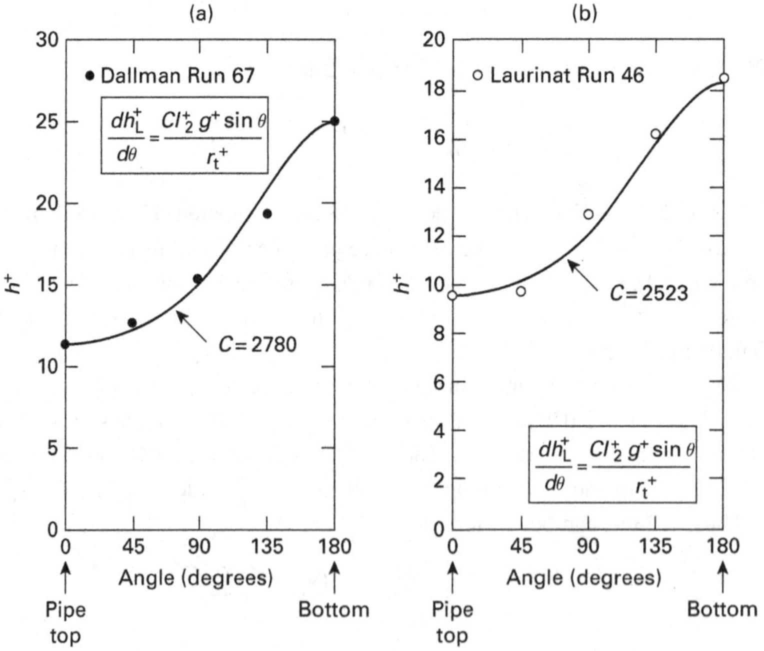

Figure 12.15 (a) Comparison of equation (12.70) with measurements by Dallman (1978) Run 67, $C = 2780$. (b) Comparison of (12.70) with measurements by Laurinat (1982) Run 46, $C = 2523$. Private communication.

and Run 46 of Laurinat, this can be accomplished by using the measurements at $\theta = 0°$ as the boundary condition for (12.70). A comparison of (12.70) with measurements is shown in Figure 12.15 with $C = 2523$ for Laurinat Run 46 and $C = 2780$ for Dallman Run 67.

The agreement of equation (12.66) with data supports the use of term (1) in equation (12.65), to represent the turbulent mixing in the film when disturbance waves are present. However, this is a limited test. One cannot conclude that only (1) and (5) are needed in general calculations. For example, it might be necessary also to consider using either or both of terms (2) and (3) in equation (12.65) to capture the behavior at the top of the pipe.

12.4 Entrainment in horizontal pipes

12.4.1 Entrainment theory

Entrainment for fully developed flow in vertical pipes is discussed in Section 11.7 by considering it to result from a balance between the rates of atomization of the wall layer and the rate of deposition. The enhancement of deposition in horizontal pipes because of gravitational settling can result in much smaller entrainment than would be experienced in vertical flows. Thus, the approach of using vertical flow equations to describe entrainment in horizontal pipes is not correct (see, for example, Dallman *et al.*, 1984).

The assumption is made that the local rate of atomization is given by (11.38), developed for symmetric flow:

$$R_{\mathrm{A}} = \frac{k'_{\mathrm{A}} \bar{u}_{\mathrm{G}}^2 (\rho_{\mathrm{L}} \rho_{\mathrm{G}})^{1/2}}{\sigma} (\Gamma - \Gamma_{\mathrm{C}}) \tag{12.71}$$

where Γ is the local mass flow rate in the wall layer per unit length and Γ_c is the critical film flow below which atomization does not occur. Take $\theta = 0$ as the bottom of the pipe. The integration of (12.71) around the perimeter gives

$$\langle R_{\mathrm{A}} \rangle = \frac{1}{\pi d_{\mathrm{t}}} \int_0^{\pi} d_{\mathrm{t}} R_{\mathrm{A}} d\theta \tag{12.72}$$

$$\langle R_{\mathrm{A}} \rangle = \frac{k'_{\mathrm{A}} \bar{u}_{\mathrm{G}}^2 (\rho_{\mathrm{L}} \rho_{\mathrm{G}})}{\sigma} \left(\frac{W_{\mathrm{LF}}}{P} - \Gamma_{\mathrm{C}}^* \right) \tag{12.73}$$

$$\Gamma_{\mathrm{c}}^* = \Gamma_{\mathrm{c}} - \frac{1}{\pi} \int_{\theta_{\mathrm{c}}}^{\pi} (\Gamma_{\mathrm{c}} - \Gamma) d\theta \tag{12.74}$$

where θ_c is the angular location at which $\Gamma = \Gamma_c$. Thus $\langle R_A \rangle$ is given by (12.71) with Γ_c replaced by Γ_c^*. If $\Gamma > \Gamma_c$ around the whole circumference, $\theta_c = \pi$ and $\Gamma_c^* = \Gamma_c$. If $\Gamma < \Gamma_c$ over a portion of the perimeter, from θ_c to π, $\Gamma_c^* < \Gamma_c$.

The local rate of deposition can be represented by (11.31):

$$R_{\mathrm{D}} = V_{\mathrm{W}} C_{\mathrm{W}} = V_{\mathrm{W}} C_{\mathrm{B}} \frac{C_{\mathrm{W}}}{C_{\mathrm{B}}} \tag{12.75}$$

where V_W is the average velocity of the particles striking the wall. For vertical flows, V_W is related to the component of the turbulent velocity fluctuations of the particles normal to the wall. If these are given by a Gaussian distribution with a mean-square value of $\overline{v_P^2}$,

$$p(x) = \frac{1}{(2\pi)^{1/2}\sigma_P} \exp\left[-\frac{(x-\mu)^2}{2\sigma_P^2}\right] \tag{12.76}$$

where μ is the mean velocity, $x = \mu + v_P$ and $\sigma_P = \left(\overline{V_P^2}\right)^{1/2}$. For a vertical flow, $\mu = 0$; for a horizontal flow $\mu = V_T \cos\theta$, where V_T is the free-fall velocity. Thus, at the top of the pipe $\cos\theta = -1$ and $\mu = -V_T$.

The velocity, V_W, varies around the circumference, so

$$V_W = \int_0^\infty x p(x) dx \tag{12.77}$$

By substituting (12.76) into (12.77), the following expression was obtained by Pan & Hanratty (2002a, b):

$$V_W = \int_0^\infty \frac{(x-\mu)}{\sqrt{2\pi}\sigma_P} \exp\left[-\frac{(x-\mu)^2}{2\sigma_P^2}\right] dx + \int_0^\infty \frac{\mu}{\sqrt{2\pi}\sigma_P} \exp\left[\frac{-(x-\mu)^2}{2\sigma_P^2}\right] dx \tag{12.78}$$

The integral in (12.78) considers only depositing particles, that is, positive values of $x = \mu + v_P$. (An integration from $-\infty$ to $+\infty$ would give $V_W = \mu$.) When $(V_T / \sigma_P) \to 0$, equation (12.78) gives

$$V_W = \left(\frac{\sigma_P^2}{2\pi}\right)^{0.5} \tag{12.79}$$

so that only particle turbulence is responsible for deposition. For $(V_T + \sigma_P) \to \infty$,

$$V_W \to V_T \cos\theta \tag{12.80}$$

so that gravitational settling is controlling deposition. Pan & Hanratty (2002a, b) present the integration of (12.78) shown in Figure 12.16. It can be seen that deposition is enhanced over what is predicted by (12.79) at the bottom of the pipe, where $V_T \cos\theta$ is positive, and diminished below what is predicted by (12.79) at the top, where $V_T \cos\theta$ is negative.

The averaging of R_D (defined by equation (12.75)) around the circumference gives

$$\langle R_D \rangle = \left\langle V_W \frac{C_W}{C_B} \right\rangle \frac{W_{LE}}{Q_G S} \tag{12.81}$$

Under fully developed conditions, $\langle R_D \rangle = \langle R_A \rangle$. Equations (12.73) and (12.81) give

$$\frac{(E/E_M)}{1-(E/E_M)} = \frac{k'_A d_t \bar{u}_G^3 S (\rho_L \rho_G)^{1/2}}{4\langle k_D (C_W/C_B)\rangle \sigma} \tag{12.82}$$

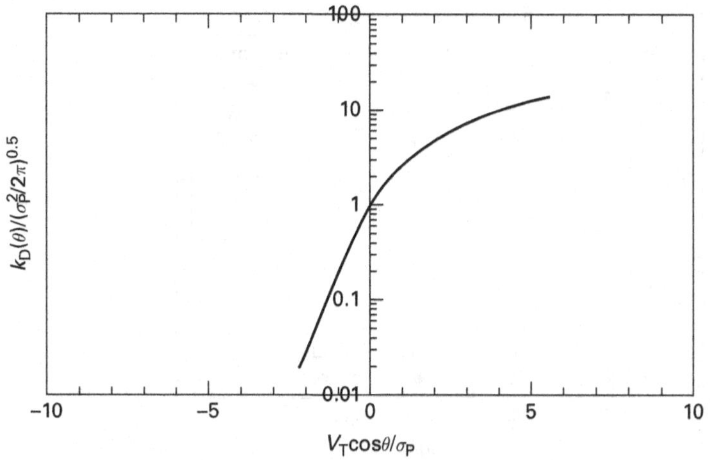

Figure 12.16 Calculated variation of k_D with θ, equation (12.78). The angle θ is zero at the bottom of the pipe. Pan & Hanratty, 2002b.

where

$$E_M = 1 - \frac{\pi d_t \Gamma_c^*}{W_L} \tag{12.83}$$

The critical flow, Γ_c, is given by Andreussi *et al.* (1985). One needs to develop a relation for $\langle k_D(C_W/C_B)\rangle$ to use (12.82).

12.4.2 Entrainment measurements

Measurements of entrainment by Williams (1986, 1990), Laurinat (1982), Dallman (1978) and by Paras & Karabelas (1991a, b) are discussed by Pan & Hanratty (2002b). They are for the air–water system and cover pipe diameters of 2.31–9.53 cm, gas velocities of 11–131 m/s, gas densities of 1.26–2.75 kg/m^3. These experiments were carried out at low enough gas velocities for gravitational settling to play a major role. A modified version of (12.82), which substitutes the terminal velocity for $\langle k_D(C_W/C_B)\rangle$ and takes S as a constant, was used by Pan & Hanratty (2002a, b) to correlate measurements at low gas velocities:

$$\frac{(E/E_M)}{(1 - E/E_M)} = A_2 \left(\frac{d_t \bar{u}_G^3 \rho_L^{1/2} \rho_G^{1/2}}{\sigma V_T} \right) \tag{12.84}$$

where A_2 is a dimensionless constant. At very high gas velocities, where particle turbulence would be the dominant mechanism for particle deposition, the equations for vertical flows can be used.

Measurements of entrainment for a horizontal 9.53 cm pipe are presented in Figure 12.17. As already noted for vertical annular flow, E/E_M is much more sensitive to changes in gas flow than to changes in liquid flow. In calculating E_M, the term Γ_c^* was set equal to Γ_c, which was calculated with the equation proposed by Andreussi *et al.* (1985) for vertical annular

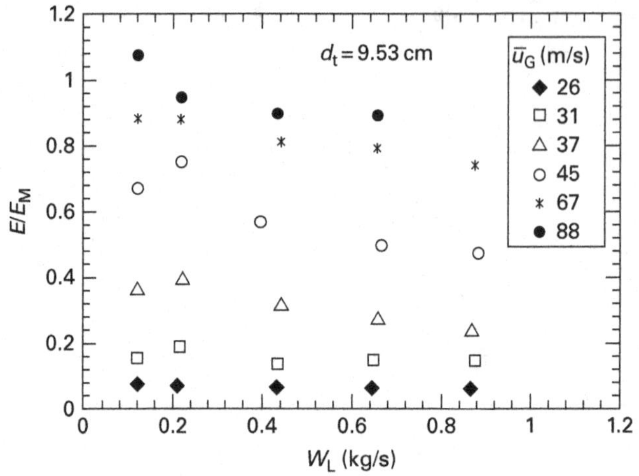

Figure 12.17 Measurements of E/E_M obtained by Williams (1986, 1990) for air and water flowing in a 9.53 cm pipe. Pan & Hanratty, 2002b.

flows. The extrapolation of these results to $E = 0$ gives a rough estimate of a critical gas velocity of $U_G \approx 21$ m/s. The measurements of E/E_M in Figure 12.17 show a slight decrease with increasing W_L at large liquid flows. This reflects inaccuracies associated with the use of linear relations for R_A and R_D. (See discussion by Pan & Hanratty (2002a, b).)

Results obtained in 9.53 cm and 2.31 cm pipes are presented in Figure 12.18 a, b. These are plots of E, rather than E/E_M. A rough estimate of a critical gas velocity of 15 m/s is obtained for the 2.31 cm pipe. Lines representing W_{LFC} are calculated with the equation proposed for vertical annular flows. Justification for the substitution of Γ_C for Γ_C^* is inconclusive.

Equation (12.84) is tested in Figure 12.19. A value of $A_2 = 9 \times 10^{-8}$ was selected to fit the data. Each point represents an average of measurements for different W_L at a fixed gas velocity. The terminal velocity is given by

$$V_T^2 = \frac{4 d_P g \rho_L}{3 C_D \rho_G} \tag{12.85}$$

with $C_D = 24/\text{Re}_P$ for $\text{Re}_P < 1$ and $C_D = 18.5/\text{Re}_P^{0.6}$ for intermediate Re_P (see Section 8.4, that is, Bird *et al.*, 1960). If Stokes law is used to calculate C_D, theory predicts that $(E/E_M)/1 - (E/E_M)$ varies as $\bar{u}_G^5$. An intermediate behavior for C_D gives a dependency of $\bar{u}_G^{4.2}$. The change of the slope of the lines calculated using (12.84) represents a change from Stokes law at large $\bar{u}_G$ to the intermediate relation for C_D at small $\bar{u}_G$. (This is because there is a decrease of drop diameter with increasing gas velocity.)

The three lines at the top part of Figure 12.19 represent the entrainment that would be observed in vertical pipes (see equation (11.43)). An important point to be made in comparing these lines with data is that correlations based on results in vertical pipes will overpredict at small gas velocities, and not capture the strong influence of gas velocity observed in horizontal pipes.

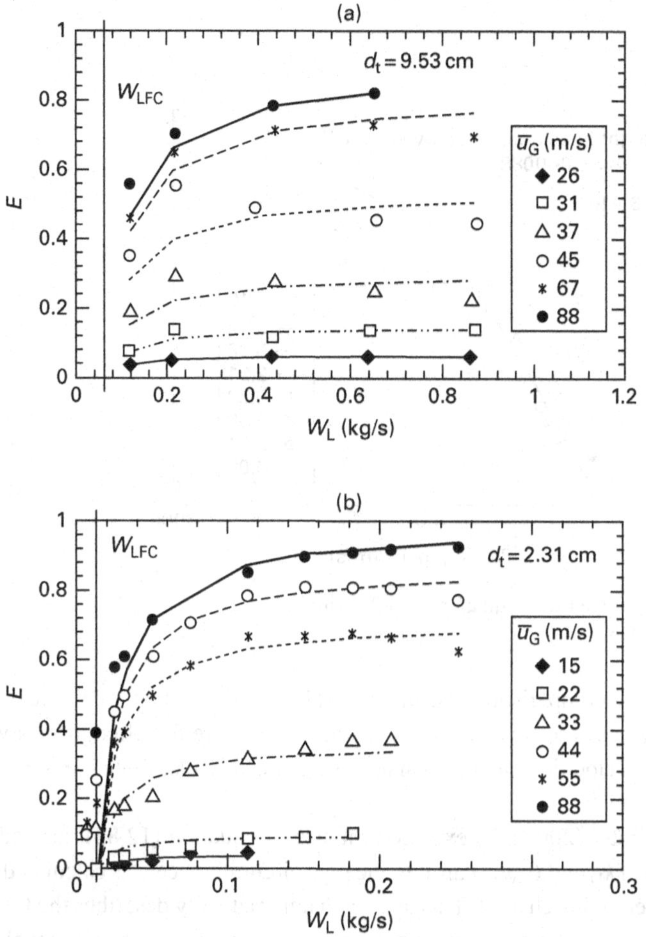

Figure 12.18 Measurements of entrainment in (a) a 9.53 cm horizontal pipe and (b) a 2.31 cm pipe. Pan & Hanratty, 2002b.

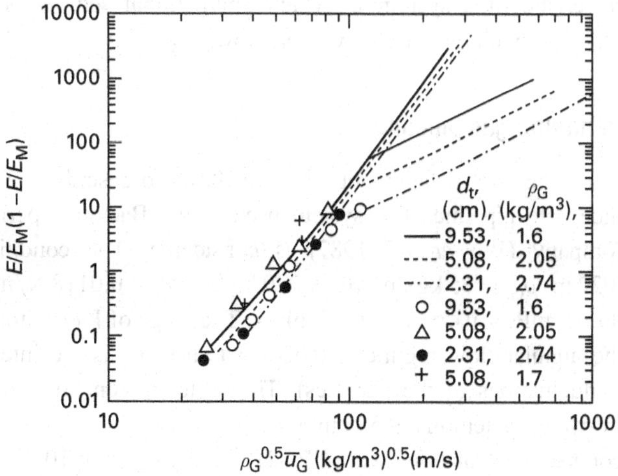

Figure 12.19 Comparison of measurements with empirical equation (12.84). The particle diameter, d_{32} is calculated with equation (11.31). Pan & Hanratty, 2002b.

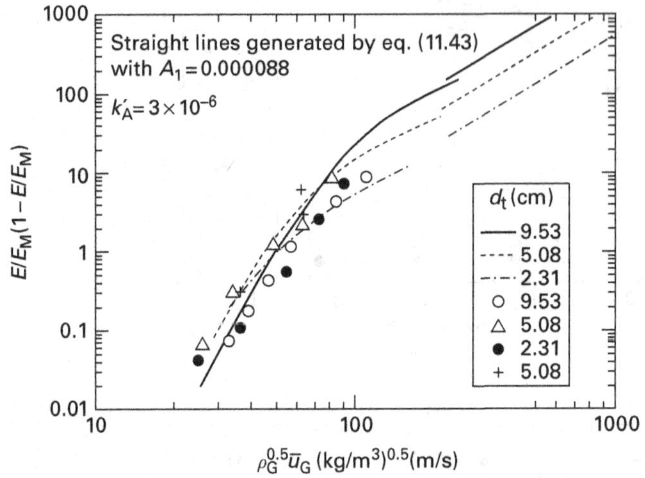

Figure 12.20 Entrainment calculated with equation (12.82) using $k'_A = 3 \times 10^{-6}$ and d_{32} given by (11.31). Pan & Hanratty, 2002b.

The favorable comparison of results with (12.84) shows that the influence of gravitational settling needs to be considered in order to capture the effect of gas velocity. For very large gas velocities, the drops would be uniformly distributed and E/E_M would be close to unity.

Pan & Hanratty (2002a, b) explored the use of equation (12.82), instead of (12.84). The variation of k_D and C_W/C_B around the pipe circumference was calculated by methods outlined earlier in this chapter. This approach automatically describes the transition from a situation where gravity is dominating to one in which gravitational effects are not important. The results of this calculation are presented in Figure 12.20. A value of $k'_A = 3 \times 10^{-6}$ was used. These theoretical calculations show a smooth transition from a low to a high velocity behavior for which $(E/E_M)/1 - (E/E_M)$ varies roughly as $\bar{u}_G^5$ and $\bar{u}_G^2$. One is characteristic of a system in which gravitational settling is controlling deposition. The other is characteristic of a vertical flow.

12.4.3 Calculations for a natural gas pipeline

It is of interest to use the interpretation for air–water flows to describe the behavior of a natural gas–condensate pipeline. The system used in the Bacton experiments done by Shell Oil Company (Wu *et al.*, 1987) is considered. The conditions are as follows: $d_t = 0.2032$ m, $\rho_G = 65$ kg/m^3, $\rho_L = 720$ kg/m^3, $\sigma = 0.0118$ N/m, pressure = 75 bars, $\mu_L = 0.00055$ mPa s, $\mu_G = 1 \times 10^{-5}$ mPa s. The range of V_T/σ_P for this natural gas pipeline in the annular flow regime is 0.02–0.5 (where σ_P is the intensity of the turbulent velocity fluctuations of the particles). This is to be compared with V_T/σ_P = 0.25–2.5 for an air–water system in a 9.53 cm pipe.

Figure 12.21 compares calculations with (12.82), using $k'_A = 3 \times 10^{-6}$ and an eddy diffusivity for droplets of $\varepsilon_P = 0.04 r_t v^*$, for both natural gas and the air–water system. The

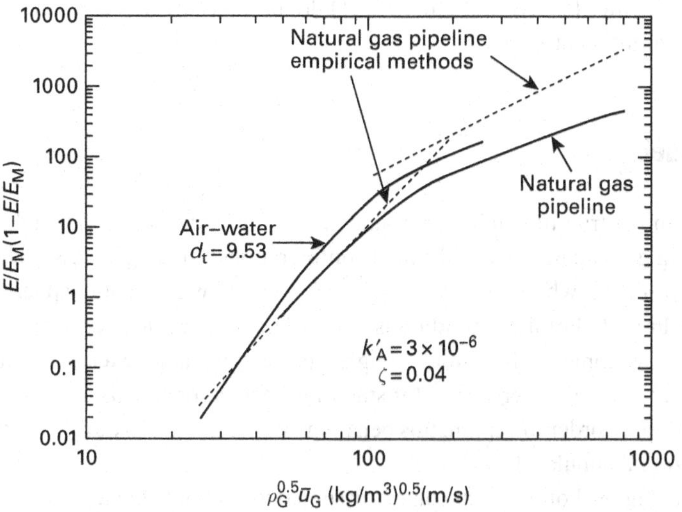

Figure 12.21. The solid curves use equations (12.82) and (11.41) to calculate entrainment for an air–water system and for a natural gas pipeline. The dashed lines are calculated for a natural gas pipeline using equations (12.84) and (11.31). Pan & Hanratty, 2002b.

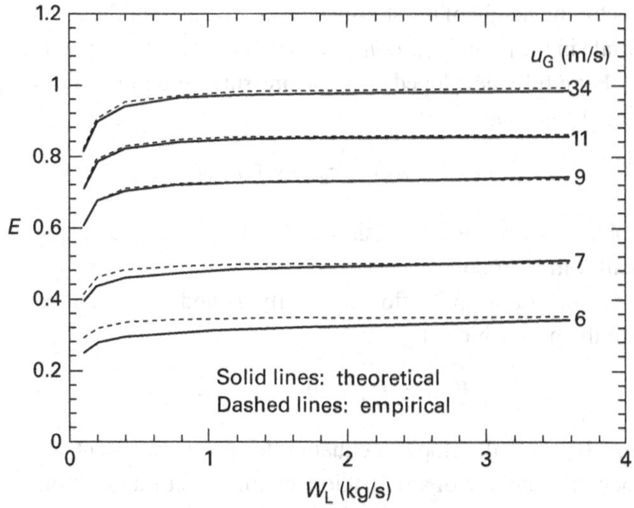

Figure 12.22. Plots of calculated E for a natural gas pipeline. The solid curves use equation (12.82) with $k'_A = 3 \times 10^{-6}$ and $\zeta = 0.04$. The dashed curves were calculated with equation (12.84). Pan & Hanratty 2002b.

dashed lines represent calculations with (12.84) using $A_2 = 9 \times 10^{-8}$ to describe the behavior of the natural gas pipeline at low gas velocities. Equation (11.43) for vertical pipes, with $A_1 = 8.8 \times 10^{-5}$, was used to represent the behavior of natural gas pipelines at large gas velocities.

Figure 12.22 compares calculations of E versus W_L for different gas velocities. The solid curves use (12.82); the dashed lines use (12.84) and empirical equation (11.43) for vertical flows. Note that large differences in predicted values of $(E/E_M)/1-(E/E_M)$ at high gas velocities (shown in Figure 12.21) do not translate into large differences in the predicted entrainment E.

12.5 Pool model

Studies of the distribution of liquid around the circumferences of horizontal pipes that are transporting gas and liquid reveal a pool configuration at small gas flows. The criterion is small $(\rho_G/\rho_L)^{1/2}$ Fr, where $\text{Fr} = u_{SG}/(gd_t)^{1/2}$. For air and water at atmospheric conditions, this yields Fr < 50. For these conditions, most of the liquid is located at the bottom of the pipe, so the asymmetric film model might not be applicable. Measurements of f_i/f_S in Figure 12.13 show smaller values for small Fr when comparisons are made at the same liquid holdup. In order to capture this behavior, Williams (1990) developed a pool model for a horizontal annular flow.

The liquid at the bottom of the pipe is assumed to resemble the idealized configuration discussed in Chapter 5 on stratified flow. The only difference is the inclusion of a thin asymmetric film on the portion of the circumference indicated by P_G and the inclusion of droplets in the gas space.

Chapter 5 describes analyses by Andritsos & Hanratty (1987) of stratified flows which provide relations for the height of the stratified layer, h_{L0}, and for the friction factor, f_{ipool}, for a given flow rate in the pool, W_{LP}. If h_{L0}/d_t is known, geometric parameters S_i and P_G can be specified. For a fully developed flow, the pressure drop is calculated by applying a force balance to the gas phase:

$$\Delta P(A - A_L) = \tau_{\text{ipool}} S + \tau_{\text{iwall}} P_G \tag{12.86}$$

where τ_{ipool} is obtained from the analysis for stratified flow and τ_{iwall} is estimated by assuming the wall film is smooth.

The use of the results on stratified flow to specify A_L and τ_{ipool} requires a prediction of the liquid flow in the pool which is given by

$$W_{LP} = W_L - W_{LE} - W_{LF} \tag{12.87}$$

Williams neglects W_{LF} and develops an equation for the entrainment, W_{LE}, by equating relations developed for the rate of atomization and the rate of deposition.

Figure 12.23 contains measurements by Williams (1990), which show that almost all of the liquid in the wall layer is contained in the "pool" at low enough gas velocities. The "pool model" pictures atomization as occurring only from an idealized layer at the bottom of the pipe. Some of the drops deposit on the walls of the pipe to be part of the film which drains downward back to the pool. Studies of the rate of atomization in a rectangular channel by Alonso (1977), by Tatterson (1975) and later work in vertical pipes suggest the following equation for small $(W_{LP} - W_{LFC})$

$$R_A = k_A'' \frac{\bar{u}_G^2 (\rho_G \rho_L)^{1/2}}{\sigma} \left(\frac{W_{LP} - W_{LFC}}{S_i} \right) \tag{12.88}$$

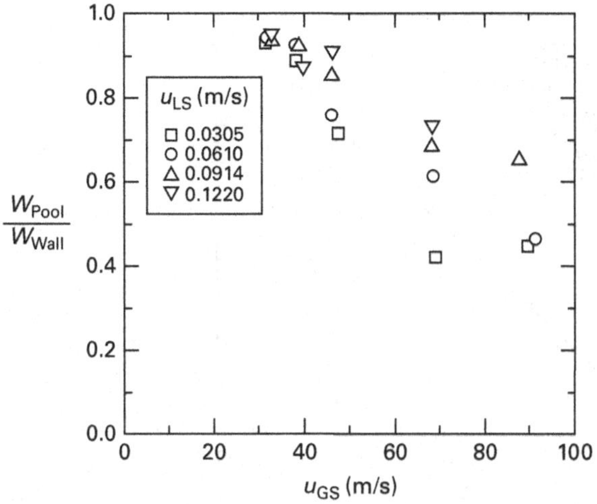

Figure 12.23 Fraction of non-entrained liquid contained in the pool for a 0.0953 m pipe. Figure 8.14 of Williams, 1990.

where R_A is the rate of atomization per unit length and W_{LP} is the weight flow in the pool. This is similar to equation (11.37), but the dimensionless constants appearing in the two equations need not be equal. Williams used A_L/h_0, rather than S_i (see notation in Chapter 5). The rate of deposition can be given as

$$(S_i + P_G)\langle R_A \rangle = C_0 S_i V_{W0} + \langle V_W C_W \rangle P_G \tag{12.89}$$

where V_{W0} and V_W are the velocities of droplets striking the stratified layer and the wall. Equation (12.88) can be written as

$$(S_i + P_G)\langle R_A \rangle = C_B \left[\frac{C_0}{C_B} S_i V_{W0} + \left\langle \frac{C_W}{C_B} V_W \right\rangle P_G \right] \tag{12.90}$$

where C_B is the bulk concentration of drops and P_G is the length of the wall above the pool (see Chapter 5). The first term on the right side is the rate of deposition on the pool. The second term is the rate of deposition on the wall layer, where $\langle \frac{C_W}{C_B} V_W \rangle$ is the average around the length P_G. The term $\langle R_D \rangle$ is the average rate of atomization over S_i and P_G. Equating the rate of atomization to the rate of deposition provides an equation for the entrainment, E, analogous to (12.82).

$$k''_A \frac{\overline{u}_G^2 (\rho_G \rho_L)^{1/2}}{\sigma} (W_{LP} - W_{LPC}) = C_B \left[\frac{C_0}{C_B} \frac{S_i}{(S_i + P_G)} V_{W0} + \left\langle \frac{C_W}{C_B} V_W \right\rangle \frac{P_G}{(S_i + P_G)} \right] \tag{12.91}$$

where V_{W0} may be assumed equal to the settling velocity, V_T. Since $E = W_{LE}/W_L$ and $C_B = W_{LE}/Q_G$, the solution of (12.91) for C_B provides information on the entrainment of drops in the gas phase.

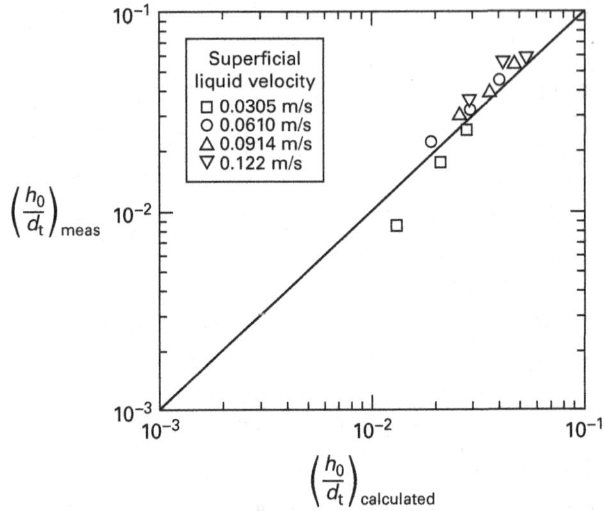

Figure 12.24 Comparison of the liquid height measured at the bottom of a pipe with results calculated with the analysis of Andritsos & Hanratty, (1987). Andritsos & Hanratty, 1987.

A knowledge of E allows a calculation of W_{LP} if W_{LF} is known since

$$W_L - W_{LE} = W_{LF} - W_{LP} \tag{12.92}$$

The flow in the film, W_{LF}, may be estimated with methods outlined in previous sections. However, a good first approximation is to neglect W_{LF} in (12.92).

The approach outlined above has not been exploited so its validity and usefulness have not been tested. However, Williams (1990) compared measurements of the layer height at the bottom of a 9.53 cm pipe carrying air and water with the analysis of Andritsos & Hanratty (1987) for a stratified flow. To simplify the calculation, the assumption was made that all of the non-entrained liquid is flowing in a pool at the bottom of the pipe. The results for Froude numbers below 50, shown in Figure 12.24, are encouraging.

12.6 Transition to the annular regime

12.6.1 Air–water flow

Two mechanisms have been presented for the transition from stratified flow to annular flow in horizontal pipes: One suggests that wetting of the top of the pipe is initiated by large-amplitude waves (or flow surges) which wrap around the circumference. The other suggests that annular flow is initiated by an entrainment–deposition mechanism. Lin & Hanratty (1987) carried out a study on flow patterns for the air–water system and concluded that the transition in a horizontal 9.53 cm pipe occurs primarily through the deposition of droplets. This mechanism is dominant in a 2.54 cm pipe only for superficial liquid velocities smaller than 0.015 m/s. The initial supposition for systems in which droplet wetting is creating an annular flow was that transition to annular flow is associated with the initiation of atomization. However, observations show that transition occurs at higher gas velocities than that needed to create droplets.

Baik & Hanratty (2003) point out that droplets must migrate in sufficient quantity to form a stable continuous film at the top of the pipe. They postulated that, for a smooth pipe, the film at the top of the pipe would be laminar and covered with capillary waves.

Consider a coordinate system for which y is the distance from the wall and x is the distance from the top of the pipe in the circumferential direction. A force balance on the film gives

$$\frac{d\tau_{yx}}{dy} + \rho_L g \sin\theta = 0 \tag{12.93}$$

where $\tau_{yx} = \mu_{yx}(du_x/dy)$ and θ is the angular distance from the top of the pipe.

Equation (12.93) can be integrated to obtain $\overline{u}_x$ by assuming that $\overline{u}_x = 0$ at $y = 0$ and that $\tau_{yx} = 0$ at $y = \overline{h}_L$, where τ_{yx} is a shear stress in the circumferential direction on a plane perpendicular to the y-axis. This implies that the influence of secondary flows in the gas can be neglected and that ripple waves are perpendicular to the x-axis. The mass flow per unit length is given as

$$\Gamma_x = \int_0^{\overline{h}_L} \rho_L \overline{u}_x dy \tag{12.94}$$

The change of Γ_x in the circumferential direction is obtained from the mass balance

$$\frac{1}{r_t}\frac{d\Gamma_x}{d\theta} = R_A - R_D \tag{12.95}$$

where r_t is the pipe radius, R_D is the rate of deposition of drops per unit area, R_A is the rate of atomization per unit area and $\theta = 0$ is taken as the top of the pipe. The rate of atomization at the top of the pipe would be zero at the initiation of annular flow by droplet deposition. The height of the film at the top of the pipe is then calculated as

$$\overline{h}_L = \left(\frac{3R_D d_t \upsilon_L}{2g\rho_L}\right) \tag{12.96}$$

where υ_L is the kinematic viscosity of the liquid and d_t is the pipe diameter.

The rate of deposition at the top of the pipe, R_D, is given as

$$R_D = C_W V_W \tag{12.97}$$

where the velocity of the particles striking the wall, V_W, is given by equation (12.78). The concentration of drops at the top wall, C_W, was assumed by Baik & Hanratty to be given by equation (12.27).

Figure 12.25 compares calculations by Baik & Hanratty (2003) of the thickness of the layer at the top of the pipe with measurements by Williams *et al.* (1996), which were limited to heights above 60 μm. The system being considered is the flow of air and water in a 9.53 cm pipe. The numbers in parentheses are calculated drop sizes. Note that, as

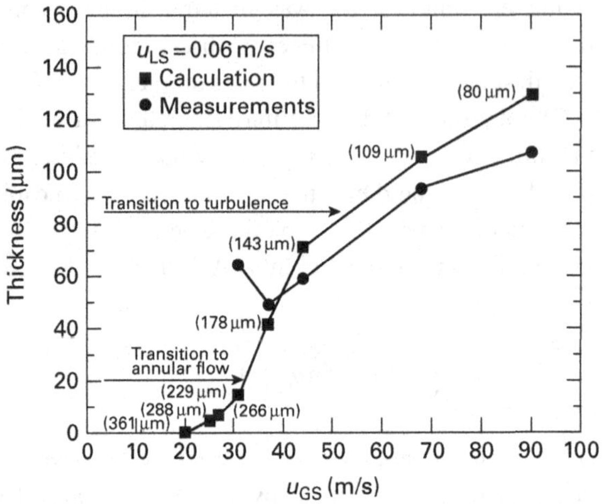

Figure 12.25 Comparison of measurements of the liquid height at the top wall with equations (12.96) and (12.97), for air–water flows in a 9.53 cm pipe. Drop sizes are shown in parentheses. Baik & Hanratty, 2003.

drop size decreases with increasing gas velocity, the concentration of drops at the top of the pipe increases, so R_D increases.

Lin & Hanratty (1987) have found that the transition to annular flow (for u_{LS} = 0.06 m/s) occurs at a superficial gas velocity slightly above 30 m/s. This is indicated by one of the arrows in Figure 12.25. Note that the film height at the top of the pipe is 20 μm. Andreussi *et al.* (1985) observed that disturbance waves (or turbulence) occur for a film Reynolds number of about 380. For the system considered in Figure 12.25, this corresponds to a thickness of about 86 μm.

Thus, according the work of Baik & Hanratty, the transition to annular flow can be related to the formation of a stable layer on the top of the pipe. Therefore, the solution of this problem would involve the development of a theoretical construct that defines the stability condition. Figure 12.26 shows transitions to annular flow defined by Lin & Hanratty, for conditions where the change is controlled by droplets wetting the top wall. The transition for a 9.53 cm pipe is indicated by the solid curve. The transition for a 2.54 cm pipe is indicated by the dotted curve at the bottom of the figure. The numbers indicate the height of the film at the top of the pipe (in μm). It is noted in Figure 12.26 that the transition defined by Lin & Hanratty, in 9.53 cm and 2.54 cm pipes, occurs when the thickness of the liquid film at the top of the pipe is about 20 μm.

Baik & Hanratty explored the possibility that the critical condition is given by dimensionless height, h_L^+. They used methods developed by Asali *et al.* (1985) (see Chapter 3) to show that the critical dimensionless height for the two air–water systems considered in Figure 12.26 is $h_L^+ = 2$.

Observations of the transition indicate that the droplets impinge on the top wall where they coalesce to form larger drops. These coalesce to form rivulets which, at large enough flows, spread out to create a continuous film. There is a similarity between

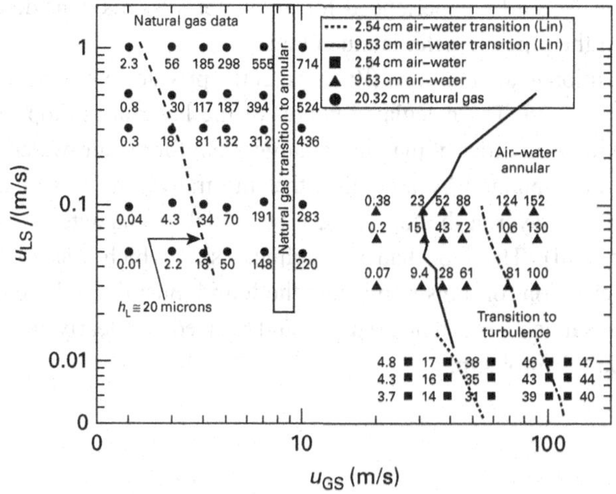

Figure 12.26 Transition from stratified flow to annular flow. Film thicknesses at the top of the pipe, calculated with equation (12.96), are given in the units of micrometers. Baik & Hanratty, 2003.

this process and the process of dryout, which has received consideration in the heat transfer community.

Therefore, it is of interest to discuss two studies carried out by Hewitt & Lacey (1965) for air flowing over a water film that was moving upward over an acrylic wall under conditions that the water film forms dry spots for h_L^+ of 1.5–2.7. In the second, a dry spot was created by blowing air at an upward-flowing annular film. If the flow of the film was large enough, the dry spot disappeared. A critical h_L^+ between 4.62 and 6.87 was determined from this second study. These results are to be compared with the critical $h_L^+ \propto 2$ that is estimated from the experiments of Lin & Hanratty (1987) on the transition to annular flow. It also should be noted that Asali *et al.* (1985) observed h_L^+ as low as about 2 for water–glycerine solutions flowing in vertical annular flows under conditions that atomization was not occurring.

12.6.2 Natural gas pipelines

Calculations of the film height at the top of a pipe for the Bacton studies, at 75 bar, described in Section 12.4.3 are presented in Figure 12.26. The properties of the fluids are quite different from air–water. The interfacial tension ($\sigma = 0.012$ N/m), the gas density ($\rho_G = 65$ kg/m^3), the liquid density ($\rho_L = 0.72$ g/cm^3) and the liquid viscosity ($\mu_L = 0.55$ cP) are to be compared with typical properties of the air–water system operating at atmospheric pressure ($\sigma = 0.072$ N/m, $\mu_L = 1$ cP, $\rho_G = 1.2$ kg/m^3, $\rho_L = 1$ g/cm^3). Also, the pipe diameter, $d_t = 20.32$ cm, is larger than the values of $d_t = 9.53$ cm and $d_t = 2.54$ cm considered in Figure 12.26 for air–water flows.

From Section 11.3 it is seen that drop sizes are much smaller for the natural gas system than for the air–water system, primarily because of the smaller surface tension. Thus, mixing of drops is faster and the calculated thicknesses of the wall layer at the top of the pipe are larger than would be experienced for air–water systems considered in Figure 12.26, even though the pipe diameter is much larger.

Data for the transition to annular flow (Zabaras, 2002, private communication) are also presented in Figure 12.26. The calculated height of the layer at the top of the pipe at transition is found to be an order of magnitude larger than for the air–water system. This difference cannot be explained by assuming that the transition is associated with an instability of the liquid layer at the top of the pipe, which is characterized by a critical h_L^+ (for a smooth wall). The pipe had a roughness size which was of the order of 100–250 μm. If a criterion for transition is that the liquid must flood these roughnesses, then film thicknesses much larger than 20 μm would be needed. Clearly, the last word has not been written on this problem.

References

Alonso, G. F. 1977 The effect of liquid viscosity on the rate of atomization in concurrent gas–liquid flow. M.Sc. thesis, University of Illinois.

Anderson, G. H. & Mantzouranis, B. G. 1960 Two-phase (gas–liquid) flow phenomena I. *Chem. Eng. Sci.* 12, 109–126.

Anderson, R. J. & Russell, T. W. F. 1970 Film formation in two-phase annular flow. *AIChE Jl* 16, 626–632.

Andreussi, P., Asali, J. C. & Hanratty, T. J. 1985 Initiation of small waves in gas–liquid flows. *AIChE Jl* 31, 119–126.

Andritsos, N. & Hanratty, T. J. 1987 Influence of interfacial waves on hold-up and frictional pressure drop in stratified gas–liquid flows. *AIChE Jl* 33, 444–454.

Asali, J. C., Hanratty, T. J., Andreussi, P. 1985 Interfacial drag and film height for vertical annular flow. *AIChE Jl* 31, 895–902.

Baik, S. & Hanratty, T. J. 2003 Concentration profiles of droplets and prediction of the transition from stratified to annular flow in horizontal pipes. *Int. J. Multiphase Flow* 20, 329–338.

Bird, R. B., Stewart, W. E. & Lightfoot, E. N. 1960 *Transport Phenomena*. New York: Wiley.

Butterworth, D. 1972 Air– water annular flow in a horizontal tube. *Prog. Heat Mass* 6, 239.

Butterworth, D. & Pulling, D. J. 1973 Film flow and film thickness measurements for horizontal annular air–water flow. AERE Report R7576.

Dallman, J. C., 1978 Investigation of separated flow model in annular gas–liquid two-phase flows. Ph.D. thesis, University of Illinois.

Dallman, J. C., Laurinat, J. E. & Hanratty, T. J. 1984 Entrainment for horizontal gas–liquid flow. *Int. J. Multiphase Flow* 10, 677–690.

Darling, R. S. & McManus, H. N. 1969 Flow patterns in circular ducts with circumferential variation of roughness: a two-phase analog. In *Developments in Mechanics*, Vol. 5, *Proceedings of the 11th Midwestern Mechanics Conference*, ed. H. J. Weiss, D.F. Young, W.F. Riley & T.R. Rogge. Ames: Iowa State University Press, pp. 153–163.

Dykhno, L. A., 1996 Waves generated on thin liquid films during transition to annular flow in a large diameter pipe: visual observation and measurements. Internal document.

Dykhno, L. A., Williams, L. R. & Hanratty, T. J. 1994 Maps of mean gas velocity for stratified flow with and without atomization. *Int. J. Multiphase Flow* 20, 691–702.

Fisher, S. A. & Pearce, A. 1979 A theoretical model for describing horizontal annular flow. In *Two-Phase Momentum, Heat and Mass Transfer in Chemical Process, and Energy Engineering Systems*, ed. F. Durst, G.V. Tsiklauri & N. Afgan. Washington, DC: Hemisphere.

Fukano, T. & Ousaka, A. 1989 Prediction of the circumferential distribution of film thickness in horizontal and near-horizontal gas–liquid annular flows. *Int. J. Multiphase Flow* 15, 403–419.

Hanratty, T. J. 1956 Heat transfer through a homogeneous isotropic turbulent field. *AIChE Jl* 2, 42–45.

Henstock, W. H. & Hanratty, T. J. 1976 The interfacial drag and the height of the wall layer in annular flows. *AIChE Jl* 22, 990–1000.

Hewitt, G. F. & Hall-Taylor, N. S. 1970 *Annular Two-phase Flow*. Oxford: Pergamon Press.

Hewitt, G. F. & Lacey, P. M. C. 1965 The breakdown of the liquid film in annular two-phase flow. *Int. J. Heat Mass Transfer* 8, 781–791.

Hinze, J. O. 1967 Secondary currents in wall turbulence. *Phys. Fluids Suppl.* 10, 112.

Hurlburt, E. T. & Newell, T. A. 2000 Prediction of the circumferential film thickness distribution in horizontal annular gas–liquid flow *Trans. ASME* 122, 396–402.

Hutchinson, P., Butterworth, D. & Owen, R. G. 1974 Development of a model for horizontal annular flow. AERE Report. R7789.

James, P.W. & Burns, A. 1979 Further developments in the modeling of horizontal annular flow. AERE Report R9373.

Laurinat, J. E. 1982 Studies of the effect of pipe size on horizontal two-phase flows. Ph.D. thesis, University of Illinois.

Laurinat, J. E., Hanratty, T. J. & Dallman, J. C. 1984 Pressure drop and film height measurements for annular gas–liquid flow. *Int. J. Multiphase Flow* 10, 341–356.

Laurinat, J. E., Hanratty, T. J. & Jepson, W. P. 1985 Film thickness distribution for gas–liquid annular flow in a horizontal pipe. *Phys.-Chem. Hydrodynamics* 6, 179–195.

Lin, P. G. & Hanratty, T. J. 1987 Effect of pipe diameter on flow patterns for air–water flow in horizontal pipes. *Int. J. Multiphase Flow* 13, 549–563.

Lin, T. F., Jones, O. C., Lahey, R. T., Block, R. S. & Mirase, M. 1985 Film thickness measurements and modelling in horizontal annular flows. *Phys.-Chem. Hydrodynamics* 6, 197–205.

McManus, H. N. 1961 Local liquid distribution and pressure drops in annular two-phase flow. Paper presented at the ASME-EIC Hydraulic Conference, Montreal, paper 61-HYD-20.

Mito, Y. & Hanratty, T. J. 2005 A stochastic description of wall sources in a turbulent field. Part 3: Effect of gravitational settling on the concentration profiles. *Int. J. Multiphase Flow* 31, 155–178.

O'Brien, M. P. 1930 Review of the theory turbulent flow and its relation to sediment transport. *Trans. Am. Geophysical Union* 14, 487–491.

Pan, L. & Hanratty, T. J. 2002a Correlation of entrainment for annular flow in vertical pipes. *Int. J. Multiphase Flow* 28, 363–384.

Pan, L. & Hanratty, T. J. 2002b Correlation of entrainment in horizontal pipes. *Int. J. Multiphase Flow* 28, 385–408.

Paras, S. V. & Karabelas, A. J. 1991a Properties of the liquid layer in horizontal annular flow. *Int. J. Multiphase Flow* 17, 439–454.

Paras, S. V. & Karabelas, A. J. 1991b Droplet entrainment and deposition in annular flow. *Int. J. Multiphase Flow* 17, 455–468.

Pletcher, R. H. & McManus, H. N. 1965 The fluid dynamics of three-dimensional liquid films with free-surface shear: a finite difference approach. In *Developments in Mechanics*, Vol. 3: *Proceedings of the 9th Midwestern Mechanics Conference*, ed. T. C. Huang & M. W Johnson. New York: John Wiley.

Prandtl, L. 1927 Turbulent flow. NASA TM 435.

Rouse, H. 1937 Modern conceptions of the mechanics of fluid turbulence. *Trans. Am. Soc. Civil Engrs* 102, 463–505; Discussion 506–543.

Swanson, R. W. 1966 Characteristics of the gas–liquid interface in two-phase annular flow. Ph.D. thesis, University of Delaware.

Tatterson, D. F. 1975 Rates of atomization and drop size in annular two-phase flow. Ph.D. thesis, University of Illinois.

Wilkes, N. S., Conkie, W. & James, P. W. 1980 A model for the droplet deposition rate for annular two-phase flow. AERE Report R9691.

Williams, L.R. 1986 Entrainment measurements in a 4-inch horizontal tube. M.Sc. thesis, University of Illinois.

Williams, L. R. 1990 Effect of pipe diameter on annular two-phase flow. Ph.D. thesis, University of Illinois.

Williams, L. R., Dykhno, L. A. & Hanratty, T. J. 1996 Droplet flux distribution and entrainment in horizontal gas–liquid flows. *Int. J. Multiphase Flow* 22, 1–18.

Wu, H. L., Pots, B. F. M., Hollenberg, J. F. & Meerhof, R. 1987 Flow pattern transitions in two-phase gas/condensation flow at high pressures in a 8-inch horizontal pipe. Paper presented at the 3rd International Conference on Multiphase Flow, The Hague, The Netherlands.

Index

Printed in the United States
by Baker & Taylor Publisher Services

Printed in the United States
by Baker & Taylor Publisher Services